D0717260

037403

Ludus Coventriæ

OR

The Plaie called Corpus Christi

COTTON MS. VESPASIAN D. VIII.

EARLY ENGLISH TEXT SOCIETY

Extra Series, No. CXX.

1922 (*for* 1917; reprinted 1960)

PRICE 30/-

Ludus Coventriæ

or

The Plaie called Corpus Christi

COTTON MS. VESPASIAN D. VIII

EDITED BY

K. S. BLOCK

Published for
THE EARLY ENGLISH TEXT SOCIETY
by the
OXFORD UNIVERSITY PRESS
LONDON NEW YORK TORONTO

HOMERTON
COLLEGE
LIBRARY

FIRST PUBLISHED 1922
REPRINTED 1960

YCH
22223

Originally printed at the Oxford University Press by Frederick Hall and reprinted by Richard Clay and Company, Ltd., Bungay, Suffolk.

PREFACE

THE *Ludus Coventriæ*, unlike the other extant mystery cycles, is not connected with the trade guilds of a town, and the nature and purpose of this collection have been recognized as a problem of special interest. Investigation has been hampered by the fact that the only edition of the plays (J. O. Halliwell, printed for the Shakespeare Society, 1841), though admirably accurate as to the text, obscures, by changes in arrangement and numbering, certain significant features of the MS. It was therefore desirable that the MS. should be re-edited in order that the bibliographical factors of the problem should be clear.

The plays have, moreover, in themselves some claim to be made more accessible. They illustrate the advance in dramatic representation that was taking place in the fifteenth century. Plays like Nos. 18 and 20 (*Adoration of the Magi* and *Massacre of the Innocents*, with the *Death of Herod*), or No. 24 (*The Woman taken in Adultery*), indicate an improvement in the art of the religious drama; passages like the satiric address of Demon to the audience, p. 225, indicate a widening of its range. The *Passion* plays, which differ from those of the other cycles in being acted, not on a sequence of pageants in procession, but on a standing group, show for this reason more attempt at construction; and there are suggestions here and in some of the other plays of specially effective acting (cf. *Introduction*, p. lvi). In the stage directions of these *Passion* plays (cf. p. 267 and p. 273) we have the most graphic picture of a fifteenth-century performance that has come down to us.

The title *Ludus Coventriæ* has been retained partly because it has been associated with these plays since the first cataloguing of the MS. by Richard James; but partly also because there is, in the opinion of the present editor, some ground for thinking that the title was used generically by James, and has therefore some importance in the history of the drama (cf. *Introduction*, p. xl).

This edition was begun in 1913, and marks of protracted and intermittent work are, I fear, obvious. My thanks are due to Sir Israel Gollancz for constant encouragement, and to the Oxford University Press for much consideration and assistance. I am indebted to my colleague Miss G. D. Willcock for help in compiling the glossary.

K. S. BLOCK.

ROYAL HOLLOWAY COLLEGE.
March, 1922.

CONTENTS

	PAGE
Introduction	xi
Ludus Coventriæ	1–377
List of Proper Names	379
Glossary	383

PLAYS AS NUMBERED IN THE MS., IN THE PROCLAMATION, AND IN HALLIWELL'S EDITION

	MS. No.	Pageant No.	Halliwell.	Page.
The Creation of Heaven and the Angels	1	1	1	16
Fall of Lucifer				17
The Creation of the World and Man	2	2		19
Fall of Man			2	21
Cain and Abel	3	3	3	29
Noah and	4	4	4	35
The Death of Lamech		No Pag^{t}.		
Abraham and Isaac	5	5	5	43
Moses	6	6	6	51
The Prophets	7	7	7	57
The Conception of Mary (Barrenness of Anna)	8	No Pag^{t}.	8	62
Mary in the Temple	9	No Pag^{t}.	9	71
The Betrothal of Mary	10	8, 9 (corr. to 10)	10	83
The Parliament of Heaven	11	No Pag^{t}.	11	97
The Salutation and Conception		10 (corr. to 11)		
Joseph's Return	12	11 (corr. to 12)	12	109

	MS. No.	Pageant No.	Halliwell.	Page.
The Visit to Elizabeth	13	No Pag^{t}.	13	115
Prologue of Summoner	None	No Pag^{t}.		123
The Trial of Joseph and Mary	14	12 (corr. to 14)	14	124
The Birth of Christ	15	13 (corr. to 15)	15	135
The Adoration of the Shepherds	16	14 (corr. to 16)	16	146
The Adoration of the Magi	18	15	17	151
The Purification	19	No Pag^{t}.	18	162
The Massacre of the Innocents	20	16	19	169
The Death of Herod		17		174
Christ and the Doctors	21	18	20	178
The Baptism	None	19	21	188
The Temptation	23	20	22	193
The Woman taken in Adultery	24	21	23	200
The Raising of Lazarus	25	22	24	210
Passion Play. I				
Prologues of Demon and of John the Baptist		No Pag^{t}.	25	225
The Council of the Jews	26	No Pag^{t}.		230
The Entry into Jerusalem		23	26	237
The Last Supper and The Conspiracy of the Jews and Judas	27	24	27	242
The Betrayal	28	25	28	262
[Prologue of the Doctors	None	No Pag^{t}.	29]	269
Passion Play. II				
King Herod		No Pag^{t}.		271
The Trial of Christ before Annas and Caiaphas	29	26		274
Peter's Denial				277
The Death of Judas		28	30	278
The Trial of Christ before Pilate	30	27		280
The Trial of Christ before Herod		No Pag^{t}.		283
Pilate's Wife's Dream	31	27, 29	31	287
The Trial of Christ and the Thieves before Pilate		27, 29		289
The Condemnation and Scourging		30		294
The Procession to Calvary	32	No Pag^{t}.	32	294
The Crucifixion		30		296
The Descent into Hell of Anima Christi	33	31	33	305

	MS. No.	Pageant No.	Halliwell.	Page.
The Embassy to Pilate of Joseph of Arimathea		32		307
The Episode of Longeus	34	31	34	310
The Descent from the Cross and Burial		32		311
The Guarding of the Sepulchre				312
The Harrowing of Hell		23 (scribal slip)	35	318
The Resurrection and Appearance to the Virgin	35			320
The Compact of the Soldiers and Pilate		No Pag^{t}.		322
The Announcement to the Three Maries	36	34	36	327
The Appearance to Mary Magdalen	37	35	37	333
The Appearance on the way to Emmaus	38	36	38	337
The Appearance to Thomas		37		346
The Ascension and	39	38	39	349
The Choice of Matthias		No Pag^{t}.		351
The Day of Pentecost	40	39	40	352
The Assumption of the Virgin	41	No Pag^{t}.	41	354
Doomsday	42	40	42	373

ERRATA

Some lines in the second Passion play between lines 985 and 1035, and again between lines 1295 and 1355, are wrongly numbered. The total number of lines in the play is correctly given.

INTRODUCTION

THE MS. Vespasian D. viii of the Cotton Collection in the British Museum is a small thick volume containing 225 leaves 8″ × 5½″. In the process of binding the leaves have all been cut down and remounted, and their breadth varies slightly: folios 91 and 92, containing the opening of the *Magi* play, are noticeably narrower than the rest. Description of MS.

A modern hand has numbered the folios. The reverse sides are numbered by tens only, 9ᵛ, 19ᵛ . . . 49ᵛ being numbered 10, 20 . . . 50; 60ᵛ, 70ᵛ . . . 110ᵛ, 60, 70 . . . 110; 121ᵛ, 131ᵛ . . . 161ᵛ, 120, 130 . . . 160; 172ᵛ, 182ᵛ . . . 212ᵛ, 170, 180 . . . 210; and 223ᵛ, 220. Pagination.

The lettering of the quires, A to W, is of about the same date. In four cases the original catchwords remain: on fo. 40ᵛ, the last page of quire B; on fo. 148ᵛ, the last page of quire N; on fo. 179ᵛ, the last page of quire S; on fo. 189ᵛ, the last page of quire T. In other cases the quire letters correspond with a change of paper. Quires.

Seven different kinds of paper are found in this MS. Quires A and B (ff. 1–40ᵛ) consist of paper of the *Bunch of Grapes* watermark (*Les Filigranes*, C. M. Briquet, Paris 1907, No. 3055). Quires C to M (ff. 41–135ᵛ), with the exception of quire E (ff. 51 and 52) and of the interpolated folios 95, 96, and 112, consist of paper of the *YHS in a Sun* watermark (*Les Filigranes*, No. 9477). The interpolated quire E has the *Pitcher*[1] watermark of the *pot d'étain* type (*Les Filigranes*, No. 12498 or 12501), not found, according to M. Briquet, before the last decade of the fifteenth century. The interpolated folios 95, 96 have the *Hand* watermark of the gloved type marked with a *ʒ* on the palm and surmounted by a pentagon.[2] Unfortunately not enough of the Paper.

[1] *Mind, Will and Understanding* in the Macro MS. is written on paper bearing a *Pot* watermark. E.E.T.S., E.S. xci, p. xxx.

[2] The three plays in the Digby MS., *The Conversion of St. Paul*, *The*

lower part of the mark on fo. 95 remains to make it clear whether the wrist is laced. No example exactly corresponding with this mark is given by M. Briquet. It belongs to the general class described by him as *Main aux quatre doigts serrés, le pouce seul écarté.* Of the subdivision *lacée au poignet,* M. Briquet gives 1526 as the date of the earliest example known to him. If the lines which can just be discerned across the gauntlet could be taken as lacing, these folios, according to the evidence of the handwriting, would be earlier. Fo. 112, also an interpolation, has no watermark, but the wire lines correspond with those of folios 95, 96. Quire N (ff. 136–48^{v}), with the exception of fo. 143, and quires P, Q, R (ff. 152–63^{v}), have the *Bull's Head* watermark (*Les Filigranes,* No. 14184). Folio 143 has no watermark, but the wire lines correspond with those of the *Bunch of Grapes* or of the *YHS* watermark. Quire O (ff. 149–51^{v}) has the *Bunch of Grapes* watermark. Quires S and T (ff. 164–89^{v}), with the exception of folios 184 and 185, have the *Two Crossed Keys* watermark (*Les Filigranes,* No. 3887). Folios 184 and 185 have again unfortunately no watermark, but the wire lines are not those of the *Two Crossed Keys* paper, but correspond with those of the paper having the *Bunch of Keys* or the *YHS* watermark. Quires V and W (ff. 190–225), with the exception of folios 213–22, have the *YHS* watermark. The interpolated folios containing the play of the *Assumption* have the *Two-Wheeled Cart* watermark (*Les Filigranes,* No. 3528).

Besides these seven watermarks in the MS. itself, another—a shield bearing a crook—is found on the fly-leaf on which Dr. James, Sir Robert Cotton's first librarian, has written a note of the contents. Only a small portion is shown, and it does not correspond exactly with any of M. Briquet's examples. It comes nearest to what he describes as *La crosse de Bâle dans un écu,* a sixteenth-century mark. Paper of apparently the same watermark is used for part (ff. 31–107) of the 1621 catalogue of the Cottonian library.

Evidence of composi-

This variety in the actual material of the MS. corresponds

Massacre, and *Mind, Will and Understanding,* are written on paper bearing different forms of the *Gloved Hand* watermark, but none exactly corresponding with that in the Vesp. D. viii. MS. The paper of *Mankind* in the Macro MS. has a *Gloved Hand* watermark. Cf. E.E.T.S., E.S. xci, p. xxviii.

in an interesting way with differences and discrepancies in the plays themselves, and throws light on the process of compilation. Some of the interpolations, folios 51 and 52, 95 and 96 and 112, are in a different handwriting from the body of the MS. and are later additions to the collection, in the case of 95 and 96 and 112 certainly, in the case of 51 and 52 possibly, replacing original material; and the play of the *Assumption*, though part of the set made up by the original compiler of this MS., is also in a different handwriting.

tion of MS as to method of compilation.

In the case, however, of fo. 143 and quire O, and of folios 184 and 185, the handwriting is the same as that of the quires N, P, Q, R and S, T in which they are interpolated, and the added passages indicate the methods of the compiler.

The quires N, P, Q, R contain a separate group of plays, those that are described later in Contemplacio's Speech (fo. 165) as having been *shewyd the last ȝere*; the subject-matter is complete without fo. 143 or quire O, and the interpolation of these gives rise to duplication. From a cancelled speech in the MS. at the bottom of fo. 142 it is clear that fo. 144 originally followed on fo. 142, and the insertion of fo. 143 (containing the incident of fetching the ass and the foal) leads to a duplication of the passage beginning *ffrendys be-holde*, which occurs again on fo. 145v, and to some inconsistency in the movements of Peter and John who 'abyden stylle' and yet advance towards Jerusalem. Again, a cancelled stage direction at the bottom of fo. 148v, the last page of quire N, shows that quire P originally followed on. The episode of Mary Magdalen in quire O (like that of the ass and the foal on fo. 143) is not mentioned in the proclamation, and the insertion of quire O leads to a duplication of the disciples' questioning of the Saviour as to the traitor (cf. fo. 150 and fo. 155v). A blank folio follows the last play of the group and the remainder of quire R is filled in with the Doctors' prologue—an entirely disconnected passage. Moreover, besides being of different paper from the rest of the MS. the quires N, P, Q, R show, as will be seen, certain slight differences in writing and rubrication.

The quires S and T offer some similarity in that here again we have traces of a set of plays which were acted separately; and from the appearance of fo. 164, the first of quire S, it would seem that we have also part of a separate small MS. bound up to make the series. But the case is more complicated; the interpolated folios 184 and 185 could not be removed, for the last of the Seven Words comes on fo. 184 and the speech of Anima Christi is continued from fo. 185 to fo. 186; and there is no break at the end of quire T. The writer or compiler of the set of plays beginning in the S quire clearly had in his mind at starting a set corresponding with the first *Passion* plays of the N, P, Q, R quires: 'we intendyn to procede þe matere *þat* we lefte þe last ȝere. . . . Now wold we procede how he was browth þan | be-forn annas *and* cayphas · *and* syth be-forn pylate | and so forth in his passyon · how mekely he toke it for man'; but this second group of *Passion* plays does not come to any definite end, the action being continuous until the end of the *Appearance to Mary Magdalen* play on fo. 201, the middle of quire U. The MS., however, affords some traces of the original plan; there is evidence of a change of purpose at the close of the Crucifixion scene, and it can be deduced with some certainty that the original end of the Passion sequel was discarded to make room for the material on folios 183v, 184, 184v, 185, 185v, and 186, and that the Senturyo's speech on fo. 186v belongs to a separate *Burial and Resurrection* play now dovetailed on. A change in the colour of the ink suggests that the scribe paused at the line: *heloy · heloy · lamaȝabathany* (fo. 183v). The writing also becomes smaller with the next line, and, at the bottom of the folio, is cramped as if to fit in more than was allowed for. It appears from slight indications in handwriting and ink as if the scribe went on from the line mentioned to fo. 186v: *Senturyo. A now trewly telle weyl I kan*; and that later he returned and filled up the remainder of fo. 183v, continuing on fo. 186 and fitting in the interpolated folios. That the material on the interpolated folios and fo. 186, though dovetailed in, is itself an interpolation is proved by the discrepancies it involves: Mary having passed into the Temple (fo. 185 interpolated) is yet at the foot of the Cross to receive her son's body (fo. 188); the Centurion makes his speech, recognizing the Crucified as the Son of God, twice (fo. 186 and fo. 186v), and Nicodemus seems to make a double entry (fo. 186 and fo. 188). Additional evidence from the MS. of disturbance at this point is the change in the pointing; the mid-line point, having been fairly abundant in the preceding folios of the T quire, ceases after fo. 183v. No instance of borrowing from *The Northern Passion*, otherwise so marked a feature of this group, occurs in this interpolated portion.[1] The MS. throws no additional light on the correspondence of change of paper and change of material at the junction of the T and U quires. Though there is no break in the action here the changes of metre, of the form of Pilate's name from Pylat to Pilatus, and of English for Latin stage directions,[2] the cessation of reminiscences of *The Northern Passion*,[1] as well as the double *exeunt* of the knights (fo. 189v, last of T, fo. 190, first of V), point to the use of different material, but the scribe appears to have gone on continuously. Quire T is half the size of quire S.

Hand-writing.

The bulk of the MS. is the work of one scribe, probably of the third quarter of the fifteenth century. The date 1468 is written at the close of the *Purification* play on fo. 100v, but this play is an interpolation, and it is possible that the date is specially connected with it rather than with the compilation.

The writing varies considerably, but the variations are better accounted for as due to parts of the MS. having been written at different times than as due to change of scribe. The system or rather want of system in the use of capital letters remains the same—the preferential use of the capital forms of *t* and *n* at the beginning of lines, the arbitrary use of *ff* and *f*, and of *A* and *a*, and the sporadic use of capital forms of *è* and *r*.

The most noticeable variation is found in the quires N, P, Q, R, where the writing is more angular and altogether rougher and less shapely. Exceptionally large capitals appear throughout these quires, and a few specially tall letters in the top lines. The scribe also more often writes *þe* instead of *þ*^{e} than elsewhere in the MS. The writing remains irregular, though less so, in quires S and T—that in the interpolated folios 184, 185, as in the interpolated fo. 143 in quire N, being noticeably firmer.

[1] Cf. *The Northern Passion*, edited F. A. Foster, Ph.D, E.E.T.S. 147, Part II, pp. 90–5.

[2] In quires S and T 56 English stage directions, 12 Latin, two of the latter on the interpolated folios; in quires V and W (excluding *Assumption* play) 2 English, 21 Latin. It may be noted that English and Latin stage directions are mingled in the latter part of the Digby MS. *Mary Magdalen* play.

It recovers its earlier form in quire V. There are minor irregularities in the earlier quires. It is sometimes much more cramped than at others, and once, on the other hand—folios 61–6, the end of quire F—it assumes a larger and more widely spaced form. The genealogies filling the bottoms of some of the folios (ff. 16v et seq., 21 et seq., 37) and the Psalms and Canticles quoted in the *Mary in the Temple* and *Visit to Elizabeth* plays are in liturgical script; the notes on the five Annas (fo. 37v) and on the April Calendar (fo. 74v) are in an ornamental book-hand.

Additions. There are three other handwritings to be found in the MS.: (1) that of the interpolated quire E (ff. 51, 52) containing the conversation between Joseph and his kinsmen in the *Betrothal of Mary* play; (2) that of the interpolated folios 95, 96 in the *Magi* play, and of the interpolated fo. 112 containing the opening of the *Baptism* play; (3) that of the *Assumption* play.[1] The first is a rough cursive hand of the close of the century decidedly later than that of the body of the MS. This writer uses the full *th* form, and neither the *þ* nor the *ȝ* appears in this passage. The second is also a cursive hand, but of a less untidy character. Halliwell speaks of it as 'a more recent hand', and this writer was, as will be seen, at work as a corrector of the MS. after its compilation; but the opinion of authorities is that the hand cannot be said to belong to a later type than the original scribe's. This writer uses the *þ* (in the later *y* form) and the full *th* form in about the same proportion as the chief scribe; like him he uses the *ȝ* in the word *ȝe*, but he uses the forms *you* and *your* where the other invariably uses *ȝow* (*ȝou*) and *ȝour*. This writer uses a curious form of final *s*.

The third hand was also thought by Halliwell to be later, but is now also assigned to the same period as the chief hand. It is of a different character, more cursive and less clear. This writer uses many fewer contractions than the other contributors to the MS. He does not use the *ȝ* at all; the only time it occurs

[1] Miss Swenson (*An Enquiry into the Composition of the Ludus Coventriæ*, University of Minnesota, Studies in Language and Literature) finds two other hands in the Prologue of the Doctors and the Prologue of Contemplacio in the *Passion* play respectively. The slight variation is better accounted for as suggested above.

in the play (fo. 217v)—*That ȝe schuld ben absent*—it comes in a line which was omitted, and written in the margin apparently in the hand of the chief scribe of the MS. He uses as a rule the full *th* form, but four examples of the *þ* (written *y*) are found. Fo. 217v *þu* (*thou*), fo. 219v *þi*, fo. 220v *þowth* (*thought*), fo. 221 *þi*. In this play (fo. 218) occurs the only example, in the MS. proper, of the *þ* in the older form, the two other examples occurring in the notes of some later reviser (fo. 144v, fo. 145).

Corrections of revisers.

The MS. has also been worked over by other hands. Corrections other than those of the original scribe are found scattered in many plays, and range from the alteration of a letter to the rewriting of two or three consecutive lines. The most important are to be found in the *Shepherd*, the *Magi*, the *Harrowing*, and the *Three Maries* plays, on folios 88v, 89, 89v, 93, 93v, 97, 101, and 185v, 191v, 192, 197v, 198. These appear all to be in the hand of the scribe of folios 95, 96, and 112.

Those in the *Shepherd* and *Magi* plays appear to be changes for the purpose of removing archaic or dialectal (Northern) words, phrases, or forms: *selkowth* > *mervelus*, *shene* > *bryght*, *carpynge* > *spekyng*, *barne* > *child*, *bale* > *sorow*, *buske* > *go*, *tholyn* > *suffyr*, *myrke* > *thyke*; a line is altered (fo. 89) so that *þat hattyht* (*that is called*) may be read as *that hateth*, and *xaln* is altered to *xalle*. The phrases erased and written over in Herod's ranting speech (ff. 93, 93v) are for the most part indecipherable, but the alterations that can be read in this passage suggest that they would all fall under the same heading: *paphawk* > *paddok*, *shaftys* > *sperys*. The corrections in the *Harrowing* play consist of indications of additions to the text; they take the form of references to fresh characters not found in the play as it stands —a*nima* latron*is* (fo. 185v), *þe* devyll*e*, a*nima* caym (fo. 191v, fo. 192)—and the manner of the reference—*and þan Cayme xalle sey his spech* (fo. 192)—suggests that the reviser is drawing on some other fuller version of the play known to the users of the MS. The revision in the *Three Maries* play consists in the substitution of eight new lines for four lines in the speech of Mary Magdalen and four in the speech of Mary Jacobi—the whole forming a continuous passage—announcing the resurrection to Peter and John. The purpose of the alteration is to harmonize the account of the resurrection, which in the original follows the narrative of St. Mark (that accepted by Tatian) and represents an angel as announcing the Resurrection to the women, with the version given by St. Matthew, according to which the women see the risen Lord himself before they meet the disciples. A few corrections have been made in the text for the same purpose, as if the marginal rewriting of the lines were a second thought.

A few added marginal stage directions appear also to be in the hand of this reviser: *go homwardys* (fo. 102), various *Incipit hic*, *nota hic* (ff. 189, 189v), and it is possible that some of the slighter corrections found singly here and there are due to him. Many of these consist of one word written above the line or over the original word in the text, and some even of the addition or change of a letter. In such cases it is impossible to form a definite opinion as to the handwriting, but a comparison of one or two typical letters leads to the conclusion that most of these corrections are not from the hand of the scribe of folios 95, 96, 112, nor from that of the scribe of fo. 51. They occur singly in the following plays: *Cain and Abel*, *Mary in the Temple*, *The Trial of Joseph and Mary*, *The Birth*, *The Purification*; three are found in the *Salutation and Conception* play, ten in the *Disputation*, nine in the *Lazarus* play. They consist partly of the insertion of omitted words, partly of changes of apparently archaic or dialectal forms: thus, *beth* (imperative) > *be* (fo. 43v), *evy* > *hevy* (fo. 135 and fo. 110v), *dede* > *dyde* (fo. 134), *fende* > *fynde* (fo. 98v), *glathe* > *gladd* (fo. 134v), *thei* > *they* (fo. 135), *perysche* (pierce) > *pers* (fo. 107v), *blysse* > *comfort* (fo. 110v); *ffor syknes and sorwe* (fo. 58v) is changed to *bothe eve and morewe* to avoid an identical rhyme. Some of the corrections show a misunderstanding of the original: the line *A mayd milke haue never man dyde se* is 'corrected' by the deletion of *haue* and the addition of the possessive *is* to *mayd*.

There are also a few added stage directions in different hands, e.g. *here goth he his way*, written against the speech of quartus consolator and nuncius in the *Lazarus* play (fo. 129). Once or twice names of characters about to appear are roughly written in the margin, e.g. *pastores* in the *Barrenness of Anna* play (fo. 39v). This is perhaps the explanation of the mysterious *Vade Worlych* which is obviously written with some purpose on fo. 206v and again on fo. 207, where the meeting of Luke and Cleophas with the other disciples occurs.

Of special interest are two notes on fo. 144v and fo. 145: *here entrith þe fyrst prophete*, and *here entrith þe parte of þe ij^de prophete*. No prophets appear in this play (*The Entry into Jerusalem*), and these notes again suggest a reference to another version of the play, or perhaps rather the use of this play in another combination than that of this compilation. A note on fo. 196 opposite the stage directions at the beginning of the *Three Maries* play. apparently in the hand of the scribe of folios 95, 96, 112—*finem 1^a die · Nota*—seems also to refer to a division of the performance of some group of the plays.

Corrections of the scribe.

The corrections of the scribe are made generally in black and red ink. Letters or words marked with deleting dot and cancelling strokes are almost always crossed through in red ink. Some of the corrections of the MS. by the original scribe are important for the light they throw on the compilation of the MS. The insertion of a line in the *Assumption* play has already been noticed (p. xvii). If this is accepted as in the hand of the chief scribe it clinches all the other arguments in favour of that play's having formed part of the original compilation, not indeed as conceived by the composer of the prologue but as carried out in the present MS. Mr. Gayley in his *Plays of our Forefathers* (1908) assigns this play to a date later than 1482, on the ground that it is based on Caxton's translation of the *Legenda aurea*—'practically', he says, 'a transcription from it'. But a comparison of the play with Caxton's translation and the original Latin, so far from giving any evidence in support of this view, proves that the writer of the play used the Latin and not the translation. The only instances of verbal coincidence are found in the translation by both writers of *ammirabile et magnum* by *grete and merueylous*, where both words and order may be fairly said to be inevitable, and of *vasculum vite* by *vessel of lyf*, which again could not easily be avoided. Moreover, in two instances passages of the original Latin not translated by Caxton are used in the play of the *Assumption*.[1]

1. Of the *Assumption* play.

[Evidence as to date of the *Assumption* play.]

Particularly significant are the corrections of the numbers

2. Of the numbering

[1] See Note A.

of the pageants in the Proclamation.

assigned to the pageants in the Proclamation (ff. 2v, 3, 3v, 4). These, being in figures, cannot be positively attributed to a certain hand, but the intimate connexion of the alterations with certain features in the arrangement of the subject-matter, and, in one case, the rubrication of the correction are sufficient evidence that they were made by the original scribe. It is, moreover, to be noted that for the first ten pageants and originally for the one now numbered xii, i.e. *Joseph's Return*, the numbers were given in words; for the pageants now x and xi and from the one now numbered xiv to the end Roman figures were used. This change of system indicates a break in the compilation, and the use of the word 'hellenthe' for the *Joseph's Return* pageant connects it with the original scheme. The corrections affect the plays *Mary's Betrothal*, *The Salutation and Conception*, *Joseph's Return*, *The Trial of Joseph and Mary*, *The Birth of Christ*, *The Shepherds*. These were originally numbered: *Mary's Betrothal*, pageants 8 and 9; *The Salutation and Conception*, pageant 10; *Joseph's Return*, pageant 11; *The Trial of Joseph and Mary*, pageant 12; *The Birth*, pageant 13; *The Shepherds*, pageant 14. In the renumbering both the *Betrothal* pageants are numbered 10, *The Salutation and Conception* 11, *Joseph's Return* 12, *The Trial* 14, *The Birth* 15, *The Shepherds* 16. The numbers 8, 9, 13 are thus left for the plays of *The Conception of Mary*, *Mary in the Temple*, and *The Visit to Elizabeth*, which actually fill those places in the compilation. But the additional stanzas necessary for the Proclamation were never written, and the correction of the numbering of the pageants was not carried beyond the *Shepherds* play, with the result that the numbers 15 and 16 are duplicated.

This renumbering affects the group of plays on the Virgin Mary—the first *Contemplacio* group, which has long been recognized by critics as forming a separate whole; and it suggests what there is much evidence to confirm, that the compiler is here combining two series of plays, one consisting of the *Betrothal*, *Salutation and Conception*, and *Return of Joseph*, as announced in the Proclamation, the other consisting of the *Conception of Mary*, *Mary in the Temple*, *Betrothal*, *Salutation and Conception*, and *Visit to Elizabeth*, as announced in the prologue of *Contemplacio*.

This matere here mad · is of þe modyr of mercy
how be joachym and anne · was here concepcion
Sythe offred into þe temple · compiled breffly
than maryed to joseph · and so folwyng þe salutacion
metyng *with* Elizabeth · *and þer with* a conclusyon.

Evidence in support of this is found in certain indications in the arrangement of the plays in the MS., in discrepancies between the *Betrothal* and *Salutation and Conception* plays and the descriptions of them in the Proclamation, and in inconsistencies and incongruities—marks of imperfect amalgamation—in the plays themselves.

That the group of Mary plays from *The Conception of Mary* to *The Trial of Joseph and Mary* is not homogeneous is apparent upon examination. The *Return of Joseph* and the *Trial* plays differ in tone from the others, and they form no part of the Proclamation of *Contemplacio.* A suggestion that the *Return of Joseph* is an interpolation between the *Salutation and Conception* and the *Visit to Elizabeth* plays can be found in the cancelled stage direction at the close of the *Salutation and Conception* play: *And þan Mary seyth,* which would lead on to the *Visit to Elizabeth* play in which Mary is the first speaker, but not to the *Return of Joseph* play, which begins with a dialogue between Joseph and Susanna. In the *Protevangelion of St. James,* and in Tatian's *Diatessaron,* the visit to Elizabeth follows at once on the Salutation and Conception, and Joseph's suspicions are only subsequently aroused, and this order is followed in the *Meditationes* of Bonaventura and in Lydgate's *Lyf of oure Lady.*

Again, though the *Betrothal* play does not appear exactly in the form in which it is described in the Proclamation, where two pageants (and three stanzas) are assigned to it, there are significant discrepancies between it and the *Contemplacio* plays which precede and follow. In the *Mary in the Temple* play the Virgin is represented as left by her parents at the Temple and dwelling there. This also seems to be implied in the speech of *Contemplacio* at the close: *Lo sofreynes here ȝe haue seyn | in þe temple of oure ladyes presentacion | she was nevyr occapyed in thyngys veyn | but Evyr besy in holy ocupacyon.* In Bonaventura's narrative, as in the *Protevangelion* and *Nativity of Mary* gospel, Mary is said to have lived in the Temple from her 3rd to her

14th year. At the opening of the *Betrothal* play, however, Mary is living at home with her parents, and this is the situation conveyed in the words of the Proclamation: *Than Joachym and anne so mylde | þei brynge forthe mary þat blyssyd chylde.* The speech of *Contemplacio* referred to, which appears to introduce the *Betrothal* play, is separated from it in the MS. by a blank folio, and the play begins on a fresh quire (D), which indicates at least a pause—possibly some reconsideration—in the process of compilation. And the name of the bishop Abysakar in the opening stage direction of the *Betrothal* play (fo. 49) represents rather the Abizachar of the *Trial* (fo. 76) and the Abyacar of the Proclamation (fo. 2^{v}) than the Ysakar of the *Conception of Mary* and the *Mary in the Temple* plays.[1]

There are in this play (*The Betrothal*) two interpolated passages in the metre of the preceding *Contemplacio* plays, one of them introducing a character *Minister* not otherwise found in the *Betrothal* play, who appears in the *Mary in the Temple* play. These passages appear to have been fitted in from the rejected *Betrothal* play of the *Contemplacio* series.

On the other hand, the *Salutation and Conception* play is not the play that was in the mind of the writer of the Proclamation, which evidently dealt simply with Gabriel's annunciation. No *parlement of hefne* is mentioned, and special reference is made to the overhearing of Gabriel's speech by the three maidens who dwelt with the Virgin. These maidens are mentioned in the pageant of the *Betrothal* play, and they appear in that and—two of them—in the *Return of Joseph* play, but they do not appear in the *Salutation and Conception* play.

It is not, however, clear that the *Salutation and Conception* play formed part of the original *Contemplacio* series. The *parlement of hefne* is first announced at the end of the second *Contemplacio* play before the *Betrothal* which, as has been seen, does not

[1] Both these names are connected with the Mary story in the pseudo gospels. Abiathar, Abiacar, Isacar, Zacharias appear in different versions of these gospels as the names of the high priest of the Temple during Mary's youth. Isaschar is in one version given as the name of the high priest who rejected Joachym. In some versions a distinction appears to be made between Abiathar sacerdos, who seeks Mary in marriage for his son, and Isacar pontifex, who presides at her Betrothal. This distinction is observed by Lydgate in his *Lyf of oure Lady*.

belong to the series. In *Contemplacio's* first prologue a simple annunciation play only is implied in the line: *Than maryed to joseph • and so folwyng þe salutacion.* This becomes in the later announcement:

> The parlement of hefne sone xal ȝe se
> And how goddys sone come man xal he
> And how þe salutacion aftere xal be.

The *Salutation and Conception* play is again closely connected—in that they are both based on Bonaventura's *Meditationes Vitæ Christi*[1]—with the *Purification* play—a play not included in the Proclamation, manifestly an interpolation where it occurs, and the play to which the date 1468 is affixed. Dr. Greg in his *Bibliographical and Textual Problems of the English Miracle Cycles* calls attention to the use in the *Purification* play and in certain passages in the *Return of Joseph* play of a metre not found elsewhere in the collection as evidence of borrowing from an independent source. This source is fairly certainly that from which the *Salutation and Conception*, though not in this metre,[2] is drawn. The *Salutation and Conception* play as we have it would therefore be the result of a further revision of the amalgamated series of Mary

[1] The connexion of the *Ludus Coventriæ* with Bonaventura's *Meditationes* was early pointed out by Hone in his *Ancient Mysteries*, and Halliwell gives the reference to the *Speculum Vitæ Christi* in his note on the *Salutation and Conception* play. The subject has been treated more recently by Mr. S. B. Hemingway in his *English Nativity Plays*, Yale Studies, No. 38. An examination of the borrowing shows that the writer had recourse to Nicholas Love's version of the Meditationes, *The Mirrour of the Blessed Lyf of Jesus Christ.* Not only do words and phrases used by Love in translating occur, but, in one or two cases, phrases or even sentences from Love, for which there is no equivalent in the original. [See Note B.]

[2] Though the different metres in these plays, as has been seen by most critics, undoubtedly give a clue to the stages of revision or sources of compilation, it must be premised that a different metre need not necessarily imply a different stage or source. Some of the writers engaged on these plays were not without power of changing their numbers to suit their theme (cf. the drowsy measures in the *Magi* and the *Resurrection* scenes, the couplet dialogue in the *Trial* scenes), and the changes from one stanza to another (e. g. in the *Magi* and *Shepherd* plays, or in the second *Passion* group, where the change from short to long octaves seems due to a natural movement towards a more emphatic style) may serve an artistic purpose. And on the other hand the 13-lined prologue stanza does not appear to be always of the same type and may not therefore always represent the same stage or source.

plays whereby the *Salutation and Conception* play based on Bonaventura's work was substituted for both alternatives, that of the Proclamation and that of *Contemplacio's* first prologue.

The *Visit to Elizabeth* belongs clearly to the *Contemplacio* series, but there is a curious confusion at the close of this play which suggests revision. As the play ends at present Joseph and Mary are first said to depart, and later Mary is spoken of as remaining with Elizabeth till the birth of John. An alternative ending is, however, suggested. In place of Joseph's speech beginning *Of ȝour dissese thynkys no greff*, a note at the bottom of the folio offers 'si placet' a continuation of Elizabeth's speech: *come I pray ȝow specialy | I-wys ȝe Are welcome mary | ffor þis comfortabelest comynge good god gramercy*, followed by *Contemplacio* as the name of the next speaker. By the adoption of this version all reference to the departure of Joseph and Mary would disappear. It would seem as if the play originally ended with the first and last stanzas of *Contemplacio's* speech, and that the remainder of this speech (which gives Mary's continued abode with Elizabeth) was intended only to be used as part of the alternative ending. In these additional stanzas the writer appears again to be drawing on Nicholas Love,[1] so that here also we have a trace of this further revision.

On the whole it does not appear rash to conclude that the theory suggested by the renumbering of the pageants is established[2]: namely, that the compiler is here grafting the plays of the 'Contemplacio series'—*The Conception of Mary, Mary in the Temple*, and *The Visit to Elizabeth* with a *Dissponsacion* or *Betrothal* and a *Salutation and Conception* play not used, plays of an ecclesiastical character based at least in part on the *Legenda aurea*,—on to another series of Mary plays described in the Proclamation—*The Betrothal, Return of Joseph, Trial of Joseph and Mary, Birth of Christ*, with a *Salutation and Conception* not used, plays of a simpler and more popular character based directly, as will be shown,[2] on the pseudo gospel of Matthew[3]; and that a later revision or

[1] Cf. Note B.

[2] For additional evidence see p. xxvi and pp. xliv to xlviii.

[3] In both the *Pseudo-Matthaei evangelium de ortu Beatæ Mariæ* and the *Historia de nativitate Mariæ* versions of Mary's story the visit to Elizabeth is omitted.

alteration of purpose led to the substitution of a more elaborate *Salutation and Conception* and to some additions, based on Bonaventura's *Meditationes Vitæ Christi.*

Rubrication, &c.

Red ink is used in this MS. (1) for the numbers of the plays and the numbers of the Commandments in the *Moses* play; (2) for paragraph signs marking stanza divisions, looped lines enclosing names of speakers, lines under stage directions, for a few miscellaneous signs—m̂ (ff. 10ᵛ and 16 marking stage directions, fo. 177ᵛ marking a couplet), ꝯ in quires S and T (ff. 168ᵛ–79 and 183ᵛ) and the *Assumption* play to mark couplets, ℣ on fo. 185ᵛ and in the *Assumption* play to mark Latin versicles; (3) for the division of two lines written in one to save space here and there when a short perpendicular stroke is made over two similar black strokes or over the point ⁄ (4) for corrections; (5) for initial capitals for most of the plays, the Commandments, the Psalms in the *Mary in the Temple* play, and a few scattered instances here and there, and for one marginal genealogy (fo. 37ᵛ), and finally red strokes are given to the initial letters of a selection of the first words in the lines of each stanza.[1]

The rubrication is of a rough character; the initial capitals are ungainly, and the various signs and underlinings, &c., are roughly and carelessly made. An exception is found in the *Doctors' Prologue* standing, as has been shown, by itself, preceded and followed by blank folios. Here each stanza has a shapely red capital, and the lines connecting the rhymes, which elsewhere in the MS. with one exception are black, are here neatly drawn in red. The red initial capitals of the Psalms and versicles in the *Mary in the Temple* and *Visit to Elizabeth* plays (for which liturgical script is used) are also done with more care.

With the exception of the interpolated folios 51, 52, 95, 96, 112, which are unrubricated, all the MS. is rubricated on the same general plan. There are, however, slight variations which seem, like those in the handwriting, to indicate rather that the work was done in different portions at intervals than that another hand was employed. Thus in this, as in the

[1] In this edition all such rubricated letters are printed as capitals as this enables the variation in rubrication, which is more significant than the variation in use of capital forms, to be noted.

writing, quires N, P, Q, R show more carelessness. A more sloping stroke is used for marking the initial letters of the lines, and whereas elsewhere in the MS. three or four of the more important words seem to be picked out, in these quires there are many pages in which every word (*ands* included) has a stroke.

The lines dividing the speeches of the different characters and the lines joining the rhymes (with the exception noted above) are black. In quires N, P, Q, R, and S and T, ¶ in black ink stands before stage directions.

Pointing. One or two instances of the use of the point in the line for emphasis or for clearness, to mark an antithesis, or to separate two similar words, or to mark an internal rhyme—(*hevyn and Erth · foulle and best* (fo. 12), *what ȝe xal sey · whan þat ȝe cum* (fo. 223v), *Ofte ȝoughe is ffowthe with · with sueche vexacion* (fo. 44v), *þat I haue wretyn · wretyn it is* (fo. 183v), *And þat he is now · I knowe wel how* (fo. 188)—can be found in most of the plays; but in certain plays or parts of plays the use of the mid-line point becomes more frequent. In a few passages of a specially declamatory character it is used systematically; such passages are the detached *Doctors' Prologue* (ff. 163, 163v), the *Prologue of the Demon* (which has a specially varied assortment of stops), and that of *John the Baptist*, and the first speech of Annas in the opening *Passion* play (ff. 136–9), Peter's address to the inhabitants of Jerusalem (fo. 143v), and *Contemplacio's Prologue* (fo. 165). In others, though quite irregular, it still occurs with too great frequency to be considered accidental, and the variations correspond with other differences which indicate the use of material from two sources.

The first passage in the collection in which the pointing becomes noticeable is the episode of Lamech in the *Noah* play (ff. 23, 23v, 24).[1] This episode is not mentioned in the Proclamation, and may be accepted as an addition to the *Noah* play. The mid-line point next becomes abundant in the group of Mary plays, and here the variations in its use support the division of this group already suggested on other grounds. Pointing occurs to a significant extent in the *Conception of*

[1] The metre of the latter part of the *Noah* play is the same as the metre of this passage, yet the pointing is confined to the Lamech passage.

Mary, the *Mary in the Temple*, the *Salutation and Conception*, and the *Visit to Elizabeth* plays, i. e. the *Contemplacio* group. It is only used in one passage in the *Betrothal* play (fo. 53), and then for the special purpose of marking internal rhymes. It does not occur in the *Return of Joseph* or *Trial of Joseph and Mary* plays. In the *Birth* play again, as in the *Noah* play, the variation in the pointing corresponds with an addition to the original. The only part of the *Birth* play in which pointing occurs with any regularity is the opening episode of the cherry-tree, which is not mentioned in the Proclamation. Of the remaining plays of the first part of the collection the *Purification* alone shows traces of systematic pointing. This play is not included in the Proclamation, and is an obvious interpolation breaking the continuity of the *Magi* and *Massacre of the Innocents* plays. In this play, as verbal reminiscences prove, the writer is again drawing from Bonaventura's *Meditationes* as in parts of the first *Contemplacio* group. It is in the *Passion* plays that most use is made of the mid-line point. Quire N opens, as has been shown, with seven pages (ff. 136-9) in which all the lines are so divided, and though this regularity is not kept up, the point continues to be fairly abundant throughout this quire. The second *Passion* group begins also with systematic pointing in the speech of *Contemplacio*, and the pointing is better maintained than in the former group, continuing to be a marked feature up to the interpolated folio 184, where it ceases abruptly, not being resumed in the remaining folios of quire T, i. e. not in the *Harrowing*, *Burial*, and *Resurrection* plays. The bearing of this on the question of the composition of the S and T quires has been pointed out.[1]

Numbering of the plays.

Of special significance as throwing light on the problem of the connexion of the Proclamation and the collection is the numbering of the plays. This is done in bold red figures in the right-hand margin—in the case of 29 only the number stands at the top of the page—by the original rubricator. The erasure and rewriting of one of the red-lettered names of the apostles at the opening of the *Pentecost* play (fo. 212) in order to leave room for the number disposes of doubt on this point. The numbers 4, 5, and 7 have the earlier Arabic form, which they changed in the course of the

[1] Cf. p. xv.

fifteenth century, and the earlier form is also used in the numbering of the Commandments. The numbers run from 1 to 42 with a duplication of 10, which is written against the closing speech of *Contemplacio* in the *Mary in the Temple* play and against the opening speech of Abysakar in the *Betrothal* play,[1] and with the omission of 17 and 22.

The latter number belongs to the *Baptism* play, and it is not found in the MS. because the original folio containing the opening has been removed at some period subsequent to the compilation to make room for the interpolated folio 112. The omission of 17 is less easy to account for. It occurs in the following sequence: *The Shepherds* play 16, *The Magi* play 18, *The Purification* play 19, *The Massacre of the Innocents* play 20. An erased 1 before the 2 of the 20 in the number of the *Massacre* play is probably connected with the irregularity. It is possible that the compiler's first intention was to put the *Purification* play—which, as has been said, is clearly an interpolation—between the *Shepherds* and the *Magi* plays, the alternative—which he finally chose—being to break up very awkwardly by its insertion the continuous action of the *Magi* and *Massacre* plays. In Tatian's *Diatessaron* the Purification comes between the visit of the Shepherds and of the Magi and in the pseudo gospel, *de Nativitate Mariæ et de infantia salvatoris*, the Purification precedes the visit of the Magi. The order chosen by the compiler is that followed by Bonaventura and Lydgate. There is now only one blank page between the *Shepherds* and *Magi* plays, but there may originally have been two folios corresponding with those removed to make room for the interpolated folios 95 and 96. The correspondence of watermarks in this quire supports this.

The significance of the numbering, however, chiefly appears in the two parts of the *Passion* plays and the *Resurrection* and *Harrowing* plays. Here the rubricator in his numbering, like the writer of the Proclamation in his enumeration of the pageants, is apparently dividing an action, continuous and for practical acting purposes indivisible, into separate sections. It has been held that the Proclamation was written for a pro-

[1] This is also evidence that the numbering was done in connexion with the compilation.

cessional performance,[1] and the form suggests this, the pageants being described as if they would come forward in order. But a closer examination shows that, in some cases at least, the writer must have had in his mind a group of pageants that could be used simultaneously—that he had in fact a standing and not a processional play in his mind. Pageants 15, 16, and 17 form such a group, and pageants 27 and 29, with the coming and going of Pilate's wife, and pageants 31 and 33 (MS. 23), with the descent and return of *Anima Christi*, could not have been presented successively in a sequence of pageants.[2] But the writer of the Proclamation divides the various episodes between the pageants to form a series more or less corresponding with the usual *Corpus Christi* cycle, and in this the numberer agrees with him. The disregard of the structure of the plays in the numbering is best illustrated by the numbers 28 and 30. No. 28 (fo. 158), dividing the *Betrayal* from the *Last Supper*, is placed after a stage direction which comes in the middle of a speech, so that the speech is actually divided between the two plays. No. 30 (fo. 169v), dividing the *Trial before Caiaphas* from the *Trial before Pilate*, is placed against Caiaphas's words to the messenger after the episode of the Denial of Peter, and is followed by the coming and going of the messenger and the episode of Judas's return of the money to Caiaphas and Annas, all of which forms part of the action of the preceding play.

That the rubricator had the Proclamation in mind in numbering the plays seems likely in itself. It is supported by the partial correction of the Proclamation, and additional evidence may be found in the numbering of the sections 33–5 (ff. 185, 186, 191). The unequal and arbitrary division of the continuous action of the *Burial*, *Harrowing*, and *Resurrection* plays as it is divided by these numbers in the MS. (33, the pescent of *Anima Christi*—six stanzas only; 34, the burial and the setting of the sepulchre guard; 35, the return of *Anima*

[1] Prof. Hardin Craig's article in the *Athenaeum*, Aug. 16, 1913.

[2] The opening of the *Visit of Elizabeth* with the stage direction *Et sic transient circa placeam* and the speech of *Contemplacio* to occupy the time of the supposed journey of 'myles two and ffyfty' marks a play belonging to a standing group. Note also the phrase *locum interludii* in the stage directions (fo. 23).

Christi with the delivered souls, the appearance to the Virgin, the awakening of the soldiers and the making of the compact between them and Pilate, Annas, and Caiaphas) would seem clearly to be made to indicate as far as possible the correspondence of the plays with the Proclamation.

Evidence of the numbering, &c., as to connexion of Proclamation and compilation.

The close connexion between the numbering and the Proclamation has been hitherto disguised by the alterations made in the former by Halliwell in his edition of the MS. By making the second play begin after instead of before the creation of man, by giving the preliminary matter of the two groups of *Passion* plays, which is not included in the Proclamation, as separate plays—*The Council of the Jews* and *King Herod*, by making a separate play of Pilate's wife's dream, and by making a more convenient division of the *Burial*, *Resurrection*, and *Harrowing* plays—moving No. 35 so as to include the setting of the guard with the Harrowing instead of with the Burial—by all these rearrangements Halliwell disturbs the agreement of the numbering and the Proclamation; and this has led to the opinion that there is more discrepancy between the Proclamation and the compilation as we have it than is in fact the case.[1]

Most instances of non-correspondence between the two can be shown to be accompanied in the MS. by indications of interpolation or substitution which suggest an explanation. The first instance is the omission in the Proclamation of any mention of the Lamech episode in the description of the fourth pageant, and it has been shown that this bears signs of being an addition to the original *Noah* play. The next discrepancies are found in connexion with the first group of *Contemplacio* plays (Nos. 8–13), and these have been shown to be due to the amalgamation of a new series of Mary plays with those described in the Proclamation and some revision of this amalgamation; and with this is connected the insertion into the middle of the *Magi-Massacre* play of the *Purification* play, which is also (like Nos. viii, ix, xiii) not included in the Proclamation—a play from the same source, Love's version of the *Meditationes Vitæ Christi*, as parts of the *Contemplacio* group.

[1] Miss Swenson in her *Enquiry into the Composition of the Ludus Coventriæ* makes a strong case for the close connexion of the Proclamation and the *Ludus Coventriæ* as we have it.

The omission from the Proclamation of the preliminary matter of the two *Passion* groups is explained by the fact that the compiler seems here to have abandoned his plan of arranging a cycle of separate plays and to have incorporated two sets of plays on the Passion which lay to his hand as they had been used for separate yearly performances, using apparently copies already made. With regard to the first *Passion* play the series used corresponds fairly with the Proclamation; the incidents of the ass and the foal and of Mary Magdalen and the box of ointment, which are not mentioned in the Proclamation, are, as has been shown, interpolations in the MS. The Proclamation itself is, however, disturbed at this point, and the use, for the first time, of one stanza for two pageants suggests that the compiler is reconsidering his plan. With the second *Passion* play the divergence is marked. Here there are discrepancies which seem to be fundamental between the Proclamation and the plays.

As regards the preceding portion of the MS., what the MS. shows to be due to interpolation or addition or revision is either absent from the Proclamation or added in manifest revision. The writer had, if not a cycle already in existence, at least a collection of plays or groups of plays ready for compilation in his mind. At this point, however, the plays diverge from the Proclamation; the material of the S and T quires belongs to a different version of the Trial and Death than that described in the Proclamation, and it is material that the MS. shows to be due to interpolation—the descent from the Cross of *Anima Christi*—that corresponds with the Proclamation. The Trial, according to the Proclamation, is before Annas, Caiaphas, and Pilate only; no mention is made of Herod, who does not appear in the *Contemplacio* prologue either. The preliminary scene with Herod can be dispensed with, but the subsequent examination before Herod is an integral part of the action; and the pageants 27 and 28 represent a different treatment of the incidents of the Trial from that of the play. According to the Proclamation the three thieves appear in the first trial before Pilate, at which Pilate's wife is also present, and the remorse and death of Judas has a pageant to itself; in the play the three thieves and Pilate's wife appear in the second scene with Pilate, and the Judas incident occupies only eight lines with two stage directions. There is no mention in the Proclamation of the Veronica incident, and the Proclamation puts the Longeus episode before the descent of *Anima Christi*—i. e. at the close of the Crucifixion as in the York, Towneley, and Chester cycles—instead of after it, at the beginning of the Burial scene, as it stands in the play. On the other hand, the material which the evidence of the MS. shows to be interpolation on folios 183v–6—the removal of the Virgin by John and the descent of *Anima Christi*—corresponds with the descriptions of pageants 30 and 31 in the Proclamation. With the setting of the sepulchre guards the divergence comes to an end. This coincides with the beginning of a new quire (U), where the metre changes to that used for the interpolated portions on ff. 183v et seq., and there are, as has been seen, various indications of change of material.

The discrepancies in the case of the *Ascension* and *Pentecost* plays are accompanied by obvious signs of haste and incompleteness in the compilation. The *Ascension* play, to which only the latter part of a stanza is allotted in the Proclamation, breaks off short after the speech of the first of the two angels announced in the Proclamation and given in the stage direction, and the incident of the choice of Matthias, not mentioned in the Proclamation, follows briefly treated. A gap is left in the MS. after the angel's speech, and the speech proposing the choice of Judas's successor is left without any name of speaker. Similarly the *Pentecost* play, a single folio, is manifestly a fragment. The omission of the *Assumption* play from the Proclamation is accounted for by its being, as the MS. shows, an interpolation.

In all cases, therefore, except in the second *Passion* group in the S and T quires, the variation of Proclamation and text is accompanied by some sign of disturbance in the text or, in the case of the first *Passion* group and the *Ascension* play, in the Proclamation.

Conclusions to be drawn from examination of MS. 1. Compilation.

The general evidence of the various features of the MS. that have been examined shows that the collection contains parts or the whole of four separate groups: (1) the composite *Contemplacio* group (viii to xiii); (2) the first *Passion* group (xxvi to xxviii); (3) the second *Passion* group (xxix to xxxii), dovetailed on by means of the *Descent into Hell* (xxxiii), of different style, to a *Burial* play (xxxiv) of similar style, which in its turn is joined (p. 314, beginning of U quire) to a (4) *Resurrection and Harrowing* play connected in style with xxxiii and forming a group with the *Three Maries* and, as it stands in the compilation, with the *Mary Magdalen* play. An examination of the text gives two[1] more groups showing that (5) the first three plays (*Creation*, *Fall*, and *Cain and Abel*) and (6) the *Visit of the Magi* and *Massacre of the Innocents* and *Death of Herod* (xviii and xx) form respectively continuous sets, diction and metre connecting also the *Shepherds* play with the latter suite.[2] There are also two interpolated separate plays: *The Purification* and *The Assumption.*

The evidence as to the composition of the series to be drawn from the characteristics of the MS. is complicated and often ambiguous, but the following points emerge:

1. MS. Vesp. D. viii is the compiler's book, not a transcript of another MS.
2. It contains a collection of plays made according to a plan which was subject to alteration as it proceeded.
3. Some of the plays and groups of plays had had a separate existence, having been acted as separate plays or groups.
4. One portion of the MS. certainly, and probably two, quires N, P, Q, R, and quires S, T, have also had a separate existence.

In support of 3, besides the evidence already adduced, may be noted the preservation of their distinctive character by the groups in the series, by the *Contemplacio* group, and especially by the two *Passion* groups the stage directions of which are conspicuously different from any others in the collection; the variety of the headings of the single plays—*introitus* (ff. 20v, 25v, 31), *modo de* . . . (ff. 106, 212), *hic incipit* . . . (ff. 127v, 201, 210, 223v), *hic intrabit pagetum de* . . . (fo. 75); the conclusion of the *Disputation* play addressed to *Att þat hath herd þis consummacion of* | *þis pagent* as to the audience of a separate performance, and to a less extent the sermon-like conclusions of the *Temptation*

[1] Or three—the *Baptism* and *Temptation* plays might be taken as continuous.

[2] The general use of stock alliterative phrases makes the evidence to be drawn from diction unreliable, but the following phrases which occur both in the *Shepherds* and *Magi* plays are sufficiently individual to be allowed to count: 'his (*þat*) bryght blood'; 'in (by) a bestys bynne'; 'Heyl blome on bedde'; and 'To *þe* blosme upon his bedde'.

and the *Woman taken in Adultery* plays; and the fact that—as the modernizing revision of certain plays, the added stage directions (e. g. p. 327), and the worn condition of the *Magi* portion of the MS. show—certain plays continued to be so acted.

5. The compiler had command of other versions of plays or groups of plays from which he drew.

The evidence for this is mainly to be deduced from the variation of the plays from the description of them in the Proclamation, chiefly in the first *Contemplacio* group and in the two *Passion* groups, but it is supported by the fact that the later users of the MS. obviously had access to such other versions (cf. the references in the later hand in the *Harrowing* play to *anima latronis*, fo. 185v, to extra speeches of the devil and of *anima caym*), and by the presence in the MS. of the detached *Doctors' Prologue* written on a blank folio at the end of quire R. This introduces characters not found together in any play or group of plays in this collection—the eleven apostles, John the Baptist, and St. Paul, who only appears in the *Assumption* play. It would seem to be part of another group in the possession of the performers of the *Passion* groups, copied for convenience on a blank sheet of the MS. of the latter. So the allusion to the 1st and 2nd prophets in notes (pp. 240–1) appears to refer to some other combination of plays than that of the text.

The evidence of the MS. supports the view that a compiler is putting together parts to make a whole rather than the view suggested by Dr. Foster (*Northern Passion*, E.E.T.S. 147, p. 99) that a reviser has separated a whole into groups, though he drew apparently on a cycle—or the remains of a cycle—of plays in thirteen-lined stanzas as well as on single plays and on groups of plays. To unravel the tissue of compilation and revision in these plays demands a full study of literary and linguistic characteristics. The examination of the MS. affords no conclusive evidence on such questions, but it gives much to support the theory that the MS. represents a selection from the repertory of a body of ecclesiastical actors. As Dr. Smith says: 'Videntur olim coram populo sive ad instruendum sive ad placendum *a Fratribus mendicantibus* repraesentata.' Nor does the evidence from the MS. throw clear light on the question of the identity of the compiler with any of the writers or revisers of the plays, but some inferences can be drawn.

The relation of the plays to the Proclamation shows, as has been said, that the Proclamation was composed in its present form in close connexion with the compiling of this series. The relation is particularly interesting at the point of divergence in the second *Passion* group. Here the interpolated portion (ff. 184, 185, 186) in the text corresponds with the Proclamation, from which the rest of the text diverges, and that with regard to a feature—the division into two scenes of the Harrowing—that is peculiar to this treatment of the theme. The inference suggests itself that the compiler was himself the writer of this link passage, which again is connected with the latter part of the *Resurrection* play and through that, by the evidence of metre and more elusive evidence of diction, with the *Magi* plays. The conclusion to be drawn from this agrees with that of Mr. Greg (*Problems of English Miracle Cycles*) in that it associates the composition of the *rime couée* portions of the cycle with the process of compilation. Dr. Greg postulates a further overworking by the writer of the *Contemplacio* prologues and possibly of the long octave plays and passages, who would therefore be the actual compiler of the Vesp. D. viii MS. But there is evidence[1] that the long octave *Contemplacio* plays underwent a further revision by a writer drawing on Bonaventura's *Meditationes Vitæ Christi*, to whom, as well as the *Salutation and Conception*, the *Purification* play[2] must be attributed. Now it would appear unlikely, had this reviser been the compiler, that the *Purification* play should not be mentioned in the Proclamation, and that the description of the *Salutation and Conception* in the Proclamation should not have been made to refer to the compiler's own work. The theory that the writer of the *rime couée* parts, with which the corresponding passages in the Proclamation agree closely, is the compiler seems to present the stronger case. This writer draws also on Bonaventura,[3] and both revisions are probably connected with the compilation. 2. Compiler.

This MS. has known the turn of fortune described by Robert Hegge, its first recorded owner, when after the invention of Marginalia.

[1] Cf. pp. xxi to xxv; xlv, xlvii.

[2] It is to this play that the date 1468 is attached.

[3] Cf. p. xlix.

printing 'old MSS. were stright bequeath'd to the Moths: and pigeons and Jack daws became the only students in church libraries'[1]; and many of its blank pages have been used for idle scribbling—attempts at copying the MS. writing, signatures, stray phrases, mostly in sixteenth-century hands. It would appear that it fell early into irreverent schoolboy hands, for some of the scribbled copying of lines of the text seems to be rather rough contemporary work than later imitation of an older script. The first or last line of a folio is often imitated, and in one instance a passage of some length has been carelessly and roughly copied on the blank page opposite (fo. 201). A jumble of ill-formed Greek and fancy letters on fo. 119v with what appear to be notes on a fraction sum on the opposite folio suggests the hand of a schoolboy. The names that occur among these scribblings are Wyll*iam* Dere (91v, 136), Polerd (91v), Hollond, Joh*an* & Hary (151v, 152v, 153v, 155v), H Kinge the yownger (111v), John Hasycham (91v), and on the same page (91v) 'John Taylphott of parish Bedonson' with the motto 'wee that will not when we paie [*sic*] when we would we shall find (? or saie) nay'. The oddness of the personal name throws doubt on that of the parish, and no record of a parish of Bedonson has been found.

By a coincidence that is possibly nothing more the names William Kinge the younger his booke 1656, John King his brother and John Holland of Brabant occur among the scribblings in the Chester plays (MS. Add. 10305, ff. 55 and 111); the motto quoted above is also found (fo. 124) in a four-lined form, of which, however, unfortunately the first line is missing.

Of a different character are the two signatures of R. Hegge, Dunelmensis, the title of the plays, and some annotations of the original scribe. The first signature comes on fo. 10 above the opening of the first play. It consisted of the full name Robert Hegge, Dunelmensis, written in large Roman hand, but, having been cut away with the margin, the Christian name is now indecipherable. The second, on fo. 164, the outside sheet of the S and T quires, is more elaborate. At the top of the page in Roman type is written: 'In no*mjne* Dei. Amen'; in the middle of the page, in a slender cursive hand, 'ego R. H.

[1] *The Legend of St. Cuthbert*, by R. Hegge.

Dunelmensis possideo'; and immediately beneath, οὐ κτῆσις[1] ἀλλὰ χρῆσις.

The title 'The plaie called Corpus Christi' in a small Elizabethan hand stands at the top of fo. 1 above the speech of the first vexillator. Mr. Hemingway in his *English Nativity Plays* speaks of this title as being in Hegge's handwriting, but this is not so. It is writing of an earlier and altogether different type.

The notes forming part of the original MS. consist of three genealogies, of Adam to Noah (ff. 16ᵛ et seq.), of Noah to Abraham (ff. 21 et seq.), of the Virgin Mary (fo. 37); the measurements of Noah's ark (fo. 24); a list of the five Annas of scriptural importance (fo. 37ᵛ) and three dates of the ecclesiastical calendar (fo. 74ᵛ). The genealogies are in liturgical script, the note on the dates and on the Annas in smaller bookhand, and the genealogies are handsomely rubricated, one part of that of Mary being entirely in red. These marginal additions give support to the conclusion to be drawn from the general characteristics of the compilation, that it is of ecclesiastical and not of civic origin.

History of the MS. and its title.

The MS. forms part of the Cottonian collection. It is not included in the 1621 catalogue of the collection (MS. Harl. 6018), and is supposed to have been acquired in 1629 by Sir Robert Bruce Cotton's first librarian, Richard James, on the death in that year at Oxford of Robert Hegge, a member of James's own college, Corpus Christi, in whose possession it had been. Richard James became Sir Robert Cotton's librarian before 1628[2] and a number of letters exist (MS. Cott. Julius C. III, ff. 212, 214, 217, 219) written by him to his patron from Oxford —unfortunately undated as to the year but clearly belonging to this period—referring to transactions in books, coins, and other rarities. The note on the fly-leaf of the MS. giving the contents is in James's handwriting. The MS. is included in the 1696 catalogue of the Cotton collection drawn up by Dr. Thomas Smith.

R. James.

The problem of the *Ludus Coventriæ* begins with James's note on the fly-leaf: 'Elenchus contentorum in hoc codice [Vespasian D. viii, added in a later hand] Contenta novi testamenti scenicè expressa et actitata olim per monachos sive fratres

[1] The word appears to have been first written κησις, and the τ to have been inserted. The οὐ is written ȣ, and the accents on κτησι and χρησις are omitted.

[2] *Dict. of Nat. Biog.*

mendicantes · vulgò dicitur hic liber Ludus Coventriæ · sive ludus corporis Chr*isti* · scribitur metris Anglicanis.'

The description of the contents is inaccurate in omitting the Old Testament plays,[1] and the connexion of the collection with Coventry is not supported by any evidence.

T. Smith. In the description of the MS. in the catalogue drawn up by Dr. Thomas Smith in 1696 the reference to Coventry is omitted, and account is taken of the Old Testament material: 'Vespasianus D. viii. A collection of plays in old English metre, i.e. Dramata sacra in quibus exhibentur historiae veteris et N. Testamenti, introductis quasi in scenam personis illic memoratis quas secum invicem colloquentes pro ingenio fingit Poeta. Videntur olim coram populo sive ad instruendum sive ad placendum a Fratribus mendicantibus repraesentata.'

W. Dugdale. James's inscription had, however, in the meantime been accepted by Dugdale, and the passage in his *Antiquities of Warwickshire Illustrated*, in which he describes the Corpus Christi pageants of the Grey Friars of Coventry, referring to this MS. as giving the text of the performance, became the authority for many subsequent writers on the subject of these plays.[2] It runs as follows: 'Before the suppression of the Monasteries their City [i.e. Coventry] was very famous for the pageants that were play'd therein, upon Corpus Christi day; which occasioning very great confluence of people thither from far and near was of no small benefit thereto; which pageants being acted with mighty state and reverence by the friers of this house [i.e. the Grey Friars] had Theaters for several scenes, very large and high, placed upon wheels and drawn to all the eminent parts of the City for the better advantage of Spectators: And contain'd the story of the New Testament, composed into old English Rithme, as appeareth by an antient MS. p intituled Ludus Corporis Christi or Ludus Coventriæ [a note in the margin here gives the reference: "p In bibl. Cotton. sub effigie Vesp.

[1] Is it possible that this mistake has any connexion with the fact that R. Hegge's second signature is found on a blank folio preceding the second *Passion* groûp which shows signs of having been at some time an outside leaf?

[2] Especially perhaps after the publication in 1722 of Stevens's additional volumes to Dugdale's *Monasticon Anglicanum* in which he prints the first five plays of Vesp. D. viii. Cf. Thomas Pennant, *Journey from Chester to London*, 1782, and 'Q' in *Gentleman's Magazine*, Feb. 1784.

D. 9" (so apparently by a slip[1] for viii)]. I have been told by some old people who in their younger days were eye-witnesses of these pageants so acted that the yearly confluence of people to see that show was extraordinary great.'[2]

The fact that Dugdale follows James in omitting the Old Testament plays shows that he had not examined the MS. Vesp. D. viii for himself. He appears to be accepting the description of the fly-leaf from which the words '*composed into old English Rithme*' seem to be translated. Thomas Sharp in his *Dissertation on the Coventry Mysteries*, 1825, was the first to point out that the Coventry civic cycle was distinct from the *Ludus Coventriæ* preserved in MS. Cott. Vesp. D. viii. The two surviving plays of this cycle and the information concerning it to be derived from Sharp's (and Halliwell's) extracts from MS. sources destroyed in the Birmingham Library fire in 1879, and from other Coventry civic records still preserved, are accessible in the E.E.T.S. edition of Prof. Hardin Craig, and it is now recognized that there is no connexion between *The plaie called Corpus Christi* and the Coventry civic cycle; and the comparison, where it is possible, notably between the two *Shepherds* plays or between the two *Disputation* plays, serves to bring out strongly the more ecclesiastical character of the present collection.

Thomas Sharp.

Coventry civic cycle.

Thomas Sharp accepts Dugdale's attribution of the *Ludus Coventriæ* to the Grey Friars of Coventry on the general ground of his local knowledge. Later scholars have pointed out that the *old people* to whose reminiscences Dugdale refers—speaking evidently of information he has gathered as an adult, not of stories heard as a child—must have been recalling the civic pageants which were only finally 'laid down' in 1580,[3] and not the performances of the Grey Friars whose house was dissolved in 1538. Sharp himself speaks somewhat dubiously of the evidence for the acting of the Grey Friars: '*a solitary mention in one MS. not older than the beginning of Charles I's reign of Henry VII's visit to the city in 1492 "to see the Plays acted by the Grey Friers"*'[4]; and this piece of evidence has lately been

Acting 'by the Grey friers,' Coventry.

[1] Unless this is again connected with the separate existence of the *Passion* plays.

[2] The *Antiquities of Warwickshire Illustrated*, &c., by William Dugdale, 1656.

[3] Sharp, *Dissertation*, p. 12. [4] Ibid., p. 5.

weakened if not destroyed by the discovery that in a compilation of the MS. annals of Coventry (Harl. 6388) made by Humphrey Wanley—himself a Coventry man—and said by him to be 'taken out' of a collection of eight MSS., the entry given by Sharp under 1492 is found under 1493 in the following form: '*The King and Queen came to see the playes at ye greyfriers and much commended them*', where the reference is clearly to the place and not the actors. Yet it might perhaps still be questioned whether Dugdale, a careful antiquary, who would have access to the MSS. used by Wanley, having also a personal connexion with Coventry, and living nearly three-quarters of a century nearer the time, could have been misled, by an ambiguous entry and confused local tradition, into the invention of the acting of the Grey Friars.

In any case his association of the MS. Vesp. D. viii with Coventry seems to have been secondhand—a repetition of the statement of James's note.

In the difficulty of explaining the note Halliwell has recourse to the theory of a lost colophon which supplied James with the information. The suggestion might be hazarded that James in the sentence '*vulgò dicitur hic liber Ludus Coventriæ, sive ludus corporis Christi*' is using the words *Ludus Coventriæ* generically, taking the name of the best-known example as the name of a type of dramatic performance. The words *vulgo dicitur* suggest this. It may be gathered from the anecdote in the 'Hundred Merry Tales', quoted by Halliwell, of the Warwickshire village priest who referred his parishioners for confirmation of his doctrine to the Coventry Corpus Christi play, that Coventry was the accepted background for a story concerning the Corpus Christi pageants; Heywood's allusion in the *Four P's* proves that Coventry was recognized as the local habitation of the Mystery devil; and there is evidence that the name Corpus Christi had come to be used for the collective mysteries apart from the occasion of their performance. Weaver, in a passage[1] also quoted by Halliwell, says of a play the subject of which was 'the sacred scriptures from the creation of the world', 'They call this Corpus Christi play in my country'. The title written in a sixteenth-century

[1] John Weaver, *Funeral Monuments*, 1631, p. 405.

hand on the first page of the present collection, 'The plaie called Corpus Christi', is apparently used in this wider sense, and James in his note on the fly-leaf may be expanding this title: 'vulgò dicitur hic liber Ludus Coventriæ sive ludus corporis Christi', i.e. 'This book is commonly called the Coventry or Corpus Christi play'.

Printing of the MS. J. Stevens.

The printing of plays from MS. Vesp. D. viii began in 1722 in John Stevens's additional volumes to Dodsworth's and Dugdale's *Monasticon Anglicanum* (vol. i, pp. 139–53 under Grey Friars of Coventry). Stevens prints the Proclamation and the first five plays, using Gothic type, translating the stage directions, and including the marginal genealogies.

W. Hone.

A century later in 1823 Hone in his *Ancient Mysteries described* gives a summary, illustrated by full quotations, of plays viii to xv (incl.) with parallel passages from the Apocryphal Gospels and various illustrations including the 'parliament of heaven' from an English translation (not Love's) of the *Speculum Vitæ Christi*, and the *Cherry-tree* carol.

J. Payne Collier.

In 1836 J. Payne Collier included *The Marriage of the Virgin Now first printed from MS. Cott. Vesp. D. viii* in *Five Miracle Plays or Scriptural Dramas* privately printed.

W. Marriott.

In 1838 William Marriott, Ph.D., in his *A Collection of English Miracle Plays or Mysteries containing ten dramas from the Chester, Coventry, and Towneley series with two of later date*, published at Basle, included two of the *Ludus Coventriæ* plays, *Joseph's Jealousy* and *The Trial of Joseph and Mary*, with one of the Coventry civic cycle, *The Pageant of the Company of Shearmen and Tailors*.

J. O. Halliwell.

In 1841 the whole MS. was edited by J. O. Halliwell for the Shakespeare Society, with introduction, notes, and glossary. Various plays have since appeared in representative collections.

A. W. Pollard; J. M. Manly; S. B. Hemingway.

The opening of the *Salutation and Conception* play is included in Mr. A. W. Pollard's *English Miracle Plays* (1890); this play and *Noah and Lamech* are included in Mr. J. M. Manly's *Specimens of Pre-Shakespearian Drama* (1900); and Mr. S. B. Hemingway in *English Nativity Plays* (1909) gives plays 11, 12, 13, 15, and 16 from this series.

W. W. Greg.

In 1915 the *Assumption* play was edited with an exact reproduction of the features of the MS. by Dr. Greg.

The present edition.

Such an exact reproduction has not been attempted in the present edition. Stops have been added at the end of stanzas or speeches, though otherwise the text is left unpunctuated as in the MS., the stops which occur, and are reproduced, being metrical except in the case of a few stage directions. A few medial capital letters have been omitted, a few words run together by the scribe have been separated, a few disconnected syllables have been hyphened, where the exact reproduction appeared disproportionately disturbing to the reader. It has not been thought necessary to record all the scribe's slips of the pen. In any special case a note has been appended. *I* and *J*, used indifferently in the MS., have been distinguished according to modern usage and the capital form substituted for the small in the case of the personal pronoun; *z* and *ȝ* have also been distinguished. The lines between the speeches and the lines connecting the rhymes have been omitted. In the MS. the tail-verses of the thirteen-lined stanzas and of the *rime couée* stanzas are sometimes written in the margin, outside the rhyme brackets of the rest of the stanza; such lines are printed as short lines in the stanza. Lines written continuously for reasons of space have been arranged in stanzas, with a note recording the change (cf. p. 120, ll. 125 et seq.); otherwise the irregularities of the MS. have been preserved.

Some of the contraction marks used have clearly lost their significance. This is notably the case with *n͡*. It is the normal form of *n* for the scribes of fo. 51 and of folios 95, 96, 112. The scribe of the MS. writes *tron͡e* (fo. 9), *gon͡e* (fo. 69), *somown͡e* (fo. 74ᵛ), and the carelessness with which the stroke is added or not to a series of rhyme-words suggests that it had become a mere flourish. This is also true of *d̄*, which occurs here and there as it were accidentally, though the form *virid̄* for *viridi* (fo. 25, *cum ramo virid̄*) shows that it retained its significance at need. The scribes of fo. 51 and of folios 95, 96, 112 use *d̄* for all final *d*'s, and write *de* in full. The case is less clear with regard to *t̄t*, though *at̄te* and *hat̄te* are found (fo. 81). On fo. 33 (p. 55, l. 132) a final *e* seems to have been erased in *kyt̄t*; *knowlac̄he* (fo. 110ᵛ) is found, but also *sc̄h* for *sche* (fo. 78ᵛ and elsewhere). These signs of doubtful significance, *d̄*, *ḡ*, *h̄*, *t̄t*, *m͡*, *n͡*, have been retained in the text.

The sign ⁀, not reproduced in the text, is abundantly used in the usual ways: over an *n* after *u* or *w*, e.g. *grou⁀dyd*, *geaw⁀t*; over *u* or *w* in the combination *ous*, *ows*, e.g. *gracyous*, *leprow⁀s*; over *u* or *n* occasionally in other cases, e.g. *secund⁀s*, *opy⁀d*; as sign of a contracted *n* or *m*, e.g. *þ⁀g*, *h⁀*; exceptional instances of this use are *⁀d* (*and*), fo. 193, and on the same folio *⁀resyn* (*am resyn*); for other contractions in *plentev⁀s* (*plentevous*), fo. 137, *c⁀nawnt* (*counawnt*), fo. 170, *do⁀* (*done*), fo. 69ᵛ, p. 114, l. 165, and *do⁀* (*down*), fo. 126. Against *a⁀*, fo. 134, *non* has been written as a correction in the margin; *⁀*, fo. 63ᵛ, p. 105, l. 244, stands for *no*. Occasionally this sign serves no purpose, as in fo. 27ᵛ, p. 47, l. 129, *hev⁀ne*; fo. 212ᵛ *crepp⁀*.

The contraction ꝯ for the plural has been expanded as *ys*, as in the fully written plural nouns this form predominates. *Es* and *is* occur, though less frequently. On fo. 202ᵛ *pilgrimes* and *pilgrimys* are found in following lines, elsewhere *sowles* and *sowlys*, &c. On the other hand, on fo. 79, p. 131, l. 234, *legges* has apparently been corrected to *leggys*; and on fo. 209 *eretykis* has been corrected to *eretykys*. There is one instance of the use of the ꝯ for *com* on fo. 112—com*pany*, p. 188, l. 19—and it is possible that the unintelligible *hese juge*, p. 116, l. 34, may be due to the scribe's misreading of this contraction in *consider-ynge*; the corresponding Latin is: *considerans sui senectutem*. Other contractions occurring once or exceptionally are *mayde* (*mayden*), fo. 90ᵛ, p. 149, l. 102; *hꝯ* (*his*) used by the scribe of folios 95, 96, 112; *wepon⁀*, fo. 160 (stage direction) (? *weponys*); *bᵗ* (*but*), fo. 162. The recognized contractions for *er*, *ur*, *us*, *n*, *m*; *i* in *ion*; final *e* after *r*, *p*, *ȝ*, *t*; *p* + *ra*, *re*, &c., and those in the Latin passages, have been expanded in italics. Overwritten as well as omitted letters are italized, e.g. *þⁱ* = *þi*, *þᵘ* = *þou*, *þᵉ* = *þe*.[1]

Notes on sources and literary relations.[2]

The *Ludus Coventriæ* differs from the other Mystery cycles in having, in its New Testament plays at least, a closer dependence on known literary sources. Its borrowings have been

[1] By inadvertence the two forms '*þat*' and '*þat*' have been used to represent the MS. form '*þᵗ*'. In a few instances also '*þer*', '*our*', and '*ȝour*' have been given for '*þer*', '*our*', and '*ȝour*'.

[2] I have not been able to consult *Sprache und Heimath des sogenannten Ludus Coventriæ* by Dr. Max Kramer (1892); nor *Die Quellen des sogenannten Ludus Coventriæ* by E. Falke (1908).

investigated by various writers. Halliwell gives references to the Apocryphal Gospels, to Lydgate's *Lyf of oure Lady*, and his anagrams on the name of the Virgin (cf. *Ludus Coventriæ*, fo. 47, and the *Minor Poems* of J. Lydgate, ed. McCracken, p. 303), to the *Speculum* (or *Meditationes*) *Vitæ Christi* of Bonaventura, and to the thirteenth-century *Harrowing of Hell* (Harl. 2253) (cf. *Ludus Coventriæ*, fo. 192^{v}, p. 320, ll. 1416 et seq., and lines 43, 44, 31, 55 of the *Harrowing*, *Altenglische Dichtungen*, ed. Böddeker, p. 271). Hone in *Ancient Mysteries described* gives in more detail the parallels between the Mary plays and the Apocryphal Gospels, and quotes from an English translation of the *Speculum Vitæ Christi* to illustrate the *Parlement of Hefne*. Miss Hope Traver in *The Four Daughters of God* (Bryn Mawr Monographs, No. 6), has discussed the relation of the *Parlement of Hefne* to other English versions of the theme. The use of the *Legenda aurea* for the *Assumption* play has been shown by Mr. Gayley in his *Plays of our Forefathers*, and by Dr. Greg, who in his edition of the *Assumption* prints the chapter from the *Legenda* used by the writer. The most recent discovery has been that of Dr. Foster, who has demonstrated the indebtedness of the writer of the *Passion* plays to the *Northern Passion*. The coincidence of the names of three out of the four Knights of Pilate in the *Ludus Coventriæ*, fo. 189, and the poem on the Resurrection, MS. Ashmole 61 (fo. 138), has been pointed out by E. Falke.

So much of the material used by the writer of Mysteries was common to different possible sources that it is difficult to determine which was the one actually used. Thus the chapter in the *Legenda aurea*: '*De Nativitate Beatæ Mariæ virginis*' (ed. Graesse, 1846, ch. 131) is in great part a transcription of the Apocryphal Gospel *De Nativitate S. Mariæ*, and the material that is versified in the *Conception of Mary* play (the description of Joachim, ll. 25–8; his rejection, ll. 76–80; the angel's speech, ll. 149–74, with the exception of the reference to *joys fyff*) might equally well have been taken from the one or the other. The connexion of the fifteen steps of the Temple with the fifteen Psalms of degree (*Mary in the Temple*, ll. 84 et seq.) is common to the two sources, and found also in both is the original of the interpolated passage in the *Betrothal of*

Mary on fo. 48v, p. 82, ll. 92 et seq. (cf. *Tunc anxiatus est pontifex eo quod neque contra scripturam quae dicit: vovite et reddite, votum infringendum putaret nec morem genti insuetum introducere auderet*); both give the answer to *Vox*; cf. *The Betrothal*, p. 89, l. 230.

It can, however, be shown that the *Legenda aurea* was the actual authority used for the *Conception of Mary* at least. The marginal genealogies on the folios preceding the *Conception of Mary* are taken from chapter 131 of the *Legenda aurea*; the phrase *regale sacerdocium* (cf. the *Conception of Mary*, l. 15) occurs in it and not in the gospel *De Nativitate*; and the words used of the rejection of Joachim's offering in the *Legenda aurea, cum indignatione nimia repulit*, seem rather to be echoed in '*with grett indygnacion þin offeryng I refuse*', than the words in the gospel: *despexit eum et munera eius sprevit.*

The same difficulty occurs in attempting to distinguish the claims of Bonaventura's *Meditationes* and of Lydgate's *Lyf of oure Lady* to be the original of parts of these plays, though the writer's use of Love's translation of the former in certain cases (cf. Note B, p. lviii) gives a clue, and the *Salutation and Conception* and the *Purification* plays may safely be attributed to the influence of the *Meditationes*. To this is due the added passage in a different metre in *Joseph's Return* to which Dr. Greg draws attention (*Bibliographical and Textual Problems of the English Miracle Cycles*, p. 129). The altered tone is consonant with the refined treatment of the theme in the *Meditationes*, and two slight verbal coincidences indicate this source. Cf. '*For on the tone side he sawh hir lyf so holy and no tokene of synne in hir*' and '*I knew never with here so god me spede | tokyn of thynge in word or dede | þat towchyd velany | nevyr þe les*', et seq.; and '*this tribulacion and this desese*' and '*for vnknowlage he is desesyd*' in the same context. From the *Meditationes* comes also the interview between the risen Lord and his Mother. That our Lord appeared first of all to the Virgin is a theory discussed in the *De Resurrectione* chapter of the *Legenda aurea*. Bonaventura describes the appearance: *Sodeynly oure lord Jesu came and aperede to her, and in alther whitest clothes with a glad and louely chere gretynge hir on side half in these wordes: Salue sancta parens that is to say Haile holy moder. . . . My dere moder I am.*

Ego sum: resurrexi et adhuc tecum sum I have uprisen and loo ȝit I am with thee . . . and deth and sorwe and alle peynes and angwische I haue ouercome. . . . And so thei bothe louely and lykyngly talkynge togidre maden a grete joyful feste. This seems to be the original of the passage *Salue sancta parens: my modyr dere*, fo. 192v, and it is recalled also by the words in the Proclamation, l. 431, *with suche cher* and *comforth his modyr he doth indew* | *þat joy it is to here* · *þer speche for to devyse.*

Less clear is the source of the material used in the *Mary in the Temple* and *Visit to Elizabeth* plays which is found alike in the *Meditationes*, the *Lyf of oure Lady*, and in part in the *Legenda aurea* and the gospel *De Nativitate.*

The name of Lydgate was early associated with the *Ludus Coventriæ.* In Bishop Tanner's *Bibliotheca Britannico-Hibernica* (1748) these plays are doubtfully identified with the 'pageants' attributed to Lydgate.[1] Halliwell, as has been said, draws attention to the likeness of the passage on the Virgin's name to poems of Lydgate ; the hymn sung by the shepherds—*Stella celi extirpauit*, fo. 90—is one that Lydgate translated (*Minor Poems*, ed. McCracken, p. 294). There is also in the *Christ with the Doctors* play an echo of a verse in the *Lyf of oure Lady*, which might however be dismissed as due to the use of a commonplace of theological illustration: cf. *L. C.*, p. 181, ll. 97 et seq., and *Lyf of oure Lady*: ed. 1531, Ch. xix, *Ryght as the sonne perissheth thorowe the glasse* | *Thorowe the cristall, berall or spectacle* | *Withoute harme, right so by myracle* | *Into her closet the father's sapyence* | *Entred is withouten violence.*

It seems at first sight in favour of the *Lyf of oure Lady* as a source for *Mary in the Temple* that it includes the story of the fifteen steps not found in the *Meditationes*, *L. C.*, p. 74, and the seven petitions, *L. C.*, p. 79, not found in the *Legenda aurea* or the gospel *De Nativitate.* The imitation of Lydgate's manner in

[1] By a fatality attaching to references to Vesp. D. viii his description contains an error. He apparently noticed the change of writing with the *Assumption* play and did not look further:

'A procession of pageants from the creation. In MS. Cotton. Vespas. D. viii. Scripto circa Lydgati tempora sunt collectanea (in 212 paginis in folio) 40 pageants sive ludorum incipiendo a creatione et finiendo tempore descensus Spiritus S. Jamesius existimat hunc fuisse Ludum C.C.C. (sic) sive Ludum Coventriae'.

the anagram (p. 80) favours this view. Against all this there is, however, one piece of evidence that is conclusively in favour of the *Meditationes* as against the *Lyf*. In this work the three precepts observed by Mary are distinct from the seven petitions, though they correspond with the first three. Lydgate gives only the seven petitions. The writer of *Mary in the Temple* follows Bonaventura, and gives the three precepts in the form of the charge of *Episcopus*, p. 77, ll. 152 et seq. Moreover, lines 168, 169, which correspond with the third precept: *Thu xalte hate no thynge but þe devyl* and *synne* | *god byddyth the lovyn þi bodyly enmy* appear to have some reference to the misleading form in which the third precept is given in the *Meditationes*: *and the thridde is*: *Thou schalt hate thyn enemy* (*Habebis odio inimicum tuum*) only explained later in the text: *his enemyes that is to saie vices* and *synnes*. The order of the seven petitions in *Mary in the Temple* differs from that in the *Meditationes* and the *Lyf*, but the change—putting the most significant last instead of fifth—is one that suggests itself; the phraseology of the passage is closer to Love's than to Lydgate's.

The opening speech of *Contemplacio* in the *Visit to Elizabeth* is taken verbally from the *Legenda aurea*, Ch. 86, 'De Nativitate S. Iohannis baptistae': cf. *L. C.*, p. 116, ll. 23–7, and *David enim rex . . . volens cultum dei ampliare xxiv summos sacerdotes instituit quorum tamen unus maior erat qui princeps sacerdotum dicebatur*. The phrase *The plage of dompnesse*, l. 35, corresponds with *plaga taciturnitatis*. The close of this play has been shown (Note B) to be taken from the *Meditationes*. The matter of the whole play is found there as well as in the *Legenda aurea* and the *Lyf of oure Lady*.

For the *Betrothal of Mary* the *Meditationes* offer no material. The story of the flowering rod and the dove is found in the *Legenda aurea* and in the gospel *De Nativitate S. Mariæ*; in the *Pseudo-Matthew* and the *Lyf of oure Lady* the dove only appears on the rod. In these also the high priest is represented as neglecting Joseph's little rod, whereas in the former as in the play Joseph himself is the delinquent and endeavours to avoid offering the rod. On the other hand, the names of the Virgin's maidens (five in each case, seven unnamed in *De Nativitate*) are given only in the *Pseudo-Matthew* and the *Lyf of oure Lady*, and the forms used for the three in the play, Rebecca, Sephora, Susanna, only in the *Pseudo-Matthew*, the second name in Lydgate being Scaphea. Here, therefore, the *Legenda* and *Pseudo-Matthew* appear to have been used.

The use of the *Pseudo-Matthew* Gospel is clearest in the *Birth* play, where the incident of the midwives follows this original very closely. Cf. *Ludus Coventriæ*, p. 141, stage direction and ll. 181–4 and 265–76 and: *Audiens autem haec Maria subrisit. Cui Ioseph dixit, Noli subridere sed cauta esto ne forte indigeas medicina*; and *Domine tu nosti quia semper te timui et omnes pauperes sine retributione acceptionis curavi, de vidua et orphano nihil accepi et inopem vacuum a me ire nunquam dimisi. Et ecce misera facta sum propter incredulitatem meam quia ausa fui temptare virginem suam.* The suggestion of the cherry-tree incident comes also from this gospel where a similar story is told of a palm-tree passed on the journey into Egypt. *The Trial of Joseph and Mary* follows less closely this gospel, which does not include the incident of the drinking of the draught by the detractor.

For plays viii to xv there is thus verbal evidence of direct borrowing from three sources: the *Legenda aurea*, the *Meditationes* or *Speculum Vitæ Christi*, and the *Pseudo-Matthew* Gospel, and the mingling of material in the plays corresponds with other evidence (cf. pp. xx to xxiv) to support a theory of dovetailing and revision. The plays of the next group (*Shepherds, Magi, Massacre*, and *Death of Herod*) are of a different character from any of the preceding ones, of higher literary quality, more lyrical, and in parts more dramatic, and in this closer to the *Passion* plays, and no source has been yet indicated for them. By similarities of metre and diction they are connected with the latter part of the *Resurrection* play.

The use made of the *Northern Passion* by the writer of the *Passion* plays differs from the use of the *Legenda aurea* in the plays derived from that source or from the use of the *Meditationes* in the *Salutation and Conception* and *Purification* plays. There is no such slavish transcription of whole passages as we find in the former, nor even such close following as we find in the latter. The parallel passages collected by Dr. Foster do not suggest that the writer set out to dramatize the *Northern*

Passion, but rather that he knew the story well in that form and its phrases recur, as Biblical phrases in a scriptural story would (or should) to-day. There is much in the *Passion* plays (the sermons of Peter and John, our Lord's exposition of the Sacrament, the Virgin's lament on his capture, the sentencing of the two thieves, and the dialogue between the Virgin and John at the foot of the Cross) which is not in the *Northern Passion*; the order of the incidents of the Trial before Pilate is different, and the dramatic handling of the Trial scenes with the accelerated dialogues in couplets is original.

In the *Passion* group, as in the earlier *Contemplacio* group, the layers of superimposed material correspond with the use of different sources. The influence of the *Northern Passion* is crossed by the influence of the *Meditationes*. A possible verbal reminiscence of Love in the scene between the Virgin and her risen Son[1] in the latter part of the *Resurrection* play gives a clue, and though there is no further verbal coincidence it is not rash to assume that the place given to the Virgin at the betrayal, at the death, and at the resurrection, the treatment of the harrowing as well perhaps as the completion of the seven words—matter found at the conclusion of the first *Passion* play, in the interpolated portions of the second, and in the latter part (Quire V) of the *Resurrection* play[2]—is due to the *Meditationes*. The chapter in the *Meditationes* on the taking of our Lord ends with a lament of the Virgin 'in Magdeleyns hous'; cf. *L. C.*, pp. 267, 268; special stress is laid on the 'seuene notable words' in the chapter on the death, and in the account of the descent into hell the actual harrowing following immediately on the death is separated from the appearance of the risen Lord to the rescued souls after the resurrection of his body. This does not exactly tally with the representation in the *Passion* play where the rescued souls are brought forth only on the resurrection morning, but it gives some suggestion for the division of the harrowing into

[1] Cf. p. xlv.

[2] Evidence of metre and diction also, as has been said, connects the interpolated matter on folios 183ᵛ–6 and the end of the *Resurrection* play on quire V. Cf. 'For mannys helpe my body is bred (fo. 185ᵛ) and 'For man I haue mad my body in bred' (fo. 192ᵛ), and in similar contexts 'chare awey þe crowe' (fo. 183ᵛ), and 'chare awey þe ravyn' (fo. 194ᵛ).

two parts—the descent at the end of the *Crucifixion* play and the rising with the redeemed souls in the middle of the *Resurrection* play—which is peculiar to the *L. C.* play. The borrowings from the *Northern Passion* cease [1] where the influence of the *Meditationes* comes in—in the interpolated portion of the second *Passion* play whereby it is dovetailed on to the *Resurrection* play and in the conclusion (Quire V) of the *Resurrection* play.[2]

This cycle is later in its compilation than the civic cycles, but it seems to have borrowed little from them beyond the general choice and treatment of material, and even in these respects there are noticeable divergencies. There are more plays peculiar to this cycle than to any other, and the absence of farcical passages in the *Cain, Noah,* and *Shepherds* plays distinguishes it from all the others.

The opening of the speech of God the Father in the first play, *Ego sum α et ω*, is found in all cases, and the further likeness of a few phrases in the Towneley play (cf. 'Aꝑ maner thyng is in my thoght' and 'aꝑ þat evyr xal haue beynge | it is closyd in my mende') would seem due to the same natural coincidence. An interpolated passage in the *Shepherds* play—the solitary attempt at humour—p. 148, ll. 78–89, may be a reminiscence of a similar farcical discussion of the angels' song in the Chester *Shepherds* play; a line in the scourging scene recalls a line in the same context in the Towneley play (cf. p. 277, ll. 168–71, and 'we shaꝑ teche hym I wote a new play of yoyꝑ') and there are coincidences in the Towneley and *L. C. Passion* plays due to the use in both of the *Northern Passion,* but otherwise there is nothing that may not be accounted for as due to similarity of theme and form.

Besides the plays which have no counterpart in the other cycles (*The Conception of Mary, Mary in the Temple, The Betrothal*

[1] There is perhaps a later reminiscence of the *Northern Passion* (though there are many possible sources) in the line in the *Assumption*: *Sere and ye slept on christis brest · seyng aꝑ celestly,* said by Peter of John. Cf. *Northern Passion,* ll. 276 et seq. The corresponding Latin words in the *Legenda aurea* from which this play is directly taken do not give so clearly the reference to the legend of John's vision here implied: *Tu insuper pectus domini recumbere meruisti et exinde sapientiae et gratiae plus ceteris fluenta potasti.*

[2] Cf also p. xiv and p. xxxi for the composition of the second *Passion* play.

of Mary, The Trial of Joseph and Mary) the following episodes are found only in the *L. C.*—the death of Cain at the hands of Lamech (play 4), the testifying of the kings (Matthew i. 6–10) with the prophets (play 7), the 'parliament of heaven' (play 11),[1] the miracle of the cherry-tree (play 15); Veronica appears only in this series, but her action is performed by one of the Maries in the York plays; in no other cycle is the crucifying of our Lord assigned to the highborn Jews ('No man xal towche ȝ*our* kyng | but yf he be knyght · or gentylman born; cf. also fo. 181, stage direction)[2]; characteristic of this collection again are the pithy short sermons put into the mouth of different characters: two from John the Baptist, one at the close of play 22 (p. 192) on repentance, and one as prologue to the *Passion* play (p. 229) on the dangers of overconfidence and of despair; one from Peter in the *Entry into Jerusalem*, scene (p. 238) on the spiritually blind, lame, and dumb; one from Christ expounding the Eucharist (p. 254). To these may be added the ironic discourse of Lucifer, a 'bountevous lord' to 'reward synners', at the opening of the *Passion* play (p. 225), and the exposition of the Commandments in the sixth play, which there is nothing to match in the Chester play in which the Commandments are recited.

In this exposition of the Commandments[3] the writer is drawing on the usual homiletic interpretation of the time. A treatise of the Commandments in a MS. at St. John's Coll.. Oxford, ed. by J. F. Royster (University of N. Carolina, Studies in Philology, vol. 6, 1910), who dates it between 1420 and 1434, contains much the same material (cf. *L. C.*, p. 55, ll. 134–7, and 'Bot agaynes þis commaundement doos he þat slaes with hond or with word or will'; and p. 57, ll. 179–82, and 'In the siȝste commaundement is forboden þe dede of lechere . . . and in þis commandement þe wille of þe syn is forboden'; cf. also *Court of Sapience*).

[1] This occurs in a different connexion in *The Castle of Perseverance*, cf. p. liv, and in a still more different in the 'Proces de Paradis' of the *Mystère du Vieil Testament*, in which, moreover, only Mercy and Justice are concerned.

[2] In the *Dialogus B. Mariæ et Anselmi de Passione Domini* a distinction is made between the *populus*, who are sympathetic after the Crucifixion, and the *Iudæi*, who continue their persecution of Christ's friends.

[3] For the curious confusion of the burning bush (*rubrus* by mistake for *rubus*) and Mount Sinai, cf. Isidore: 'Interpretatur autem Sinai rubus quod significat ecclesiam.'

The discourse of Lucifer is specially interesting as helping to date the compilation.[1] Attacks on extravagant fashions are common in the fifteenth century; the one in the text seems to be dealing with the same objectionable features in costume as the poem in MS. Harl. 372, printed by J. Payne Collier for the Percy Society, vol. xv, p. 55, and assigned to the mid-century —'not later than 1467' (cf. *L. C.*, p. 227, ll. 69 et seq., and 'Ye prowd galonttys hertlesse | W*ith* yo*ur* high cappis witlesse | And youre schort gownys thriftlesse | Haue brought this lond in gret heuynesse | W*ith* yo*ur* long peked schone . . . *And* yo*ur* long here in to yo*ur* eyen . . . With yo*ur* wyde furryd hodes . . . leue yo*ur* short stuffide doublett*ys and* yo*ur* playtid gownys'). Similar fashions, however, seem to be attacked later in the century in the morality *Nature* (the 'side hair', the 'short gown', and 'wide sleeves' that 'would make a doublet and coat for some lad in this town').

The mingling of the kings of Jesse's race with the prophets in the seventh play is not found in any other play of this type that has come down to us, but Sepet in *Les prophètes du Christ*[2] takes it as the surviving example of a class: 'L'idée de faire paraître à côté des prophètes proprement dits la ligne de Jesse, les rois de Juda, fils de David et ancêtres du Messie, n'est pas particulière au Ludus Coventriæ. La scène a certainement eu ce caractère dans les mystères français.' He quotes in support of this view a description of a similar combination in a Corpus Christi procession still held at Mayenne in the seventeenth century.[3]

The Lamech episode is found in *Le Mystère du Vieil Testament*, but here the boy who guides Lamech and whom he slays is his own son, Tubal-cain. The writer of the *L. C.* interpolation seems to have drawn straight from P. Comestor, cf. pp. 39–41, including stage direction and: 'Lamech vero vir sagittarius diu vivendo caliginem oculorum incurrit, et habens adolescentem

[1] Lucifer comes in as a 'prowde galaunt' in *Mind, Will and Understanding*, and there is a reference in *The Castle of Perseverance* to 'longe Crakows on þi schos'.

[2] Bibliothèque de l'École des Chartes, vol. 38, 1877.

[3] Prof. Hardin Craig in a Note appended to Miss Swenson's *Enquiry* points out that apparently a number of kings played a part in the Lincoln pageants, which may, in his opinion, give a clue to the connexion of the *Ludus Coventriæ* with Lincoln.

ducem dum exerceret venationem pro delectatione . . . casu interfecit Cain inter fructeta aestimans feram quia ad indicium iuvenis dirigens sagittam interfecit. Et cum experiretur quod hominem scilicet Cain interfecisset iratus illic cum arcu ad mortem verberavit eum. Et ideo cum peccatum Cain punitum esset septuplum ut diximus suum punitum est septuagies septies id est septuaginta animae et septem egressae de Lamech in diluvio perierunt.'

The independence of this cycle is perhaps most clearly illustrated in the treatment of a theme common to all the cycles—Christ and the Doctors. Five plays on the subject have come down to us,[1] and, of these, four—those in the York, Towneley, Chester, and Coventry civic cycles—go back to one original; the fifth, that in the *Ludus Coventriæ*, stands alone; and the play with the learned brag of the opening speeches and the theological problems and paradoxes introduced into the discussion is characteristic of the more ecclesiastical and perhaps also of the more sophisticated and 'stagy' cast of this compilation. The questions raised—(*a*) the unity of the Trinity, (*b*) the distinction of the three persons, (*c*) the manner of the incarnation, (*d*) the reason for the choice of the second person, (*e*) the 'double birth' of Jesus, (*f*) the reasons for the marriage of Mary—were all theological commonplaces,[2] but

[1] Another is mentioned in the Epilogue to the *Candlemas* play: 'And the next yeer as we be purposid in our mynde | The disputacion of the doctours to shew in your presens'.

[2] (*a*) The comparison of the Trinity to *ignis, splendor, calor* is found in V. de Beauvais and goes back to St. Augustine. Cf. also similar comparison in *Candlemas* play of the humanity, soul, and godhead of Christ to the wax, wick, and light of a taper, a comparison used in *Piers Plowman* for the Trinity (C. xx. 168 and Skeat's note on the passage).

(*b*) Cf. *L. C.*, p. 17, &c.

(*c*) Cf. Lydgate, *Lyf of oure Lady*, v. s. p. xlvi.

(*d*) Cf. *L. C.* p. 103, and Love: 'Man forfeted by vnwitte and foly'—an addition to Bonaventura; also *Court of Sapience*, where this is the second of three reasons.

(*e*) 'bis genitus dicitur sive quia Pater eum genuit sine Matre in æternitate sive quia Mater sine Patre in tempore'.—Isidore (*Etymologiarum* lib. vii, vol. 2, p. 265 Migne).

(*f*) These are two of the three reasons given by Bonaventura and Comestor. V. de Beauvais gives four, referring to Jerome and Ignatius.

the writer deserves some credit for original ingenuity in collecting them.

A connexion has been often suggested between this cycle and other religious plays of the fifteenth century.[1] There are obvious similarities in the stage directions of the *L. C. Passion* plays and those of such plays as *Mind, Will and Understanding*, the *Candlemas* play, *Mary Magdalene, The Croxton Play of the Sacrament*, which suggest similar methods of staging and perhaps similar business on the part of the actors.[2] There are also similarities in diction and in metre and in the handling of metrical form—the varying of the metre to suit the speaker or the situation—and in other characteristics of treatment which deserve further investigation but which lie outside the scope of this introduction. *The Castle of Perseverance* offers one or two special points of contact. It has, like the *L. C.*, a prologue in thirteen-lined stanzas for vexillatores in which the name of the town at which the play is to be performed is left to be filled up. It introduces the debate of the Four Daughters of God (*The Parliament of Heaven*) over the fate of the soul of *Humanum Genus*, though in this case it is the salvation of the individual and not the redemption of the race that is in question, and the debate does not lead up to the Incarnation. It introduces also the figure of Death; the entry of this character has not the quality of dramatic irony[3] that makes the entry in the *Death of Herod* play so impressive (*Humanum Genus*, an old man, has just remarked 'on Coveytyse is al my lay | *And* schal; tyl deth me ouer-throw'—Herod: 'I was nevyr meryer here beforn | Sythe þat I was fyrst born | Than I am now ryght in þis morn')—but the

[1] Mr. H. R. Patch ('The *Ludus Coventriæ* and the Digby *Massacre*', Publications of the M. L. A. of America, xxxv), basing his conclusions on an examination of the metres of the plays, makes the suggestion that the Dublin *Abraham and Isaac* and the Digby *Massacre* (or *Candlemas* play) might have formed part of the cycles or groups from which the compiler of *L. C.* made his selection.

[2] The Chester and Towneley cycles have a few, the York cycle a very few, brief directions in Latin; the Coventry civic cycle has fuller directions in English but not of the descriptive kind found in the *L. C. Passion* plays. This similarity as regards *Mind, Will and Understanding* has been noted by Mr. Pollard in his edition of the *Macro Moralities* (E.E.T.S., Extra Series 91).

[3] Cf. also *Everyman*: 'Full lytell he thynketh on my comynge | His mind is on flesshely lustes and his treasure.'

general effect is similar as are (naturally) the two speeches of Death. In *The Castle of Perseverance* is found also one of the geographical lists of which there are examples in the Towneley play *Herod the Great* (considered by Mr. Pollard to be one of the group of later plays), in the *L. C. Temptation* play,[1] and in *The Croxton Play of the Sacrament.* The arrangement of names of countries and towns in lists, often alphabetical, in the geographical treatises of the time afforded convenient material for the alliterative artist (cf. *Rel. Ant.*, vol. i, p. 271, and MS. Arundel 123). Though some of the names are common, the lists seem to be independent, but they testify doubtless to some kinship, as do perhaps the attacks in all the *Macro Moralities* and the *Ludus Coventriæ* on the extravagant fashions in clothing of the time.

The *Ludus Coventriæ* has generally been considered inferior in literary merit to the other cycles. It lacks the genuine feeling that dignifies the York cycle (cf. *Birth* and *Flight into Egypt* plays); it has nothing to compare with the pathos of the Chester *Isaac* play or the humour of the Towneley second *Shepherds* play. Though Mary plays such an important part, no speech of hers in the *L. C.* has the poetic quality of the Lament, in the Towneley *Crucifixion* play, of the Mother at the foot of the Cross. Apart from the greetings of the Shepherds, the speech of Death, the lament of Adam and Eve, and a few *rime couée* passages there is very little poetry in the pages of Vesp. D. viii. But the plays of this collection offered their audience compensations for deficiencies—more singing,[2] more 'devices',[3] more processions,[4] more harangues, and, above all,

[1] The names are in all cases chosen for reasons other than geographical, but no other writer trades so boldly on the ignorance of his audience as the writer of the *Temptation* play in including *Zebee and Salmana* among the countries to be seen from the Mount (cf. Ps. lxxxiii. 11).

[2] Notably in the *Contemplacio* group (viii to xiii) and the *Assumption*, but also in the *Creation, Noah, Shepherds, Purification,* and *Entry into Jerusalem* plays.

[3] Cf. Proclamation: '*With* whiche devys as we best may | The holy gost xal ovyr him on'. None of the other cycles attempt the spectacular representation of the Incarnation described in the stage direction, p. 107.

[4] The *Moses* play included apparently a procession of the Commandments. Cf. Proclamation: '*þe* ten comaundement*ys* alle be dene | in our*e* play ȝe xal hem sene'.

in some cases, more acting. The passages of couplet dialogue in the Trial scenes (ff. 168v, 169, 171v, 177, 178, 178v, 179) mark an important advance from recitation and declamation to acting, and in several of the plays the writer or writers show command of stage effect, and understanding of the impressiveness of significant gesture and movement. A comparison of the *Woman taken in Adultery* in this series and in the York and Chester cycles illustrates the superior skill of at least one of the *Ludus Coventriæ* writers in developing a dramatic situation. The preliminary sermon of Christ on forgiveness, the angry comments of the Jews, the arrival of Accusator with his welcome scandal 'a ryght good sporte I kan yow telle', the breaking open of the door of the house, the escape of the man 'If any man my wey doth stoppe . . . I xal þis daggere putt in his croppe'; the pleading of the woman 'I pray ȝow kylle me prevely . . . lete not þe pepyl up-on me crye'—all these dramatic features are found only in the *Ludus Coventriæ* example. The most dramatic passage in the series is, perhaps, the unnoted entrance of Death in the midst of the revelry of Herod and his knights, where, as has been said, the style rises above the usual level. The effect of silent stealthy movement followed by sudden outcry on the imagination of an audience is again recognized in the stage direction for Pilate's wife's dream: 'her xal þe devyl gon to Pilatys wyf · þe corteyn drawyn as she lyth in bedde *and* he xal no dene make but she xal sone after þ*at* he is come in · makyn a rewly noyse · comyng and rennyng of þe schaffald . . . leke a mad woman'; and this direction 'he xal no dene make' is the more significant as Pilate's wife afterwards describes the 'sounds unheard': 'As wylde fyre *and* thondyr blast | he cam cryeng on to me'. Very effective, too, must have been Judas's secret passings to and from the scaffolds in the *Last Supper and Conspiracy* play: 'here Judas rysyth prevely' and 'here Judas goth in sotylly'. The excitement of escape is suggested in the vivid stage direction *et curret* which follows Pilate's dismissal of Barabbas, and again in the description of the flight of the young man in the *Woman taken in Adultery*—'hic iuuenis quidam extra currit in deployde calligis no*n* ligatis et braccas in manu tenens'. And we can still feel the thrill that stirred the 'gentyllys and ȝemanry' of *N.* town in the fifteenth century when we read the stage direction which opens the first Trial scene: 'here xal a massang*er* com i*n* to þ*e* place ren*nyng and cri*yng Tydyngys tydyng*ys* · *and* so round abowth þ*e* place · Jh*e*sus of nazareth is take · Jh*e*sus of nazareth is take'. Such an appeal may be, in the words of the mediaeval preacher, more to 'þe siȝt wiþoute forþ' than to 'þe feiþ wiþinne forþ', but there is no doubt of its potency, and in this effective stage representation the *Death of Herod*, the *Woman taken in Adultery*, and the *Passion* plays of the *Ludus Coventriæ* are outstanding.

NOTE A [to page xix].

1. *Legenda sanctorum aurea* (1481).	*The Golden Legend* (Caxton).	*The Play of the Assumption.*
i. |*Tu insuper super pectus domini recumbere meruisti* | et exinde sap*ient*ie ac gracie plus ceteris fluenta potasti.	And thow oughtest to bere this palme of lyghte atte the xequyes of chastyte and holynes thou that drankest of the fontayn of perdurable clenes.	Sere | *and ye slept on christis brest* · | seyng alle celestly Ye are goddis clene mayde · wyth-outyn ony nay This observaunce is most like · you to do dewly.
ii. Memor *enim* esse debes qualiter |*tibi astiti*| et qualiter te accusante ancilla hostiaria te excusavi.	Thow oughtest remember how the chambriere that was Vssher accused thee and [? I] excused thee.	In cayfas halle · when thou were seyne And of thee pet*er* a mayde acusid there | *I halpe the tho* | · now helpe me ageyne.

iii. Caxton translates all the versicles which are given in the original Latin in the play, and in one instance the writer of the play has a variant reading.

habebit fructum *in respectione*[1] animarum sanctarum.	She shalle haue fruyte *in refection* of noble sowles.	Habebit requiem *in respectu* a*nim*arum sa*n*ctar*um*.

[1] Graesse's edition *refeccione*.

NOTE B [to pages xxiii, xxiv, and xlv].

The following passages show the correspondence of the text of the plays and passages dependent on the *Meditationes* with the translation rather than with the original work.

Meditationes Vitæ Christi. Bonaventura, *Opera*, 1588.	*The Mirrour of the Blessed Lyf of Jesus Christ*, p. 29. Nicholas Love, ed. L. F. Powell.	*'Salutation and Conception' Play.*
i. Intuere hic pro deo et meditare qualiter tota Trinitas est ibi expectans responsionem et consensum huius suae filiae singularis amanter et delectabiliter aspiciens verecundiam eius et mores et verba; et etiam qualiter angelus diligenter et sapienter inducit eam et ordinat verba sua stans inclinatus et reverens coram domina sua . . .	Now take good heed and haue in mynde how fyrst all the holy trinyte is there abidynge a fynal answer and assent of his blessid douꝝter Marye takynge hede and byholdynge lykyngliche her schamfast semblaunt, her sad maneres and her wise wordes, | *and furthermore howe alle the blessid spirites of heuene, and alle the ryꝝtwis lyuynge men in erthe and alle the chosen soules that weren that tyme in helle as adam, abraham, david and alle othere desireden hir assent in the whiche stood the sauacioun of all mankynde,* | and also how the aungel gabriel stondynge with reverence byfore his lady enclynynge . . .	Mary come of and haste the and take hede in thyn entent whow þe holy gost blyssyd he be abydyth þin answere and þin assent . . . . . . | *fferthermore take hede þis space* *whow alle þe blyssyd spyrytys of vertu* *þat are in hefne by ffore goddys face* *and alle þe gode levers and trew* *That are here in þis erthely place* *thyn owyn kynrede · þe sothe ho knew* *And þe chosyn sowlys þis tyme of grace* *þat are in helle and byde rescu* *as Adam, abraham and david in fere* *and many othere of good reputacion* *þat þin answere desyre to here* *and þin assent to þe incarnacion* *in whiche þou standyst as persevere* *of alle man-kende savacion.* |

Meditationes Vitæ Christi.	*The Mirrour of the Blessed Lyf of Jesus Christ*, p. 39.	*'Visit to Elizabeth' Play.*
O qualis domus, qualis camera, qualis lectus, in qua et quo pariter commorantur et requiescunt tales matres talibus filiis foecundatae, Maria et Elisabeth, Jesus et Joannes. Sunt et ibi magnifici senes, scilicet Zacharias et Joseph . . .	| *A lord god what house was that* | or what chamber and what bedde in the whiche dwelleden togidre and resteden so worthi moderes with so noble sones that is to saie Marie and Elizabeth Jesu and John And also with them dwellynge tho worschipful olde men zacharie and Joseph.	| *A lord god what hous was þis on* | þat þese childeryn and here moderys to as mary and elizabeth jhesus and john and joseph and zakarye also.
. . . . .		
Et tunc apertum est os Zachariae, et prophetavit, dicens: *Benedictus Dominus Deus Israel*, etc. · Et sic in domo illa haec duo cantica pulcherrima, scilicet *Magnificat* et *Benedictus*, facta fuerunt . . .	And the mowthe and the tunge of the fader zacharye *by-fore closed for vntrowynge* was than opened: and so he prophecied seienge: . . .	And þan zakarye spak i-wus *þat had be dowm and his speche lorn* he and Elizabeth prophesyed as þus they mad BENEDICTUS · them be-forn and so · MAGNIFICAT · and BENEDICTUS · ffyrst in *þat* place *þer* made worn.
. . . . .		
	And so in that house thise two noble and worthy canticles that is to seyen Magnificat and benedictus weren first spoken and made.	
Tandem valefaciens Elisabeth et Zachariae, ac benedicens Joanni rediit ad domum suae habitationis in Nazareth.	And at the laste | *whan al this was done* | sche toke hir leue at Elizabeth and Zacharye and blessid the child John and so went home et seq.	| *Whan att was don* | oure lady fre toke here leve than aftere this at Elizabeth and at Zakarie and kyssyd johan and gan hym blys.

	Ch. ix, *preliminary paragraph not in original.*	*'Purification' Play.*
	After the kynges had performed her offerynges . . . ȝit stode that worthy lady of all the world in that symple herborwe . . . paciently abidyng unto the fourty day ordeyned	To be purefyed · haue ȝe no nede Ne þis on · to be offeryd · so god me spede ffor fyrst þou art ful clene *Vndefowlyd in thought* and *dede*

Meditationes Vitæ Christi.	Ch. ix. *preliminary paragraph not in original.*	*'The Purification' Play.*
	by the lawe to hir purificatioun as sche were another womman of the peple *defoyled thoruȝ synne, and as the childe Jesu were a pure man and not god nedy to kepe the observaunce of the lawe.*	and *anothyr þi son* with-*owtyn drede* *is god* and *man to mene* *Wherefore it nedyd not to bene* *but to kepe þe lawe* · on Moyses wyse.
	The Mirrour of the Blessed Lyf of Jesus Christ, pp. 62–3.	
ii. Accipite pater excellentissime unigenitum vestrum quem secundum mandatum vestrae legis vobis offero quia primogenitus matris est.	Taketh now *hiȝest fader ȝoure owne dere sone* whom I offre here to ȝow after the biddynge of ȝoure lawe: for he is the firste born of his moder.	*hyest ffadyr* god of powere ȝour *owyn dere son* I offre ȝow here as I to ȝ*our* lawe am sworn receyve *þi* childe in glad manere for he is *þe* fyrst þis childe so dere *þat* of his modyr is born.
iii. Accipite clementissime pater hanc oblationem et munusculum et primum donum quod parvulus vester hodie de sua paupertate praesentat.	*Al miȝty and merciful fader of heuene* vnderfonge ȝe this litel ȝifte and offerynge and the firste ȝifte that ȝoure litel child this day presenteth *vnto ȝoure hiȝe maieste of his symple poueite.*	*alle myghty fful fadyr* · *mercyful kynge* receyvyth now þis lytyl offerynge ffor it is *þe* fyrst in degre *þat* ȝ*our* lytyl childe so ȝynge presentyth to day be my shewyng *to* ȝour *hyȝ mageste of his sympyl poverte.*

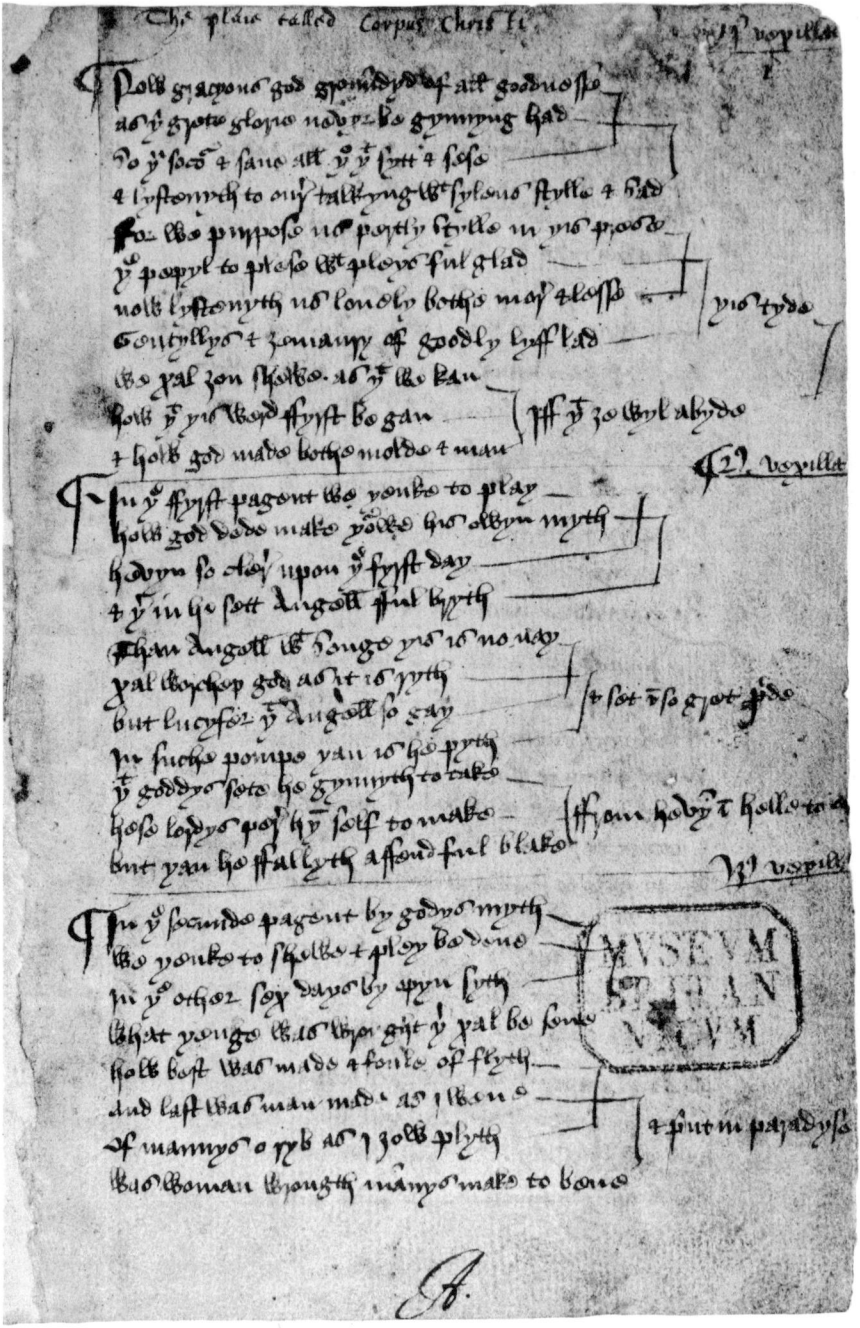

British Museum Cott. MS. Vespasian D. VIII, f. 1a

LUDUS COVENTRIÆ.

Cotton MS. Vespasian D. viii.

Fo. 1

(A quire)

The Proclamation.

The Plaie called Corpus christi.[1]

1us *vexillat[or*

¶ Now gracyous god groundyd of alł goodnesse
as þi grete glorie nevyr be-gynning had
So þ*ou* soco*ur and* saue alł þo þ*at* sytt *and* sese
and lystenyth to our*e* talkyng w*ith* sylens stylle *and* sad
Ffor we purpose us pertly stylle in þis prese 5
þe pepyl to plese w*ith* pleys ful glad
now lystenyth us louely bothe mor*e and* lesse
Gentyllys *and* ȝemanry of goodly lyff lad
þis tyde
we xal ȝou shewe · as þ*at* we kan 10
how þ*at* þis werd ffyrst be-gan
and how god made bothe molde *and* man
Iff þ*at* ȝe wyl a[byde.

¶ 2us *vexilla[tor*

¶ In þe ffyrst pagent we þenke to play
how god dede make þ*ur*owe his owyn myth 15
hevyn so cler*e* upon þ*e* fyrst day
and þ*er* in he sett Angeł fful bryth
Than Angeł w*ith* songe þis is no nay
xal worchep god as it is ryth
but lucyfer þ*at* Angeł so gay 20
in suche pompe þan is he pyth
and set i*n* so gret p*r*ide
þ*at* goddys sete he gynnyth to take
hese lordys per*e* hy*m* self to make
but þan he ffallyth a ffend ful blake 25
ffrom hevy*n* i*n* helle to a[byde.

[1] A note at the top of the folio in small Elizabethan handwriting.

HOMERTON
COLLEGE
LIBRARY

3us *vexill[ator*

¶ In þe secunde pagent by godys myth
we þenke to shewe *and* pley be-dene
in þe other sex days by opyn syth
what þenge was wrought þ*er* xal be sene
How best was made *and* foule of flyth
and last was man made as I wene
Of mannys o ryb as I ȝow plyth
was woman wrougth m*a*nnys make to bene
and put in paradyse
Fo. 1v Ther were flour*ys* bothe blewe *and* blake
of all frut*ys* þei myth þ*er* take
Saff frute of cu*n*nyng þei xulde for-sake
And towche it no wyse.

¶ The Serpent toke Eve an Appyl to byte
and Eve toke Adam a mursel of þe same
whan þei had do þus a-ȝens þe rewle of ryte
than was oure lord wroth *and* grevyd al w*ith* grame
Oure lord gan appose þem of þ*er* gret debyte[1]
both to Askuse hem of þ*at* synful blame
and þan almythy god ffor þ*at* gret dyspite
Assygned hem grevous peyn · as ȝe xal se i*n* game
In dede
Seraphyn An Angell gay
w*ith* brenny*ng* swerd þis is verray
From paradise bete hem a-way
in bybyl as we rede.

1us *vexillator*

¶ We purpose to shewe in þe thryd pagent
the story of Caym · *and* of hese brother Abelle
of here tythyng*ys* now be we bent
In þis pagent þe trewth to telle
How þe tythyng of Abel w*ith* feyr was brent
and accept to god yf ȝe wyl dwelle
We purpose to shewe as we haue ment
and how he was kyllyd of his brother so felle
And than
how Caym was cursyd in al degre

[1] Halliwell reads *delyte*. The MS. is not clear.

of godys owyn mowthe *þer* xal ȝe se
of trewe tythyng þis may wel be
exawple[1] to every man. 65

2[us] *vexilator*

¶ The iij[de] pagent is now ȝow tolde
þe ffourte pagent of Noe xal be
how god was wroth · *with* man on molde
because fro synne man dede not fle
Fo. 2 He sent to Noe An Angel bolde 70
A shyp ffor to makyn *and* swy*m*men on *þe* se
vpon *þe* water both wood *and* coolde
And viij sowles *þer* savyd xulde be
And i peyr*e* of everich best*ys* in brynge
whan xl[ti] days *þe* flode had fflowe 75
*þ*an sente Noe out a crowe
and After hym he sent a dowe
*þ*at brouth ryth good tydyng.

3[us] *vexil[lator*

¶ Of Abraham is *þe* fyfte pagent
and of ysaac his sone so fre 80
how *þ*at he xulde[2] w*ith* fere be brent
and slayn w*ith* swerd as ȝe xal se
Abraham toke w*ith* good A-tent
his sone ysaac *and* knelyd on kne
his suerd was than ful redy bent 85
and thouth his chylde *þ*er offered xuld be
Vpon An hyll full Ryff
than god toke tent to his good wyl
and sent An Angel ryth sone hy*m* tyl
and bad Abraha*m* a shep to kyl 90
And sauyd his chyldys lyff.

1[us] *vexil[lator*

¶ The sexte pagent is of Moyses
and of tweyn tabelys *þ*at god hy*m* took
in *þe* which were wrete w*ith*-out les
þe lawes of god to lerne *and* lok 95
and how god charged hym be wordys these
þe lawes to lerne al of *þ*at book
Moyses than doth nevyr mor*e* sese

[1] So in MS. [2] A letter or letters obliterated between *xulde* and *with*.

but *pre*chyth duly · bothe ȝere *and* woke
Fo. 2v The lawes as I ȝow telle
þe ten comaundement*ys* alle be-dene
In oure play ȝe xal hem sene
to alle þo þat þere wyl bene
If þat ȝe thenke to duelle.

2us *vexill*ator

¶ Off þe gentyl Jesse rote
þe sefnt pagent for sothe xal ben
out of þe which doth sprynge oure bote
as in *pro*phecye we redyn *and* sen
Kyngys *and* *pro*phet*ys* w*ith* wordys fful sote
Schuł *pro*phesye al of a qweñ
Þe which xal staunch oure stryff *and* moote
And wynnen us welthe w*ith*-outy*n* weñ
In hevyn to Abyde
they xal *pro*phecye of a mayde
Ał ffendys of here xal be Affrayde
here sone xal saue us be not dismayde
W*ith* hese woundys wyde.

3us *vexillator*

¶ Of þe grete bushop Abyacar
þe tende[1] pagent xal be w*ith*-out lesyng
þe which comaundyth men to be war
and brynge here douter*ys* to dew weddyng
Ał þat ben xiiij ȝere *and* more
to Maryage he byddyth hem bryng
wher evyr þei be he chargyth sore
Þat þei not ffayle for no lettyng
Þe lawe byddyth so thañ[2]
Than Joachym *and* anne so mylde
Þei brynge forthe mary þat blyssyd chylde
but she wold not be de-fylyde
w*ith* spot nor wem of man.

Fo. 3 ¶ In chastyte þat blysful mayde
A-vowyd there here lyff to lede

[1] A later correction. The original word is erased.

[2] The word *sertayñ* is written before *thañ* and crossed out. Halliwell retains it.

þan is þe busshop sore dysmayde
And wonderyth sore al of þis dede
he knelyd to god as it is sayde 135
and pra*y*th than for help *and* rede
þan seyth an Angel be not a-frayde
of þis dowte take þ*ou* no drede
but for þe kynrede of dauyd þ*ou* sende
lete hem come w*ith* here offryng 140
And in here handys white ȝerdys brynge
loke whose ȝerde doth ffloure *and* sprynge
And he xal wedde þ*a*t mayden hende.

1us *vexillator*

¶ In þe x^te^[1] pagent sothe to say
A masangere fforthe is sent 145
Dauyd is kynrede w*ith*-out de-lay
they come fful sone w*ith* good Entent
Whan joseph offeryd his ȝerde þ*a*t day
Anon ryth fforth in present
þe ded styk do floure fful gay 150
and þ*a*n joseph to wedlok went
Ryth as þe Angel bad
Than he plyth to his wyff
In chastyte to ledyn here lyff
þe busshop toke here iij · maydonys ryff 155
Som comforte þere she had[2].

2us *vexilla*tor

¶ In þe xi^de^[3] pagent goth Gabryell
And doth salute oure lady ffre
Than grett w*ith* chylde as I ȝow tell
þ*a*t blyssyd mayde for sothe is she 160
Fo. 3v þo iij maydenys þ*a*t w*ith* here dwelle
here gret spech but noon þei se
than they suppose þ*a*t su*m* Angell
goddys masangere þat it xuld be
And thus 165
þe holy gost in here is lyth

[1] This is corrected from an original ix^te^.
[2] First written *hadde* and corrected.
[3] This is corrected from an original x^de^; the writing becomes slightly smaller and there is a faint difference in the colour of the ink with this stanza.

and goddys sone in here is pygth
þe Au*n*geꝰ doth telle what he xal hyght
And namyth þe chylde jhesus.

3us *vexillator*

¶ In þe xij[1] pagent as I ȝow telle
Joseph comyth hom fro fer countre
Oure ladyes wombe w*ith* chylde doth swelle
and þan joseph ful hevy is he
He doth forsake here w*ith* hert ful felle
out of countre he gynnyth to fle
he nevyr more thenkyth w*ith* here to dwelle
and than oure lady ryth sore wepyth she
An Angeꝰ seyd hym ryf
god is w*ith* þi wyff sertayn[2]
þ*er*fore Joseph turne hom a-gayn[2]
þan is Joseph in herte ful fayn
And goth a-geyn onto his wyff.

1us *vexillator*

¶ The xiiij^te^[3] pagent I sey ȝow be-dene
xal be of joseph *and* mylde mary
how they were sclawndryd w*ith* trey *and* tene
and to here purgac*i*on þei must hem hy.[4]

Fo. 4 2us *vexillator*

¶ In þe xv[5] pagent shewe we xal
how joseph went w*ith*-oute varyauns
for mydwyuys to helpe oure lady at aꝰ
of childe that she had delyuerau*n*s.

3us *vexillator*

¶ In þe xvj[6] pagent Cryst xal be born
of þ*a*t joy Aungelys xul synge
and telle þe shepherdys in þ*a*t morn
the blysseful byrth of þ*a*t kyng

[1] Here the original word *hellenthe* has been crossed out in red ink, and xii written in fresher black ink above.

[2] *Serteyn* and *ageyn* appear to have been first written and corrected by scribe.

[3] This is corrected from an original xii^te^.

[4] A space is left at the bottom of Fo. 3^v^, beneath this speech, and at the top and in the middle of Fo. 4 above and beneath the next speech, as if to be filled in later.

[5] This is corrected from an original xiii.

[6] This is corrected from an original xiv.

The shepherdys xal come hy*m* be-fforn 195
w*ith* reuerens *and* w*ith* worchepyng
ffor he xal sauyn þ*at* was for-lorn
and graunt us lyf evyr mor*e* lestyng
i-wys
þis gle in gryth̄ 200
is mater of myrth̄
now crystys byrth̄
Bry*ng*[1] us to his blys.

1^ns^ *vexill*ator

¶ The xv^te^ pagent come kyng*ys* iii
Fo. 4^v^ w*ith* gold myrre *and* ffrankynsens 205
kyng herowdys styward hem doth se
And bryngyth aꝗ to his presens
The kyng*ys* of Coleyn w*ith* hert ful ffre
tolde kyng herownde her*e* dylygens
that þei south in þ*at* countre 210
A kyng of kyng*ys* ffrom fer*e* thens
A sterr*e* led hem þ*e* way
The Chylde is ȝoung *and* lyth in staꝗ
he xal be kyng of kyng*ys* aꝗ
beffor*e* hy*m* we thynk on kne to faꝗ 215
And worchep hym þis day.

2^ns^ *vexillator*

¶ In þ*e* xvi pagent as wroth as wynde
is kyng herownde þ*e* soth̄ to say
And cruel knyt*ys* *and* vn-kende
to sle male chylderyn he sendyth þ*at* day 220
But cryst jhesu þei may not ffynde
Ffor joseph hath led þ*at* childe away
Vnto Egypth as we haue mende
As Angeł to joseph dyd byd *and* say
In hyȝht 225
þ*o* chylderyn þ*at* syt in her*e* moder*ys* lap
to sowkyn ful swetly her*e* moder*ys* pap
þ*e* knyght*ys* do sle hem euy*n* at A swap
þis is a rewly syth.

[1] Some word beginning with *s* (or f) and ending with *l* is crossed out before *Bryng*.

3us *vexillator*

¶ In þe xvii pagent þe knyght*ys* be-dene
Shulł brynge dede childeryn be-for þe kyng
whan kyng herownde þ*at* syth hath sene
fful glad he is of here kyllyng
Than kyng herownde w*ith*-owty*n* wene
is sett to mete at his lykyng
in his most p*ri*de xal come gret tene
As ȝe xal se at our*e* pleyng
Fo. 5 his sorwe xal a-wake
whan he is sett at hese most pryde
Sodeyn deth xal thrylle his syde
and kylle his knytt*ys* þat w*ith* hym byde
þe devyl þer soulys xal take.

1us *vexillator*

¶ In þe xviii pagent we must purpose
to shewe whan cryst was xij ȝer of Age
how in þe temple he dede appose
and answerd doctoris ryth wyse *and* sage
The blyssyd babe w*ith*-owte glose
ouer cam olde clerk*ys* w*ith* suych langage
þ*at* þei meveylyd · ȝe xal suppose
how þ*at* he cam to suche knowlage
and in þis whyle
thre days he was oute
ffro his modyr w*ith*-out doute
wepyng she sowth hy*m* rownde aboute
jher*u*selem many a myle.

2us *vexillator*

¶ in þe xix pagent xal seynt jhon̄
baptyse cryst as I ȝow say
in þe watyr of flom jordon̄
w*ith* which devys as we best may
The holy gost xal ouyr hym on
þe ffader*ys* voys xal be herd þ*at* day
out of hevyn þ*at* blisful tron̄
þe fadyr xal be herd þis is no nay
and forth w*ith* pleyn̄
þe holy gost xal be his gyde
in to desert þ*er*-in to.A-byde

xl^ti days A terme ful wyde
and xl^ti nygtht*ys* to faste sarteyn̄[1].

3us *vexill*ator

¶ in þe xx^ti pagent aff þe deuelys of helle
they gadere a p*ar*lement as ȝe xal se
they haue gret doute þe trewth to telle
Of cryst jh*es*u whath he xulde be
Fo. 5v They Sende Ssathan þat Ffynde so Ffelle
Cryst for to tempte in fele degre
We xal ȝow shewe if ȝe wyl dwelle
How cryst was temptyd in synnys thr*e*
Of þe devyl sathan̄
And how cryst Answeryd on to Alle
and made þe fende awey to falle
As we best may þis shewe we xalle
Thorwe grace of god *and* man

1us *vexill*ator

¶ The xxi^ti pagent of a woman xal be
þe which was take in Adultrye
The pharysewys ffalsed þ*er* ȝe xal se
Cryst to convycte how they were slye
they conseyvyd þis sotylte
yf cryst þis woman dede dampne trewly
A-geyn his prechyng than dede he
Which was of pete *and* of mercy
And yf he dede here save
þan were he A-ȝens moyses lawe
þ*at* byddyth w*ith* stonys she xulde be slawe
þus they thowth vndyr þ*er* Awe
Cryst jh*es*u for to haue.

2us *vexill*ator

¶ The grettest meracle þ*at* evyr jh*es*us
In erthe wrouth be-forn his passyon̄
In xxii^ti pagent we purpose vs
to shewe in dede þe declarac*i*on
þ*at* pagent xal be of lazarus
In whos place *and* habytac*i*on
Cryst was logyd þe gospel seyth thus

[1] An original *e* in the first syllable appears to have been corrected by the scribe to an *a*.

and ofte[1] tyme toke þ*er* consolac*i*on
but ȝyt
Fo. 6 Lazarus As I ȝow say
was iiij days ded *and* beryed in clay 305
ffrom deth to lyve þ*e* iiij^te^ day
Cryst reysed hym ffrom þ*at* pyt.

3us *vexill*ator

¶ In þe xxiij^ti^ pagent palme sunday
in pley we purpose ffor to shewe
how chylderyn of Ebrew w*ith* flowr*ys* ful gay 310
þe wey þ*at* cryst went þ*ei* gu*n* to strewe

1us *vexill*ator

In þe xxiiij^ti^ pagent as þ*at* we may
cryst *and* his apostelys alle on rewe
the mawnde of god þ*er* xal they play
and sone declare it w*ith* wordys ffewe 315
And than̄
Judas þ*at* fals trayto*ur*
ffor xxx^ti^ platys of werdly treso*ur*
xal be-tray our*e* savyo*ur*
to þ*e* jewys certan̄[2]. 320

2us *vexill*ator

¶ ffor grevous peyn þis is no les
in þe xxv^ti^ pagent cryst xal pray
to þ*e* fadyr of hevyn þat peyn for to ses
his shamful deth to put Away
Judas þ*at* trayto*ur* be-for gret pres 325
xal kys his mouth̄ *and* hy*m* be-tray
All̄ his dyscyples than do dyscres
and forsake cryst þ*e* soth̄ to say
ffor doute þei do hem hede
hese dyscyplys all̄ every-chon̄ 330
do renne awey *and* leve hy*m* alon̄
they lete hym stondyn amonge his ffon
And ronne Away ffor drede.

3us *vex[illator*

Fo. 6^v^ Than in þ*e* xxvj^ti^ pagent
to Cayphas cryst xal be brouth 335

[1] The *o* is blotted and may have been corrected from another letter.
[2] The *e* is blotted and not clearly decipherable.

þo jewys fful redy *þer* xul be bent
Cryst to Acuse w*ith* worde *and* thouth
Seynt petyr doth folwe w*ith* good intent
to se w*ith* cryst what xuld be wrouth
Ffor cryst*ys* dyscyple whan he is hent 340
thryes he doth swer*e* he knew hy*m* nowth
A kok xal crowe *and* crye
Than doth petyr gret sorwe make
Ffor he his lord þus dede for-sake
But god to grace hym sone doth take 345
Whan he doth aske mercye.

1[us] *vexill*ator

¶ In þe xxvij pagent ser*e* Pylat
Is sett in sete as hy justyce
Whan he is set in his astat
thre thevys be brout of synful gyse 350
And cryst *þat* louyd nevyr stryff nor bat
but trewth *and* goodnesse on every wyse
as for a thef w*ith* ryth gret hat
is browth to stondyn at *þat* same syse
And þan as I ȝow say 355
b coveryd w*ith* clothis al of *þe* best[1]
a the Wyff of Pylat goth to rest[1]
Than for to slepe she is ful prest
Att þis we thenke to play.

2[us] *vexill*ator

¶ in *þe* xxviij[ti] pagent xal judas 360
þat was to cryst a ffals trayto*ur*
w*ith* wepyng sor*e* evyr crye Alas
þat evyr he solde our*e* savyo*ur*
Fo. 7 He xal be sory ffor his trespas
And brynge a-ȝen att his treso*ur* 365
att xxx pens to ser*e* Cayphas
he xal them brynge w*ith* gret dolowr*e*
Ffor *þe* which cryst was bowth
Ffor gret whanhope as ȝe xal se
he hangyth hym self vpon a tre 370
Ffor he noth trostyth in godys pete
to helle his sowle is browth.

[1] So, in wrong order with indication of the mistake, in MS.

3us *vexillator*

¶ In þe xxix pagent to Pylatus Wyff
in slepe aperyth þe devyl of helle
ffor to savyn cryst*ys* lyff 375
The devyl here temptyth as I ȝow telle
Sche sendyth to pylat a-non ful ryff
And prayth þ*at* cryst he xuld not qwelle
þan pylat is besy *and* ryth blyff
cryst for to savyn he ȝevyth coun*c*elle 380
ffor he dede neuyr trespas
The jewys do crye fast ffor to kylle[1]
the[2] rythful man þei aske to spylle
A thef þei saue w*ith* herty wylle
þ*at* Callyd is barrabas. 385

1us *vexill*ator

¶ In þe xxx^ti^ pagent þei bete out cryst*ys* blood
And nayle hy*m* al nakyd upon a rode tre
betwen ij thevys · i-wys they were to wood
they hyng cryst jhesu · gret shame it is[3] to se
Fo. 7v vii wurdyes cryst spekyth hangyng upon þe rode 390
þe weche ȝe xal here aft þo þ*at* wyl þ*er* be
þan doth he dye ffor oure allether good
his modyr doth se þ*at* syth gret mornyng makyth she
Ffor sorwe she gynneth to swowne
Seynt joh*a*n Evyn þ*er* as I ȝow plyth 395
doth chere oure lady w*ith* al his myth
And to þe temple anon forth ryth
he ledyth here in þ*at* stownde.

2us *vexill*ator

¶ we purpose to shewe in oure pleyn place[4]
in þe xxxj^ti^ pagent þorwe godys myth 400
how to cryst*ys* herte a spere gan pace
and rent oure lordys bryst in ruly plyth
Ffor longeus þat olde knyth blynd as he was
A ryth sharpe spere to cristys herte xal pyth

[1] The writing becomes slightly less regular with this line to the bottom of the folio.

[2] *ryff* written before the next word and crossed through by scribe.

[3] *was* first written and corrected by scribe to *is*.

[4] The writing in this speech is slightly more sprawled than heretofore.

þe blod of his wou*n*de to his eyn xal tras 405
and þorwe gret meracle þ*er* hath he syth
Than in þ*at* morn
cryst*ys* soule goth dowñ to helle
and þer ovyr comyth þ*e* fende so felle
comfortyth þ*e* soulys þ*at* þ*er* in dwelle 410
and savyth þ*at* was fforlorñ.

3us *vexillator*

¶ Joseph *and* nycodemus to cryst trew servau*n*t [1]
in þ*e* xxxij page [2] þ*e* body þei Aske to haue
Pylat ful redyly þ*e* body doth hem grau*n*t
þan þei w*ith* reverens do put it in grave 415
þe jewys mor*e* wyckyd þan ony geawnt
Ffor cryst*ys* ded body kepers do þei craue
pylat sendyth iiij knyt*ys* þ*at* be ryth hardau*n*t
to kepe þ*e* blody body in his dede co*n*claue
Fo. 8 And ȝit be his owyn myth 420
The body þ*at* was hevy as led
be þ*e* jewys nevyr so qwed
A-ryseth from grave þ*at* þ*er* lay ded
And ffrayth than every knyth.

1us *vexillator*

¶ in þ*e* xxiij [3] pagent þ*e* soule of cryst jh*es*u 425
xal brynge alł his ffrendys ffrom helle to paradyse
þ*e* soule goth þan to þ*e* graue · *and* be ryth [4] gret vertu
þ*at* body · þ*at* longe ded hath loyn · to lyf aȝen doth ryse
Than doth cryst jh*es*u on to his modyr sew
and comfortyth alł her*e* car*e* in temple þ*er* she lyse 430
w*ith* suche cher *and* co*m*forth his modyr he doth indew
þ*at* joy it is to her*e* · þ*er* spech for to devyse
and thañ
our*e* lady of hefne so cler
In herte sche hath ryth glad cher*e* 435
whan her*e* sone þus doth aper*e*
her*e* car*e* awey is tan.

2us *vexill*ator

¶ in þ*e* xxxiiii^ti^ pagent xal maryes thre

[1] The writing for the first five lines of this speech is smaller.
[2] So in MS. for *pagent*. [3] So in MS. for xxxiii.
[4] This word omitted and written over the line by the scribe.

seke cryst jh*es*u in his grave so coolde
An Au*n*gel hem tellyth þ*at* aresyn is he 440
and whan þ*at* þis tale to them is tolde
To cryst*ys* dyscyplis w*ith* wurdys fful fre
they telle these tydyng*ys* w*ith* brest ful bolde
than petyr *and* joh*an* as ȝe xal se
down renny*n* in hast ou*er* lond *and* wolde 445
The trewth of þis to haue
whan þei þer comy*n* as I ȝow say
he is gon ffrom vndyr clay
þan þai wytnesse a-noon þ*at* day
he lyth not in his grave. 450

3us *vexill*ator

Fo. 8v ¶ On to Mary Mawdelyn as we haue bent
Cryst jh*es*u xal than Aper*e*
In þ*e* xxxv^ti^ pagent
and she wenyth he be A gardener*e*
Mary be name verament 455
Whan cryst her*e* callyth w*ith* spec̄h ful cler*e*
She ffallyth to grou*n*d w*ith* good entent
to kys his fete w*ith* gladsom̄ cher*e*
but cryst byddyth her*e* do way
he byddyth his feet þ*at* sche not kys 460
tyl he haue styed to hefne blys
to cryst*ys* dyscyplys mary i-wys
than goth · þ*e* trewth̄ to say.

1us *vexillator*

¶ In þ*e* xxxvj^ti^ pagent xal cleophas
And sent luke · to a castel[1] go 465
Of cryst*ys* deth as þ*e*i fforth pas
they make gret mornyng *and* be ful wo
Than cryst þem ovyr-tok as his wyl was
and walkyd in felachep fforth w*ith* he*m* too
to them he doth expowne bothe more *and* las 470
alł þ*at* pr*o*phet*ys* spak ad[2] of hym self Also
that nyth in fay
whan þei be set · w*ith*-in þ*e* castełł

[1] The initial *c* is corrected from some other letter.
[2] So in MS.

In brekyng of bred · þei know[1] cryst weꝗ
Than sodeynly as I ȝow teꝗ
Cryste is gon his way.

2us *vexill*ator

¶ In þe xxxvij^ti^ pagent þ*a*n purpos we
to Thomas of ynde cryst xal aper*e*
And thomas euyn þ*er* as ȝe xal se
xal put his hand in his wou*n*d*ys* der*e*.

3us *vexill*ator

Fo. 9 ¶ In þ*e* xxxviij^ti^ pagent up stye xal he
in to hefne þ*at* is so cler*e*
Aꝗ hese Apostel*e* þ*er* xul be
And wou*n*der*e* sor*e* *and* haue gret dwer*e*
Of þ*at* ffer*l*y syth
Þer xal come Au*n*geꝗ tweyn
and comfforte hem þis is certeyn
and tellyn þ*a*t he xal comyn ageyn
Evyn by his owyn myth.[2]

1us *vexill*ator

¶ Than ffolwyth next sekyrly
of Wyttsunday þ*at* solempne ffest
whych pagent xal be ix *and* thretty
to þ*e* apostely*s* to aper*e* be cryst*ys* hest
In hierusalem wer*e* gaderyd xij opynly
to þ*e* Cenacle comyng ffrom west *and* Est
þ*e* holy gost apperyd fful veruently
w*ith* brennyng ffer*e* thyrlyng her*e* brest
procedyng from hevyn tron̄e
Aꝗ maner langage hem spak w*ith* tung̃
latyn grek *and* Ebrew among̃
And Affter þ*ei* departyd *and* taryed not long̃
here deth · to take ful sone.

2us *vexill*ator

¶ The xl^ti^ pagent xal be þ*e* last
and domysday þ*at* pagent xal hyth
who se þ*at* pagent may be agast
to grevyn his lord god eyther day or nyth
The erth xal qwake bothe breke *and* brast

[1] Or *knew*; the letter is partially blotted.

[2] This speech forms one stanza with the last speech, but it is marked as a separate stanza in the MS.

beryelys *and* gravys xul ope ful tyth
ded men xul rysyn *and* þat *þer* in hast
And ffast to here ansuere þei xul hem dyth 510
Fo. 9v[1] Beffore godys fface
but prente wyl þis in ȝ*our* mende
Who so to god · hath be vnkende
Ffrenchep *þer* xal he non ffynde
ne *þer* get he no grace. 515

¶ 3[us] *vexillator*

¶ Now haue we told ȝow all̃ be-dene
the hool mater *þat* we thynke to play
whan *þat* ȝe come *þer* xal ȝe sene
this game wel pleyd in good a-ray
Of holy wrytte þis game xal bene 520
and of no fablys be no way
now god þem save from trey *and* tene
ffor us *þat* prayth upon *þat* day
And qwyte them wel *þer* mede
A sunday next yf *þat* we may 525
At vj of *þe* belle we gynne our*e* play
In N. town̄ wherfor*e* we pray
That god now be ȝour*e* Spede. Amen.

Remainder of Fo. 9v—3½ inches—left blank.

Fo. 10[2] *Ego sum alpha et* oo · *principium et finis.* Deus

¶ My name is knowyn god *and* kynge
My werk for to make · now wyl I wende
in my self restyth my reynenge
it hath no gynnyng ne non ende
And all̃ *þat* evyr xal haue beynge 5
it is closyd in my mende
whan it is made at my lykynge
I may it saue I may it shende

[1] Corrected to 10 and then crossed through.

[2] At the top of this page in the MS., the bottom part of the letters of the name of the owner Hegge can be seen. The remains of the letters of the Christian name are indistinguishable, but *Hegge Dunelmensis* can still be deciphered.

[3] This line is written in larger form and underlined in red.

After my plesawns
So gret of myth is my pouste
Aſſ thyng xal be wrowth be me
I am oo god · in personys thre
knyt in oo substawns.

¶ I am þe trewe trenyte
here walkyng in þis wone
thre p*er*sonys myself I se
lokyn in me god Alone
I am þe ffadyr of powste
my sone w*ith* me gynnyth gon
my gost is grace in mageste
weldyth welthe up in hevyn tron
O god thre · I calle
I am fadyr of myth
my sone kepyth ryth
my gost hath lyth
and grace w*ith*-alle.

¶ My-self begynnyng nevyr dyd take
Fo. 10v And endeles I am thorw my*n* owyn myth
now wole I be-gynne my werke to make
Ffyrst I make hevyn w*ith* sterrys of lyth
In myrth *and* joy eu*er*mor*e* to wake
In hevyn I bylde Angeſſ fful bryth
my servau*ntys* to be *and* for my sake
w*ith* merth *and* melody worchepe my myth
I belde them in my blysse
Aungeſſ in hevyn evyr mor*e* xal be
In lyth ful cler*e* bryth as ble
W*ith* myrth *and* song to worchip me
Of joye þei may not mys.

hic cantent angeli in celo. ¶ Tibi *omn*es angeli tibi celi *et* vniu*er*se potestates · Tibi cherubyn et seraphyn incessabili voce proclamant · S*anct*us · S*anct*us · S*anct*us · D*omin*us deus sabaoth.

¶ To whos wurchipe synge ȝe þis songe
to wurchip god or reverens me
but ȝe me wurchipe ȝe do me wronge
ffor I am þe wurthyest þ*at* evyr may be.

lucifer*e*

Angeli boni

We wurchipe God of myth most stronge
whiche hath fformyd · bothe vs *and* the[1]
we may nevyr wurchyp hym to longe
Ffor he is most worthy of mageste
On knes to god we ffalle
oure lorde god wurchyp we
And in no wyse honowre we[2] the
A gretter lord may nevyr non be
than he *þat* made us alle.

lucifere

¶ A wurthyer lord forsothe am I
and worthyer than he · Euyr wyl I be
Fo. 11 In evydens *þat* I am more wurthy
I wyl go syttyn in goddys se
Above sunne *and* mone *and* sterr*ys* on sky
I am now set as ȝe may se
now wurchyp me ffor most mythy[3]
and for ȝ*our* lord honowre now me
Syttyng in my sete.

angeli mali

Goddys myth we for-sake
and for more wurthy we þe take
þe to wurchep honowre we make
and ffalle down at þi ffete.

Deus

¶ Thu lucyfere ffor þi mekyl pryde
I bydde þe ffalle from hefne to helle
And all þo þat holdyn on þi syde
in my blysse nevyr more to dwelle
At my comawndement anoon down þ*ou* slyde
w*ith* merth *and* joye nevyr more to melle
In myschyf *and* manas evyr xalt þ*ou* abyde
in byttyr brennyng *and* fyer so felle
in peyn evyr to be pyht[4].

[1] The *e* here is thickly written as if a correction, and a faint second *e* is written above.

[2] This word omitted and written over the line by the scribe.

[3] *wurthy* first written and crossed through.

[4] *pyth* first written and crossed through.

lucyfere

At thy byddyng þ*i* wyl I werke 75
and pas fro joy to peyne smerte
now I am a devyl ful derke
þ*at* was An Aungeł bryht.

¶ Now to helle þ*e* wey I take
in endeles peyn þ*er* to be pyht 80
Ffor fere of fyre a fart I crake
In helle donjoon̄ · myn dene is dyth.

Deus

Fo. 11v Now hevyn is made ffor Aungeł sake **2**
þ*e* fyrst day *and* þ*e* fyrst nyth
The secunde day watyr I make 85
The walkyn also ful fayr and b[r]yth[1]
The iij^de^ day · I parte watyr from erthe
tre *and* every growyng thyng
both erbe *and* floure of Suete smellyng
the iij^de^ day is made be my werkyng 90
now make I þ*e* day þ*at* xal be þ*e* fferthe.

¶ Sunne *and* mone *and* sterrys Also
þ*e* forthe day I make in same
þe v^te^ day · werm̄ *and* ffysch̄ þ*at* swy*m*me *and* go
byrdys *and* best*ys* · bothe wylde *and* tame 95
the sexte day my werk I do
And make þ*e* man Adam be name
In erthelech̄ paradys w*ith*-owtyn wo
I graunt þ*e* bydyng̃ lasse þ*ou* do blame.

¶ Fflesch̄ of þ*i* fflesch̄ · And bon of þ*i* bon̄ 100
Adam here is þ*i* wyf *and* make
both̄ ffysche *and* foulys þ*at* swy*m*myn *and* gon̄
to everych̄ of hem a name þ*ou* take
Bothe tre *and* frute *and* bestys echon̄
red *and* qwyte bothe blew *and* blake 105
þ*ou* ȝeve hem name be þ*i* self alon̄
Erbys *and* gresse both beet*ys* *and* brake
þ*i* wyff þ*ou* ȝeve name Also

[1] This word has been altered by erasure to *lyth*.

loke þ*at* ȝe not ses
ȝowr*e* ffrute to encres 110
þ*at* þer may be pres
me worchipe for to do.

Fo. 12 ¶ Now come Fforth Ad*a*m to paradys
ther xalt þ*ou* haue alt maner thynge
bothe flesch *and* ffysch *and* frute of prys 115
alt xal be buxu*m* at þ*i* byddyng
Her*e* is pepyr pyan *and* swete lycorys
take hem alt at þ*i* lykyng
both appel *and* per*e* *and* gentyl rys
but towche nowth þis tre þ*at* is of Cu*n*nyng 120
Alt thynge saff þis ffor þ*e* is wrought
her*e* is alt þinge þ*at* þ*e* xulde plese
Alt redy made on to þi*n* ese
Ete not þis frute ne me dysplese
ffor-þan þ*ou* deyst þ*ou* skapyst nowth. 125

¶ Now haue I made alt thynge of nowth
hevyn *and* Erth · foult *and* best
to alt thynge þ*at* my*n* hand hath wrowth
I graunt my*n* blyssyng þ*at* evyr xal lest [1]
My wey to hefne is redy sowth 130
of werkyng I wole þe [2] vij^te^ day rest
And alt my creatur*ys* þ*at* be a-bowth
my blyssyng ȝe haue both Est *and* west
of werkyng þ*e* vij^te^ day ȝe sees
and alt þ*o* þ*at* sees of laboryng her*e* 135
þ*e* vij^te^ day w*ith*-outyn dwer*e*
and wurchyp me in good maner*e*
þei xal in hefne haue endles pes.

¶ Adam go forth *and* be prynce in place
Fo. 12v ffor to hefne I sped my way 140
þ*i* wyttys wel loke þ*ou* chase
And gostly gou*e*rne þ*e* as I say.

Adam

¶ Holy ffadyr blyssyd þ*ou* be

[1] *last* first written, corrected to *lest*.
[2] A word, possibly *sefte*, written before vij^te^ and crossed through.

Ffor I may walke in welthe anow[1]
I ffynde datys gret plente 145
and many ffele frut*ys* ful every bow
Alł þis wele is ȝevyn to me
and to my wyf þ*a*t on me lowh
I haue no nede to towche ȝon tre
Aȝens my lordys wyl to werke now 150
I am a good garden*ere*
Euery frute of rythe name
I may gaderyn w*ith* gle *and* game
to breke þ*a*t bond I were to blame
þ*a*t my lord bad me kepyn here. 155

Eua

¶ We may both be blyth *and* glad
our*e* lordys comaundement to fulfyłł
w*ith* ffele frutys be we ffayr ffad
wou*n*dyr dowcet *and* nevyr on iłł
Euery tre w*ith* frute is sprad 160
of them to take as plesyth us tyłł
our*e* witte were rakyl *and* ovyr don bad
to fforfete Ageyns our*e* lordys wyłł
in ony wyse
in þis gardeyn I wyl go se 165
alł þ*e* fflour*ys* of fayr bewte
and tastyn þ*e* frut*ys* of gret ple*n*te
þ*a*t be i*n* paradyse.

Serpens

Fo. 13 ¶ Heyl Ffayr Wyff *and* comely dame
þis ffrute to Ete I þ*e* cownselle 170
take þis Appyl *and* Ete þis ssame
þis frute is best as I þ*e* telle.

Eua

That appyl to Ete I were to blame
ffrom joy our*e* lorde wold us expelle
we xuld dye *and* be put out w*ith* schame 175
in joye of paradyse nevyr more to duelle

[1] The final *e* of *welthe* and the *a* (?) of *anow* are obscured by a large blot in the MS. which may also hide some letter or mark above *he* in the first line of the stanza.

God hym self þus sayde[1]
what day of *þat* frute we Ete
with þese wurdys god dyd us threte
þat we xuld dye *our* lyff to lete 180
*þer*ffor*e* I am affrayde.

Serpens

¶ Of þis Appyl yf ȝe wyl byte
Evyn as god is so xal ȝe be
wys of Connyng as I ȝow plyte
lyke on to god in al degre 185
Su*n*ne *and* mone *and* sterrys bryth
ffysch *and* foule bo*þe* sond *and* se
at ȝo*ur* byddyng bothe day *and* nyth
aſſ thynge xal be in ȝowr*e* powste
ȝe xal be goddys per*e* 190
Take þis appyl in þin hond[2]
and to byte *þer* of *þou* ffond
take a nother to þin husbond
þer of haue *þou* no dwer*e*.

Eua

¶ So wys as god is in his gret mayn̄ 195
and ffelaw in ku*n*nyng ffayn wold I be.

Serpens

Fo. 13v Ete þis Appyl *and* in certeyn̄
þat I am trewe sone xalt *þou* se.

Eva

¶ To my*n* husbond *with* herte fful fayn̄
þis appyl I ber*e* as *þou* byddyst me 200
þis frute to Ete I xal asayn̄
so wys as god is yf we may be
And goddys per*e* of myth
To my*n* husbond I walke my way
and of þis appyl I xal a-say 205
to make hym to Ete yf *þat* I may
and of þis ffrewte to byth.

¶ *hic eua reueniet ade viro suo* et *dicet ei.*
¶ My semely spowse *and* good husbond

[1] The *a* in *sayde* has been corrected from an *e* by the scribe.
[2] Corrected by the scribe from *hand*, the *a* marked with deleting dot beneath and *o* written above the line.

lysteneth to me sere I ȝow pray
take þis ffayr appyl aꝗ in ȝ*our* hond̄ 210
þ*er* of a mursel byte *and* a-say
To Ete þis appyl loke þ*at* ȝe fonde
Goddys ffelaw[1] to be al-way
Aꝗ his wysd*a*m to vndyrstonde
and goddys per*e* to be ffor Ay 215
aꝗ thyng for to make
both̄ ffysch̄ *and* foule se *and* sond
byrd *and* best watyr *and* lond
þis appyl þ*ou* take out of my*n* hond
a bete þ*er*of þ*ou* take. 220

Adam

¶ I dar*e* not towch̄ þi*n* hand ffor dred
of our*e* lord god omnypotent
if I xuld werke aft*er* þi reed
of god our*e* makere I xuld be shent
Fo. 14 If þ*at* we do þis synful dede 225
we xal be ded by goddys jugement
out of þi*n* hand w*ith* hasty spede
cast out þat appyl a-non present
ffor fer of goddys threte.

Eva

Of þis appyl yf þ*ou* wylt byte 230
goddys per*e* þ*ou* xalt be pyht
So wys of kunnyng I þ*e* plyht
þis frute yf þ*ou* wylt Ete[2].

Adam̄

¶ If we it Ete our*e* self we kylle
as god us told we xuld be ded 235
to ete þ*at* frute · *and* my lyf to spylle
I dar not do aftyr þ*i* reed.

Eua

A ffayr Aungeꝗ þus · seyd me tylle
to Ete þ*at* appyl take nevyr no dred
so ku*n*nyng as god in hevyn hille 240
þ*ou* xalt sone be w*ith*-inne a sted
þ*er*fore þis frute þ*ou* Ete.

[1] The second *f* is doubtful here. It seems to have been faintly inserted.
[2] Final *e* blotted.

Adam

Off goddys wysdam for to lere
and in kunnyng to be his pere
of thyn hand I take it here
and xal sone tast þis mete.

Adam dicit sic

Alas Alas ffor þis fals dede
my flesly frend · my fo I fynde
Schameful synne doth us vn-hede [1]
I se vs nakyd be-fore *and* be-hynde
oure lord*ys* wurd wold we not drede
Fo. 14v þerfore we be now caytyvys vn-kynde
oure pore *preuytes* ffor to hede
Su*m*me ffygge levys fayn wolde I fynde
Ffor to hyde oure schame
Wo*m*man ley þis leff on þ*i* pryvyte
And w*ith* þis leff I xal hyde me
Gret schame it is vs nakyd to se
Oure lord god þus to grame.

Eva

¶ Alas þ*at* evyr þ*at* speche was spokyn
þ*at* þ*e* fals Aungel seyd on to me
Alas oure makers byddyng is brokyn
Ffor I haue towchyd his owyn dere tre
oure fflescly eyn [2] byn al vnlokyn
nakyd for synne oure sylf we se
þ*at* sory Appyl þ*at* we han sokyn
to deth̄ hath̄ brouth my spouse *and* me
Ryth grevous is oure synne
of mekyl shame now do we knowe
Alas þ*at* evyr þis Appyl was growe
to dredful deth now be we throwe
in peyne vs evyr to pynne.

Deus

¶ Adam þ*at* w*ith* my*n* handys I made
where art þ*ou* now · what hast þ*ou* wrought.

Adam

A lord for synne oure flour*ys* do ffade
I here þi voys · but I se þ*e* nought.

[1] *vn* corrected by scribe from *un*.
[2] An *h* crossed out between *eyn* and *byn*.

Deus

Adam why hast þ*ou* synnyd so sone
þus hastyly to breke my bone
Fo. 15 And I made þ*e* mayst*er* vndyr mone
Trewly of Euery tre 280
O tre I kept for my owe
Lyff *and* deth þ*er*-in I knowe
þ*i* synne fro lyf now þ*e* hath throwe
Ffrom deth þ*ou* mayst not fle.

Ad*a*m

¶ Lord I haue wrought a3ens þ*i* wyll 285
I sparyd nat my sylf to spylle
þ*e* woman þ*at* þ*ou* toke me tylle
sche brougth me þ*er* to
It was here cou*n*sell *and* here reed
sche bad me do þ*e* same deed 290
I walke as werm̃ w*ith*-out*y*n wede
A-wey is schrowde *and* sho.

Deus

¶ Wo*m*man þ*at* arte þis mannys wyffe
why hast þ*ou* steryd 3*our* bothers stryffe
now 3e be[1] ffrom 3*our* ffayr lyffe 295
and are demyd for to deye
Vnwys wo*m*man sey me why
þ*at* þ*ou* hast don þis fowle foly
and I made þe a gret lady
in paradys for to pleye. 300

Eua

¶ Lord whan þ*ou* wentyst from þis place
A werm w*ith* An Aungelys face
he hyth vs to be ful of grace
þ*e* frute yf þ*at* we Ete
I dyd his byddyng Alas Alas 305
now we be bowndyn in dethis las
I suppose it was sathanas
to peyne he gan vs pete.

Deus

Fo. 15v ¶ Thou werm w*ith* þ*i* wylys wyk
þ*i* fals fablis þei be ful thyk 310

[1] *be* omitted and written above the line.

why hast *þou* put dethis pryk
in Adam *and* his wyff
thow þei bothyn my byddyng haue brokyn
out of whoo ȝet art not wrokyn
in helle logge *þou* xalt be loky[1] 315
And nevyr mo lacche lyff.

Diabolus

¶ I xal *þe* sey where ffore *and* why
I dede hem all̄ þis velony
ffor I am ful of gret envy
Of wreth̄ *and* wyckyd hate 320
That man xulde leve above *þe* sky
where as su*m* tyme dwellyd I
and now I am cast to helle sty
streyte out at hevyn gate.

Deus[2]

¶ Adam ffor *þou* *þat* appyl boot 325
A-ȝens my byddyng well̄ I woot
Go teyl *þi* mete wi*th* swynk *and* swoot
in to *þi* lyvys ende
Goo nakyd vngry *and* bare ffoot
Ete both̄ erbys gres *and* root 330
thy bale hath non other boot
as wrecch̄ in werlde *þou* wende.

¶ wo*m*man *þou* sowtyst þis synnyng
And bad hym breke my*n* byddy*n*g
*þer*fore *þou* xalt ben vndyrlyng 335
to mannys byddyng bend
what he byddyth *þe* · do *þou* *þat* thynge
and bere *þi* chyldere · wi*th* gret gronynge
Fo. 16 In daungere *and* in deth dredynge
in to *þi* lyvys ende. 340

¶ Thou wyckyd worm ffull of pryde
ffowle envye syt be *þi* syde
Vpon *þi* gutt *þou* xalt glyde
As werm wyckyd in kende
tyl a mayden in medyl-erth be born̄ 345
þou ffende I warn *þe* be-forn

[1] So in MS. for *lokyn*. [2] *Ad* first written and crossed through.

thorwe here *þi* hed xal be to-torn
On wombe a-wey *þou* wende.

Diabolus

¶ At *þi* byddy*ng* ffowle I falle
I krepe hom to my stynkyng stalle 350
helle pyt *and* hevyn halle
xul do *þi* byddyng bone
I ffalle dowñ here a ffowle freke
ffor þis ffalle I gynne to qweke
w*ith* a ffart my brecħ I breke 355
my sorwe comyth ful sone.

Deus

¶ Ffor ȝ*our* synne *þat* ȝe haue do
out of þis blysse sone xal ȝe go
in erthly labo*ur* to levyn in wo
and sorwe þe[1] xal a-tast 360
Ffor ȝ*our* synne *and* mys-doyng
An Angeł w*ith* a swerd brennyng
out of þis joye he xal ȝow dyng
ȝ*our* welth awey is past

ñ *Hic recedit deus et angelus seraphicus cum gladio ffl*a*mmea verberat adam et Euam extra paradisum.*

Seraphi*m*

Fo. 16v ¶ ȝe wrecchis vnkend *and* ryht vnwyse 365
out of þis joye hyȝ ȝow in hast
w*ith* flammyng swerd ffrom paradyse
to peyn I bete ȝow of care to tast
ȝ*our* myrth is turnyd to carfułł·syse
ȝ*our* welth w*ith* synne a-wey is wast 370
ffor ȝ*our* ffalse dede of synful gyse
þis blysse I spere ffrom ȝow ryth fast
here-in come ȝe no more
Tyl a chylde of a mayd be borñ
and vpon þe rode rent *and* torñ 375
to saue ałł *þat* ȝe haue forlorn
ȝ*our* welth for to restore.

Eva

¶ Alas alas *and* wele away
þat evyr towchyd I *þe* tre
I wende as wrecch in welsom way 380

[1] So in MS. *ye* for *ȝe*.

in blake busshys my boure xal be
In paradys is plente of pleye
Ffayr frutys ryth gret plente
þe ȝatys be schet w*ith* godys keye
my husbond is lost be-cause of me 385
leve spowse now þ*ou* fonde
Now stomble we on stalk *and* ston
my wyt a-wey is fro me gon
wrythe on to my necke bon
w*ith* hardnesse of þ*in* honde. 390

Adam

Wyff þ*i* wytt is not wurth a rosch
leve woman turne þ*i* thought
Fo. 17 I wyl not sle fflescly of my fflesch
Ffor of my flesch · þi fflesch was wrought
Oure hap was hard · oure wytt was nesch 395
to paradys whan we were brought
my wepyng xal be longe ffresch
schort lykyng xal be longe bought
no more telle þ*ou* þ*at* tale
Ffor yf I xulde sle my wyff 400
I sclow my self w*ith*-owtyn knyff
in helle logge[1] to lede my lyff
w*ith* woo in wepyng dale.

¶ But lete vs walke forth in to þe londe
w*ith* ryth gret labo*ur* oure fode to fynde 405
w*ith* delvyng *and* dyggyng w*ith* my*n* hond
oure blysse to bale *and* care to-pynde
And wyff to spynne now must þ*ou* ffonde
oure nakyd bodyes in cloth to wynde
tyll sum comforth of godys sonde 410
w*ith* grace releve oure careful mynde
Now come go we hens wyff.

Eva

Alas þ*at* ever we wrought þis synne
oure bodely sustena*un*s for to wynne

[1] Or lo*n*gge. There is a stroke over the *og*, but it seems to be in a different ink, higher and not quite like the ordinary mark of the contracted *n*. The words *in helle logge* occur on Fo. 15v.

ȝe must delve *and* I xal spynne 415
in care to ledyn oure lyff.

Abeel

I wolde ffayn knowe · how I xuld do
to serue my lord god to his plesyng
þer fore Caym brother lete us now go
vn-to oure ffadyr w*ith*-owte lettyng *
Fo. 17v Suenge hym in vertu and in norture 5
to com to þe hyȝe joy celestyall
remembryng to be clene *and* pure
for in mys-rewle we myth lythly fall
A-ȝens hevyn kynge
lete us now don oure dyligens 10
to come to oure fader*ys* pr*e*sens
Good brother passe we hens
to knowe ffor oure levynge.

Caym

¶ As to my fadyr lete us now tee
to knowe what xal be his talkyng 15
and þat[1] I holde it but vanyte
to go to hy*m* ffor Any spekyng
to lere of his lawe
Ffor if I haue good[2] a-now plente
I kan be mery so moty the 20
thow my fadyr I nevyr se
I ȝyf not þer of An hawe.

Abel

¶ Ryth sovereyn fadyr semely sad *and* sure
eu*er* we thank ȝow in hert body and thowth

* Ff. 16–17 at the foot of both pages crosswise in large liturgical script with rubrication :

Adam. genuit { Caym / Abel / Seth } Caym. *genuit.* Enoch. *genuit.* Iradh. *genuit* Maynael. *genuit* Matussahel. *genuit* lamech. þat slow Caym. þis lamech had 2 wyffys. Ada and[3] Sella. Of Ada com Jabel. fadere of tent*ys* and of herdme*n*. [The rest of this note on Lamech's offspring is illegible, the letters having been cut away with the margin. It is in the handwriting of the scribe of the MS.]

[1] MS. *y*[t] for *ȝet* ?

[2] The *d* is written over some other letter.

[3] Some miswritten letters (? Se) crossed out after *and*.

And alwey shuꝫ whyꝫ oure lyf may indure 25
as inwardly in hert it kan be sought
bothe my brother *and* I[1]
Ffadyr I ffalle on-to ȝ*our* kne
to knowe how we xul rewlyd be
ffor godys þ*at* fallyth bothe hym *and* me 30
I wolde ffayn wete trewly.

Adam

¶ Sonys ȝe arn to spekyn naturaly
The ffyrstffrute · of kendely engendrure
Be-fforn whom saff ȝ*our* modyr *and* I
Fo.18 were nevyr non of mannys nature 35
And ȝit were we al of a nother portature[2]
As ȝe haue me oftyn herd seyd sothly
Wherfore sonys yf ȝe wyl lyff sad *and* sure
Ffyrst I ȝow counseyꝫ most syngulerly
God ffor to loue *and* drede 40
And suche good as god hath ȝow sent
the fyrst frute offyr to hym in sacryfice brent
hym evyr be-sechyng w*ith* meke entent
In aꝫ ȝ*our* werkys to save *and* spede.

Abeeꝫ

¶ Gramercy ffadyr ffor ȝ*our* good doctrine 45
Ffor as ȝe vs techyn so xal we do
And as ffor me þ*orwe*[3] goddys grace[4] dyvyne
I wyl fforth w*ith* applye me þ*er*to.

Cayme

And þow me be lothꝫ I wyl now also
On to ȝ*our* counseꝫ ffadyr me inclyne 50
and ȝitt I say now to ȝow bothꝫ too
I had levyr gon hom weꝫ ffor to dyne.

Adam

Now god graunt good sacryfice to ȝow bothꝫ too
he vowche-saff to acceptyn ȝow *and* aꝫ myne

[1] The writing becomes noticeably smaller from this line to the bottom of the folio.

[2] *As ȝe haue me oftyn seyd sothly*, with *haue* written above the line, is written as the next line and crossed through.

[3] MS. þõ. [4] A letter y or þ obliterated after *grace*.

and ȝeve ȝow now grace to plesyn hy*m* soo 55
þ*at* ȝe may come to þ*at* blysse þ*at* hy*m* self is inne
W*ith* gostly grace
þ*at* all ȝ*our* here levyng[1]
may be to his plesyng
and at ȝ*our* hens partyng 60
to com̄ to good place.*

Abell dicit

Fo.18v ¶ Al-myhtty god and god ful of myth
be whom all þing is made of nowth
to þ*e* my*n* hert is redy dyht
for upon þ*e* is all my thought 65
O souereyn lord reygnyng in eternyte
w*ith* all þ*e* mekenesse þ*at* I kan or may
This lombe xal I offre it up to the
accept it blyssyd lord I þ*e* pray
my ȝyft is but sympyl þis is no nay 70
but my wyl is good *and* evyr xal be
þ*e* to servyn *and* worchepyn both̄ nyht *and* day
and þ*er* to þ*i* grace grawnt þ*ou* me
throwh̄ þ*i* gret mercy
which̄ in a lombys lyknes 75
þ*ou* xalt for mannys wyckydnes
Onys ben offeryd in peynfulnes
and deyn ful dolfoly.

¶ Ffor trewly lord þ*ou* art most worthy
þ*e* best to haue in eche degre 80
both̄ beste *and* werst ful certeynly
all is had þ*ur*owe grace of þ*e*
The best schep full hertyly
among*es* my flok þ*at* I kan se
I tythe it to god of gret mercy 85
And bettyr wold if bettyr myht be
Evyn here is myn offryng
I tythe to þ*e* w*ith* ryht good wylle

[1] Lines 58–61 are written in two lines divided by two black strokes.

* Ff. 17v–18. At the foot of the pages as before : Seth gen*uit* Enos. *genuit* Caynan. *genuit* Malachel. *genuit* Jared gen*uit* Enok *genuit* Matussalem. *genuit* Lamech *genuit* Noe.

of þe best þou sentyst me tylle
now gracyous god on hevyn hille
Accept now my tythyng.

Caym

Fo. 19 ¶ Amonges all folys þat gon on grownd
I holde þat þou be on of þe most
to tythe þe best þat is not sownd
and kepe-þe werst þat is nere lost
But I more wysly xal werke þis stownde
to tythe þe werst *and* make no bost
Off all my cornys þat may be fownde
In all my ffeldys both croft *and* cost
I xal lokyn on every syde
here I tythe þis vnthende sheff
lete god take it · or ellys lef
þow it be to me gret repreff
I ȝeve no ffors þis tyde.

Abell

¶ Now Caym brother þou dost ful ill
Ffor god þe sent both best *and* werst
þerfore þou shewe to hym good wyll
and tythe to god evyr of þe best.

Caym

In feyth þou shewyst now a febyll skyll
it wolde me hyndyr *and* do me greff
what were god þe bettyr þou sey me tyll
to ȝevyn hym awey my best sheff
and kepe my self þe wers
he wyll neyther ete nor drynke
Ffor he doth neyther swete nor swynke
þou shewyst a ffebyl reson me thynke
what þou fonnyst as a best I gesse.

Abell

ȝit me thynkyth my wyt is good
to god euer more sum loue to shewe[1]
Fo. 19v (marked 20) Off whom we haue oure dayly food
and ellys we had but lytyl drewe.

Caym

ȝitt me thynkeht þi wytt is wood

[1] The writing in this and the following three lines at the top of the next folio is smaller.

Ffor of þ*i* lor*e* I ffynde but ffewe
I wyłł neu*er* þ*e* mor*e* chawnge my mood
Ffor no wordys þ*at* þ*ou* dost shewe 125
I sey I wyłł tythe þ*e* werst.

Abełł

Now god þ*at* syt in hefne aboue
On whom is sett ałł my*n* hool loue
þis wyckyd wyłł from þ*e* he showe[1]
As it plesyth hym best. 130

¶ *Hic ardent decimum Abel · et Caym quo facto dicit* Caym̃

¶ herke abel brother what a-ray is þis
thy tythyng brennyth as ffyr*e* fful bryght
it is to me gret wondyr i-wys
I trow þis is now a straunge syght.

Abełł

Goddys wyłł fforsothe it is 135
þ*at* my tythyng w*ith* fyr*e* is lyth
Ffor of þ*e* best were my tythis
and of þ*e* werst þ*ou* dedyst hy*m* dyght
bad thyng þ*ou* hy*m* bede
of þ*e* best was my tythyng 140
and of þ*e* werst was þin offryng
þ*er*for god almyghty hevyn kyng
Alowyht ryht nowth þ*i* dede.

Caym̃

¶ What þ*ou* stynkyng losel *and* is it so
doth god þ*e* love *and* hatyht me 145
þ*ou* xalt be ded I xal þ*e* slo
þ*i* lord þ*i* god þ*ou* xalt nevyr se
Fo. 20 Tythyng mor*e* xalt þ*ou* nevyr do
W*ith* þis chavyl bon I xal sle þ*e*
þ*i* deth is dyht þ*i* days be go 150
out of my*n* handys xalt þ*ou* not fle
W*ith* þis strok I þ*e* kylle
Now þis boy is slayn *and* dede
Of hym I xal nevyr mor*e* han drede
He xal her*e* after nevyr ete brede 155
W*ith* þis gresse I xal hym hylle.

[1] So in MS. for *shoue*.

¶ Deus

Caym come fforth *and* answere me
a-soyle my qwestyon anon ryght
thy brother Abel wher is now he
ha don and answere me as tyght. 160

Caym

My brothers kepere ho made me
Syn whan was I his kepyng knyght
I kan not telle wher þat he be
to kepe hym was I nevyr dyght
I know not wher he is. 165

Deus

A cursyd Caym þou art vntrewe
and for þi dede þou xalt sore rewe
þi brothers blood þat þou slewe
Askyht vengeauns of þi mys.

¶ Thu xalt be cursyd on þe grounde 170
Vn-prophitable where so þou wende
both veyn *and* nowthty *and* no thyng sounde
with what þing þou medele þou xalt it shende.

Caym

Alas in whoo now am I wounde
a-cursyd of god as man vn-kende 175
of any man yf I be founde
He xal me slo I haue no ffrende
Alas and wele Away.

Deus

Fo. 20v Of what man þat þou be sclayn̄
he xal haue vij folde more payn 180
hym were bettyr to be sayn̄[1]
on lyve be nyth ne day.

Caym

¶ Alas alas whedyr may I go
I dare nevyr se man in þe vesage
I am woundyn as a wrecch in wo 185
And cursyd of god ffor my ffalfage[2]

[1] The word *nevyr* is written above the line after *bettyr* in blacker ink and a different hand. The original construction may be compared with the following on Fo. 22: *I pray to god . . . that he me kepe in such a plyght . . . I to affendyn hym day nor nyght.*

[2] So in MS. for *ffalsage.*

Vn-profytabyl and vayn also
In felde *and* town̄ in strete and stage
I may nevyr make merthis mo
I wot nevyr whedyr to take passage 190
I dare not here abyde
Now wyl I go wende my way
W*ith* sor*e* syeng *and* wel away Introit*us* Noe
to loke where þ*at* I best may
Ffrom mannys ssyht me hyde. 195

Remainder of Fo. 20ᵛ—4 inches—left blank.

¶ Noe

Fo. 21 (B quire) GOd of his goodnesse · and of grace grounde **4**
By whoys gloryous power all̄ thyng is wrought
in whom all̄ vertu plentevously is ffounde
w*ith*-owtyn whos wyl may be ryth nought
Thy seruaunt*ys* saue lord fro synful sownde 5
In wyl in werk in dede *and* in thouht
our*e* welth in woo lete nevyr be fownde
Vs help lord from synne þ*at* we be in brought
Lord god fful of myght
Noe ser*ys* my name is knowe 10
my wyff *and* my chyldere here on rowe
to god we pray w*ith* hert ful lowe
to plese hym in his syght.

¶ In me Noe þ*e* secunde age
in dede be-gynnyth as I ȝow say 15
afftyr Adam w*ith*-outy*n* langage
þ*e* secunde fadyr am I in fay
But men of levyng be so owt-rage
bothe be nyght and eke be day
þ*at* lesse þan synne þ*e* soner swage 20
god wyl be vengyd on vs su*m* way
In dede
Ther may no man go þ*er* owte
but synne regnyth in every rowte
In every place rownde a-bowte 25
Cursydnes doth sprynge *and* sprede.

Vxor Noe

¶ Alł myghty god of his gret grace
enspyre men *with* hertely wyłł
for to sese of here trespace
Ffor synfułł levyng oure sowle xal spyłł *
Fo. 21v Synne offendyth god in his face
and a-grevyth oure lorde ffułł ylle
It causyth to man ryght grett manace
and scrapyth hym out of lyvys bylle
þat blyssyd book
What man in synne doth alł wey scleppe
He xal gon to helle ful deppe
than xal he nevyr aft*er* creppe
out of *þat* brennyng brook.

¶ I am *ȝour* wyff[1] *ȝour* childeryn þese[2] be
On to us tweyn it doth longe
hem to teche in alł degre
Synne to for-sakyn *and* werkys wronge
Therfore fere[3] for loue of me
Enforme hem wele evyr amonge
Synne to for-sake *and* vanyte
and vertu to ffolwe *þat* þei ffonge
oure lord god to plese.

Noe

I warne ȝow childeryn on *and* alł
Drede oure lord god in hevy[4] hałł
and in no forfete *þat* we[5] ne fałł
oure lord for to dysplese.

Shem

¶ A dere ffadyr god for-bede
þat we xulde do in ony wyse
Ony werke of synful dede
oure lord god *þat* xulde a-gryse

* In the bottom margin of Fo. 21 as before:—Noe genuit { Sem / Cham / Japhet }

[1] *þese ch* written after *wyff* and crossed through.
[2] þese here appears to be the correction of another word.
[3] So in MS. for *sere*. [4] So in MS. for *hevyn*. [5] So in MS. for *ȝe*.

my name is Shem[1] ȝ*our* son of p*ri*se
I xal werke aftere ȝ*our* rede
and also wyff þe weyll a-wyse
Wykkyd werkys þ*at* þ*ou* non brede 60
nev*er* in no degr*e*.

Vx*or* Seem̃

For-sothe ser*e* be goddys grace *
Fo. 22 I xal me kepe from all trespace
þ*at* xulde offende goddys face
be help of þ*e* trynyte. 65

Cham̃

¶ I am cham ȝ*our* secunde son̄
and purpose me be goddys myght
nevyr suche a dede for to don
þ*at* xuld a-greve god in syght.

Vx*or* Cham̃

I pray to god me grawnt þis bon̄e 70
that he me kepe in such A plyght
mornynge hevenynge mydday *and* none
I to affendyn hym day nor nyght
lord god I þ*e* pray
Bothe wakynge *and* eke in slepe 75
gracyous god þ*ou* me keppe
þ*at* I nevyr in daunger crepe
On dredfull domys day.

Japhet

¶ Japhet þ*i* iij^de^ sone is my name
I pray to god wher so we be 80
þ*at* he vs borwe fro synfull shame
and in vertuous levynge evyr mor*e* kepe me.

Vx*or* Japhet

I am ȝ*our* wyff *and* pray þ*e* same
þ*at* god vs saue on sonde *and* se

* At the foot of Ff. 21ᵛ–2: Sem gen*uit* Arfaxat. *genuit* Sale. *genuit* Heber. *genuit* Phaleg. gen*uit* Reu. *genuit* Sarug. *genuit* Nachor. *genuit* Thare. *genuit*. Abraham, Nacor, Aran

[1] Manly gives *Chem*, but the *S* is clear in the MS. Some miswritten letter between *Shem* and ȝ*our* has been crossed out by the scribe.

W*ith* no grevauns þ*a*t we hy*m* grame 85
he grawnt vs grace synne to fle
lord god now here oure bone.

Noe

Gracyous god þ*a*t best may
w*ith* herty wyl to the we pray
þ*ou* save us sekyr bothe nyght *and* day 90
Synne þ*a*t we noon done.

Deus

Fo. 22v ¶ Ow what menyht this mys-levyng man
whiche myn hand made *and* byldyd in blysse
Synne so sore grevyht me ȝa in certayn [1]
I wol [2] be vengyd of þis grett mysse 95
Myn aungel dere þ*ou* xalt gan [3]
to noe þ*a*t my servaunt is
A shypp to make on hond to tan̄
þ*ou* byd hym swyth̄ ffor hy*m* *and* his
Ffrom drynchyng hem to save 100
Ffor as I am god off myght
I xal dystroye þis werd down̄ ryght
Here synne so sore grevyht me in syght
þei xal no mercy haue.

¶ Ffecisse ho*min*em nu*n*c penitet me 105
þ*a*t I made man sore doth me rewe
myn handwerk to sle sore grevyth me
but þ*a*t here synne here deth doth brewe
Go sey to noe as I bydde þ*e*
hy*m* self his wyf his chylderyn trewe 110
tho viij sowlys in shyp to be
thei xul not drede þ*e* flodys fflowe
þ*e* fflod xal harme them nowht
of all̄ ffowlys *and* bestys thei take a peyre
In shypp to saue both ffoule *and* ffayere [4] 115
Ffrom all̄ dowtys and gret dyspeyre
This vengeauns or it be wrought.

[1] First written *tayn* and corrected to *teyn* or vice versa.
[2] *he* written after this word and crossed through.
[3] *gon* first written and corrected by scribe to *gan*.
[4] In this line and the remaining six to the bottom of the folio the writing is thicker and slightly more cramped.

Angelus ad noe

¶ Noe · Noe · A shypp loke þ*ou* make·[1]
and many a chaumbyr þ*ou* xalt haue þ*er*inne
Of euery kyndys best a cowpyl þ*ou* take·
w*ith*-in þ*e* shypp bord · her*e* lyvys to wynne·*
Fo. 23 Ffor god is sor*e* grevyd w*ith* man for his synne
þ*at* all̄ þis wyde werd xal be dreynt w*ith* flood
saff þ*ou* *and* þi wyff xal be kept from þis gynne
and also þ*i* chylderyn w*ith* her*e* vertuys good.

Noe

¶ How xuld I haue wytt a shypp for to make
I am of ryght grett Age V. C. ȝer*e* olde
it is not for me þis werk to vndyr-take
Ffor ffeynnesse[2] of Age my leggys gyn ffolde.

Angelus

This dede ffor to do be bothe blythe *and* bolde
God xal enforme þ*e* *and* rewle þ*e* ful ryght
of byrd *and* of beste take as I þ*e* tolde
A peyr in to þ*e* shypp *and* god xal þ*e* qwyght.

Noe

¶ I am ful redy as god doth me bydde
A shypp for to make be myght of his grace
Alas þ*at* ffor synne it xal be so be-tydde
þ*at* vengeauns of flood xal werke þis manase
god is sor*e* grevyd w*ith* oure grett tresspas
þ*at* w*ith* wylde watyr þ*e* werd xal be dreynt
A shypp*e* for to make now lete us hens pas
þ*at* god a-ȝens us of synne haue no compleynt

Hic transit noe cum familia sua pro naui quo exeunte locum interludij sub intret statim lameth conductus ab adolescente et di[cit

lameth

¶ Gret mornyng I make *and* gret cause I haue
Alas now I se not · for age I am blynde
blyndenes doth make me of wytt for to rave
Whantynge of eye syght · in peyn doth me bynde
whyl I had syht þ*er* myht nevyr man fynde
my pere of Archerye · in all̄ þis werd A-boute

* At the foot of Fo. 22ᵛ. Aran. gen*uit*. loth.

[1] The stops at *make*, *take*, and *wynne* are in fresher coloured ink.

[2] MS. *ffeyynnesse* with a dot under the second *y*. The letters for *þ* and for *y* in this MS. are indistinguishable; Halliwell reads *ffeyþnnesse*.

Fo. 23v Ffor ȝitt schet I nevyr at hert Are · nere hynde
but yf þat he deyd · of þis no man haue doute.

¶ Lameth þe good archere my name was ovyr all
Ffor þe best Archere myn name dede ever sprede
Record of my boy here wytnes þis he xal
What merk þat were set me · to deth it xuld blede.

Adolescens

It is trewe mayster · þat ȝe seyn in dede
Ffor þat tyme ȝe had ȝoure bowe bent in honde
If þat ȝour prycke had be half a myle in brede[1]
ȝe wolde þe pryk han hitte if ȝe ny had stonde.

lameth

¶ I xuld nevyr affaylid · what marke þat ever were sett
whyl þat I myght loke · *and* had my clere syght
and ȝit as me thynkyht no man xuld shete bett
Than I xuld do now · if myn hand were sett a-ryght
A-spye som̄ marke boy · my bow xal I bende wyght
and sett myn hand euyn to shete at som̄ best
and I dar ley a wagour his deth for to dyght
þe marke xal I hitt · my lyf do I hest.

Adolescens

¶ Vndyr ȝon grett busche mayster · A best do I se
take me þin hand swyth̄ *and* holde it ful stylle
now is þin hand evyn as euyr it may be
Drawe up þin takyll ȝon best for to kylle.

lameth

My bowe xal I drawe ryght with herty wylle
this brod arwe I shete þat best ffor to sayll
now haue at þat busch̄ · ȝon best for to spylle
A sharppe schote I shote þer of I xal not fayll.

Caym

¶ Out out *and* alas · myn hert is on sondyr
With a brod arwe I am ded *and* sclayn
Fo. 24 I dye here on grounde · myn hert is all to tundyr
With þis brod arwe it is clovyn on twayn,

lameth

Herke boy cum telle me þe trewth̄ in certeyn
what man is he þat · þis cry doth þus make.

Adolescens

Caym þou hast kyllyd I telle þe ful pleyn̄

[1] *dede* first written and crossed through.

with þ*i* sharp shetyng · his deth̃ hath he take.

lameth

¶ Haue I slayn cayme · alas what haue I done
þ*ou* stynkynge lurdeyn · what hast þ*ou* wrought
þ*ou* art þ*e*[1] why I scle hym so sone
þer fo*re* xal I kyll̃ þ*e* her*e* · þ*ou* skapyst nowght 185

Hic lameth cum arcu suo verberat Adolescentem ad mortem, dicente Adolescente

Adolescens

Out out I deye her*e* · my deth is now sought
þis theffe w*ith* his bowe hath broke my brayn̄
þ*er* may non helpe be · my dethe is me brought
Ded her*e* I synke dow̄n · as man þ*at* is sclayn.

lameth

¶ Alas what xal I do wrecch̃ wykkyd on woolde 190
God wyl be vengyd ful sadly on me
Ffor deth of Caym I xal haue vii folde
mor*e* peyn þan he had · þat abell̃ dede sle.
These to mennys deth full̃ sor*e* bought xal be
Vpon all̃ my blood god wyll̃ venge þis dede 195
wher*e* fo*re* sor*e* wepyng hens wyl I fle
and loke wher*e* I may best my hede sone heyde.

*Hic recedat lameth et statim intrat noe cum naui cantantes.**

Noe

Fo. 24v ¶ With doolful hert syenge sad and sor*e*
Grett morny*ng* I make ffor this dredful flood
of man *and* of best is dreynte many a skore 200
All̃ þis werd to spyll̃ þese flodys be ful wood
and all̃ is for synne of ma*n*nys wylde mood
þ*at* god hath ordeyned þis dredfull̃ vengeau*n*ce
In þis flood spylt is many a ma*n*nys blood
Ffor synfull̃ levynge of man we haue gret grevau*n*s. 205

¶ All̃ þis hundryd ȝer*e* ryght her*e* haue I wrought
this schypp for to make as god dede byd me

[1] So in MS.

* Note at bottom of Fo. 24 in what might be the less careful writing of the scribe of the MS. but might also be a different hand:

Noe Schyp was in lenght CCC. Cubytes. } þ*e* fflod 15 Above
In brede ffyfty. And þ*e* heyth̃. thretty. } hyest montayn̄.

of all maner best*ys* a copyll is in brought
W*ith*-in my Shypp borde · on lyve for to be
Ryght longe god hath [1] soferyd · amendy*ng* to se 210
All þis hundyrd ȝer*e* god hath shewyd grace
Alas fro gret syn man wyl not fle
God doth þis vengeau*n*s for our*e* gret trespase.

Vx*or* noe

¶ Alas for gret ruthe of þis gret vengeau*n*ce
gret doyl it is to se · þis watyr so wyde 215
but ȝit thankyd be god of þis ordenau*n*ce
þ*at* we be now savyd on lyve to abyde.

Seem̃

Ffor grett synne of lechory all þis doth betyde
Alas þ*at* evyr such synne xulde be wrought
þis fflood is so gret on every asyde 220
þ*at* all þis wyde werd to car*e* is now brought.

Vx*or* Seem̃

¶ Becawse of chylderyn of God · þ*at* weryn good [2]
dede forfete ryght sor*e* · what tyme þ*at* þ*ei* were
synfully compellyd to caymys blood
Fo. 25 Ther-fore be we now cast in ryght grett car*e*. 225

Cham

Ffor synful levynge þis werde doth for-far*e*
So grevous vengeau*n*s myght nevyr man se
ouyr all þis werd wyde þ*er* is no plot bar*e*
w*ith* watyr and w*ith* flood god vengyd wyll be.

Vx*or* Cham

¶ Rustynes of Synne is cawse of þese wawys 230
Alas in þis fflood þis werd xal be lorn
ffor offens to god · brekyng his lawys
On Rokkys ryght sharp is many a man torn.

Japhet

So grevous fflodys were nevyr ȝett be-forn̄
Alas þ*at* lechory þis vengeau*n*s doth gynne 235
it were well bettyr eu*er* to be vnborn̄
than ffor to forfetyn evyr mor*e* in þ*at* synne.

Vx*or* Japhet

¶ Our*e* lord god I thanke of his gret grace
þ*at* he doth us saue from þis dredful payn̄

[1] An *f* between *hath* and *soferyd* obliterated.
[2] *gode* first written and crossed out.

hy*m* for to wurchipe in euery stede *and* place 240
we beth gretly bownde w*ith* myght *and* w*ith* mayn.

Noe

xl^ti^ days and nyght*ys* hath lasted þis rayn
And xl^ti^ days þis grett flood be-gynnyth to slake
this Crowe xal I sende out to seke sum playn
Good tydyng*ys* to brynge þis massage I make 245

hic emittat coruum et *parum expectans iterum dicat*

¶ This crowe on su*m* careyn is fall for to ete
þer fo*re* a Newe masange*re* I wyll fforth now sende
ffly fforth þ*ou* fayr dove ovyr þese wate*r*ys wete
and aspye afftere su*m* drye lond oure mornyng to A-mend

hic euolet columba qua redeinte[1] *cum ramo virid' oliue.*

Fo. 25v Joye now may we make of myrth þat þ*at* were frende 250
A grett olyve bush þis dowe doth us brynge
Ffor joye of þis token ryght hertyly we tende
oure lord god to worchep a songe lete vs synge

Hic decantent hos versus · ¶ Mare vidit et fugit ⁝ jordanis conu*er*sus est retrorsum · Non nob*is* do*mi*ne non nob*is* ⁝ *sed* no*mini* tuo da glo*ri*am · *Et sic recedant cum naui.*

Introit*us* abrahe &c.

5

MOst myghty make*re* of Sunne and of mone
Kyng *of* kyng*ys* *and* lord ou*er* all
All myghty god in hevyn trone
I þ*e* honowr*e* *and* evyr more xal
My lord my god to þ*e* I kall 5
w*ith* herty wyll lord I þ*e* pray
In synfull lyff lete me nevyr fall
but lete me leve evyr to þ*i* pay.

¶ Abraha*m* my name is kydde
and patryarke of Age ful olde 10
And ȝit be þe grace of god is bredde
In my*n* olde age a chylde full bolde
Ysaac lo here his name is tolde
My swete sone þ*at* stondyth me by

[1] So in MS.

Amonge*s* aſſ chylderyn þ*at* walkyn on wolde 15
A louelyer chylde is non trewly.

¶ I thanke god · w*ith* hert weſſ mylde
of his gret mercy *and* of his hey grace
and pryncepaly ffor my suete chylde
þ*at* xal to me do gret solace[1] 20
Fo. 26 Now suete sone fayr*e* far*e* þ*i* fface
Fful hertyly do I love the
Ffor trewe herty love now in this place
my swete chylde com kysse now me.

Ysaac

¶ At ȝour*e* byddynge ȝo*ur* mouthe I kys 25
w*ith* lowly hert I ȝow pray
ȝour*e* fadyrly love lete me nevyr mysse
but blysse me ȝo*ur* chylde both nyght *and* day.

Abraham

Al-myghty god þ*at* best may
his der*e* blyssyng he graunt þ*e* 30
And my blyssyng þ*ou* haue aſſ way
in what place þ*at* evyr þ*ou* be.

¶ Now Isaac my sone so suete
Al-myghty god loke þ*ou* honour*e*
wich þ*at* made both drye *and* wete 35
Shynyng sunne *and* scharpe schour*e*
thu art my suete childe *and* paramour*e*
Fful wele in herte do I þ*e* loue
loke þ*at* þ*in* herte in hevyn tour*e*
be sett to serve our*e* lord god a-bove. 40

¶ In þ*i* ȝonge[2] lerne god to plese
and god xal quyte þ*e* weyl þ*i* mede
now suete sone of wordys these
w*ith* aſſ þ*in* hert þ*ou* take good hede
Now far*e* weyl sone god be þ*in* spede 45
Evyn her*e* at hom þ*ou* me a-byde
I must go walkyn ffor I haue nede
I come a-ȝen w*ith*-inne a tyde.

[1] The initial *s* (*ſ*) is thickened and may be intended to be double (*ſſ*).
[2] So for *ȝongþe* (?).

Ysaac

Fo. 26v ¶ I pray to God ffadyr of myght
þat he ȝow spede in alł ȝ*our* waye
From shame *and* shenshipp daye *and* nyght
God mote ȝow kepe in ȝ*our* jornay.

Abraham

Now fare weyłł sone I þ*e* pray
Evyr in þ*in* hert loke god þ*ou* wynde
hym to serue bothe nyght *and* day
I pray to god send þ*e* good mynde.

¶ Ther may no man love bettyr his childe
þan Isaac is lovyd of me
Almyghty god mercyful *and* mylde
Ffor my swete sone I wurchyp þ*e*
I thank þ*e* lord w*ith* hert ful fre
Ffor þis fayr frute þ*ou* hast me sent
now gracyous god wher so he be
to saue my sone evyr mor*e* be bent.

¶ Der*e* lord I p*ray* to þ*e* Also
me to saue for þ*i* seruuau*n*te
and sende me grace nevyr for to do
thyng þ*a*t xulde be to þ*i* displesaunte
Bothe ffor me *and* for my*n* infaunte
I p*ray* þ*e* lord god vs to help
thy gracyous goodnes þ*ou* us grawnt
and saue þ*i* serwau*n*t from helle qwelp.

Angelus

¶ Abrah*a*m how abrah*a*m
lyst *and* herke weyłł on to me.

Abrah*a*m

Al redy ser*e* here I am
tełł me ȝ*our* wyłł what þ*a*t it be.

Angelus

Fo. 27 Almyghty god þus doth bydde þ*e*
ysaac þ*i* sone Anon þ*ou* take
and loke hym þ*ou* slee a-noon lete se
and sacrafice to god hym make.

¶ Thy welbelouyd childe þ*ou* must now kylle
to god þ*ou* offyr hy*m* as I say

Evyn vpon ȝon hey hylle
þat I þe shewe here in þe way
Tarye not be nyght nor day 85
but smertly þi gate þou goo
Vpon ȝon hille þou knele *and* pray
to god · *and* kylle þe childe þer and scloo.

Abraham

¶ Now goddys comaundement must nedys be done
Alł his[1] wyl is wourthy to be wrought 90
but ȝitt þe fadyr to scle þe sone
grett care it causyth in my thought
In byttyr bale now am I brought
my swete childe with knyf to kylle
but ȝit my sorwe avaylith ryght nowth 95
for nedys I must werke goddys wylle.

¶ With evy hert I walke *and* wende
my childys deth now for to be
now must þe fadyr his suete son schende
Alas for ruthe it is pete 100
my swete sone come hedyr to me
how Isaac my sone dere
com to þi ffadyr my childe so fre
ffor we must wende to-gedyr in fere.

Isaac

¶ Alł redy fadyr Evyn at ȝour wyłł 105
and at ȝour byddyng I am ȝow by
Fo. 27v With ȝow to walk ovyr dale *and* hiłł[2]
At ȝoure callyng I am redy
To þe fadyr evyr most comly
It ovyth þe childe evyr buxom to be 110
I wyl obey ful hertyly
to alł thyng þat ȝe bydde me.

Abraham

¶ Now son in þi necke þis fagot þou take
and þis fyre bere in þinne honde
Ffor we must now sacrefyse go make 115
evyn aftyr þe wyłł of goddys sonde

[1] The *h* is partly effaced through stain on MS.

[2] This line is preceded by part of the last line—*Andat ȝour byddyng I am*—written again by mistake and crossed out.

Take þis brennyng bronde
my swete childe and lete us go
ther may no man þ*at* levyth in londe
haue mor*e* sorwe than I haue wo.

Ysaac

¶ Ffayr*e* fadyr ȝe go ryght stylle
I pray ȝow fadyr speke on to me.

Abrah*am*

Mi gode childe what is þ*i* wylle
telle me thyn hert I p*ray* to the.

Ysaac

Ffadyr fyr*e* *and* wood her*e* is plente
but I kan se no sacryfice
what ȝe xulde offre fayn wold I se
þ*at* it wer*e* don at þ*e* best avyse.

Abrah*am*

¶ God xal þ*at* ordeyn þ*at* sytt in hevy*n*ne
my swete sone ffor þis offryng
A derere[1] sacryfice may no man nempne
þan þis xal be my der*e* derlyng.

Ysaac

Lat be good fadyr ȝ*our* sad wepynge
ȝ*our* hevy cher agrevyth me sor*e*
tell me fadyr ȝ*our* grett morny*ng*
and I xal seke sum help þ*er*-for*e*.

Abrah*am*

Fo. 28 ¶ Alas der*e* sone for nedys must me
Evyn her*e* þ*e* kylle as god hath sent
thyn owyn fadyr þ*i* deth must be
Alas þ*at* evyr þis bowe was bent
W*ith* þis fyr*e* bryght þ*ou* must be brent
An Au*n*gelle seyd to me ryght so
Alas my chylde þ*ou* xalt[2] be shent
þ*i* careful fadyr must be þ*i* ffo.

Ysaac

¶ Al-myghty god of his grett mercye
Fful hertyly I thanke þ*e* sertayne
At goddys byddyng her*e* for to dye
I obeye me her*e* for to be sclayne

[1] Double *r* corrected in the middle of this word.
[2] *u* apparently first written and altered to *a*.

I pray ȝow fadyr be glad and fayne
trewly to werke goddys wyłł
take good comforte to ȝow agayne
and haue no dowte ȝ*our* childe to kyłł.

¶ ffor godys byddyng for sothe it[1] is
þ*at* I of ȝow my deth schulde take
Aȝens[2] god ȝe don amys
his byddyng yf ȝe xuld for-sake
ȝowr*e* owyn dampnac*i*on xulde ȝe bake
if ȝe me kepe from þis reed
w*ith* ȝ*our* swerd my deth ȝe make
and werk evyr mor*e* þ*e* wyłł of god.

Abraha*m*

¶ The wyłł of god must nedys be done
to werke his wyłł I seyd nevyr nay
but ȝit[3] þe ffadyr to[4] sle þ*e* sone
my hert doth clynge and cleue as clay.

Ysaac

Fo. 28v ȝitt werk goddys wyłł fadyr I ȝow pray
and sle me her*e* anoon forth ryght
and turne fro me ȝ*our* face a-way
myne heed whan þ*at* ȝe xul of smyght.

Abraha*m*

¶ Alas der*e* childe I may not chese
I must nedys my swete sone kylle
my der*e* derlyng now must me lese
my*n* owyn sybb blood now xal I spylle
ȝitt þis dede or I fulfylle
my swete sone þ*i* mouth I kys.

Ysaac

Al redy fadyr evyn at ȝ*our* wyłł
I do ȝ*our* byddyng as reson is.

Abraha*m*

¶ Alas der*e* sone her*e* is no grace
but nedis ded now must þ*ou* be

[1] *it* omitted and written above the line in slightly darker ink.
[2] *Aȝeng* first written and crossed through.
[3] An *e* stands above the *it* of *ȝit*.
[4] The letters *sch* between *to* and *sle* are crossed through.

*wit*h þis kerchere I kure þ*i* face
In þe tyme þat I sle the 180
Thy lovely vesage wold I not se
not for all þis werdlys good
*wit*h þis swerd þat sore grevyht me
my childe I sle *and* spylle his blood.

Ang*e*lus

¶ Abrah*a*m Abrah*a*m þ*ou* fadyr fre. 185

Abrah*a*m

I am her*e* redy what is ȝ*our* wylle.

Ang*e*lus

Extende þ*in* hand in no degre
I bydde þ*ou* hym not kylle
her*e* do I se by ryght good skylle
All myghty god þat þ*ou* dost drede 190
For þ*ou* sparyst nat þ*i* sone to spylle
God wyll aqwhyte þe well þ*i* mede.

Abrah*a*m

Fo. 29 ¶ I thank my god[1] in hevyn a-bove
and hym honowr*e* for þis grett grace
and þat my lord me þus doth prove 195
I wyll hym wurchep in every place
my childys lyff is my solace
I thank my*n* god evyr for his lyff
In sacrifice her*e* or I hens pace
I sle þis shepe w*it*h þis same knyff. 200

¶ Now þis shepe is deed *and* slayn
w*it*h þis fyr*e* it xal be brent
Of Isaac my sone I am ful fayn
þat my swete childe xal not be shent
This place I name w*it*h good entent 205
þe hill of[2] godys vesytacion
Ffor hedyr god hath to us sent
his comforte aftyr grett trybulac*i*on.

Angelus

¶ Herke Abrah*a*m *and* take good heyd
by hym self god hath þus sworn̄ 210

[1] Some miswritten word (? me) after *god* crossed through.
[2] *of* omitted and written above the line.

Ffor þ*at* þ*ou* woldyst a done þis dede
he wyll þ*e* blysse both evyn and morn
Ffor þ*i* der*e* childe þ*ou* woldyst haue lorn
at goddys byddyng as I the telle
god hath sent þ*e* word be-forn 215
þ*i* seed xal multyplye wher so þ*ou* duelle.

¶ As sterr*ys* in hevyn byn many *and* fele
so xal þ*i* seed encrese *and* growe
þ*ou* xalt ovyr-come in welth *and* wele
all þ*i* fomen reknyd be rowe 220
Fo. 29v (marked 30) As sond in þ*e* se doth ebbe and flowe
Hath cheselys many vnnum*er*abyll
so xal þi sede þ*ou* mayst me trowe
encres *and* be evyr p*ro*phytabyll.

¶ Ffor to my spech þ*ou* dedyst obeye 225
thyn enmyes port*ys* þu shalt possede
and all men on erthe as I þ*e* seye
thei xal be blyssed in þ*i* sede
Al-myghty god þus þ*e* wyll mede
Ffor þat good wyll þ*at* þ*ou* ast done 230
þ*er*fo*r*e thank god in word *and* dede
Both þ*ou* þ*i* self *and* ysaac þ*i* sone.

Abrah*a*m

¶ A my lord god to wurchep on kne now I fall
I thank þ*e* lord of þ*i* mercy
now my swete childe to god þ*ou* kall 235
and thank we þ*at* lord now hertyly.

Isaac

w*ith* lowly hert to god I crye
I am his seruuant both day *and* nyght
I thank þ*e* lord in hevyn so hyȝe
w*ith* hert w*ith* thought w*ith* mayn w*ith* myght. 240

Abrah*a*m

¶ Gramercy lord and kyng of grace
gramercy lord ou*er* lordys all
now my joye returnyth his trace
I thank þ*e* lorde in hevyn þi*n* halle.

Isaac

Ovyr all kyng*ys* crownyd kyng I þ*e* kalle 245
at þ*i* byddy*ng* to dye w*ith* knyff
I was fful buxvm evy*n* as þ*i* thralle
lord now I thank the þ*ou* grauntyst me lyff.

Abrah*am*

Fo. 30 ¶ Now we haue wurchepyd our*e* blyssyd lorde
on groun*de* knelyng upon our*e* kne 250
now lete us tweyn sone ben of on A-corde
and goo walke hom in to our*e* countr*e*.

Ysaac

Ffadyr as ȝe wyll so xal it be
I am redy w*ith* ȝow to gon
I xal ȝow folwe w*ith* hert full fre 255
all þ*at* ȝe bydde me sone xal be don.

Abrah*am*

¶ Now god all thyng of nowth þ*at* made
evyr wurcheppyd he be on watyr *and* londe
his grett honowr*e* may nevyr mor*e* fade
in feld nor town se nor on sonde 260
As althyng lord þ*ou* hast in honde
so saue us all wher so we be
whethyr we syttyn walk or stonde
Evyr on þ*in* handwerke þ*ou* haue pyte.

explicit.

Remainder of Fo. 30—3⅜ inches—and Fo. 30ᵛ left blank.

Fo. 31 (40 written by the side and crossed out)

¶ Introit*us* Moyses

HE þ*at* made all thynge of nought 6
Hevyn *and* erth both sunne *and* mone
Saue all þ*at* his hand hath wrought
All myghty god in hevyn trone
I am Moyses þ*at* make þis bone 5
I pray þe lord god w*ith* all my mende
to us incline þ*i* mercy sone
þ*i* gracyous lordchep lete us fynde.

¶ The to plesyn in all degre
Gracyous god *and* lord ovyr all 10

þou graun*te* us grace wher so we be
and saue us sownd fro synfuꝗ faꝗ
thy wyꝗ to werke to us *þi* thraꝗ
enforme *and* teche us aꝗ *þi* plesans
in purenesse put us *þat* nevyr not faꝗ
and groun*de* us in grace ffrom aꝗ grevau*ns*.

Hic moyses videns rubrum ardentem admirande dicit[1]

¶ A mercy god what menyth ȝon syte
A grene busch as fyr*e* doth flame
and kepyth his colowr*e* fayr *and* bryghte
Ffresch *and* grene w*ith*-owtyn blame
It fyguryth su*m* thynge of ryght gret fame
I kan not seyn what it may be
I wyꝗ go ner*e* in goddys name
and wysely loke þis busch to se.

Deus

¶ Moyses how Moyses
herke to me a-non þis stou*n*de.

Moyses

Fo. 31v I am her*e* lorde w*ith*-owtyn les
ȝowr*e* gracyous wyꝗ to do I am bounde.

Deus

Thu take þi schon anon ful rownde
of *þi* fete i*n* hast lete se
fful holy is *þat* place and grownde
þer *þou* dost stonde I sey to the.

Moyses

¶ Barfoot now I do me make
and puꝗ of my schon fro my fete
now haue I my schon of take
what is ȝo*ur* wyꝗ lord · fayn wold I wete.

Deus

Com ner*e* moyses w*ith* me to mete
these tabeꝗ-is I take þ*e* in þi*n* honde
w*ith* my ffynger in hem is wrete
aꝗ my lawys *þou* vndyrstonde.

¶ Loke *þat* *þou* pr*e*che aꝗ abowte[2]
hoo so wyꝗ haue frenshipp of me

[1] This is written in larger form than other stage directions.
[2] The colour of the ink changes slightly with this line.

to my lawys loke þei lowte
þat þei be kept in alł degre
Go forth̄ and preche a-non let se 45
loke þou not ses nyght nor day.

Moyses

ȝour byddyng lord alł wrought xal be
ȝour wyłł to werk I walk my way.

Custodi precepta domini dei tui · deutronomini · vj^to^.

¶ The comaundement of þi lord god man loke þou kepe
where þat þou walk · wake or slepe 50
Euery man take good hede
and to my techynge take good intent
for god hath sent me now in dede
ȝow for to enforme his comaundment
ȝow to teche god hath me sent 55
Fo. 32 his lawys of lyff þat arn ful wyse
them to lerne be dyligent
ȝour soulys may þei saue at þe last Asyse.

¶ The preceptys þat taught xal be
be wretyn in þese tablys tweyn̄ 60
In þe fyrst ben wretyn thre
that towch̄ to god þis is serteyn̄
In þe secund tabyl be wretyn ful pleyn̄
þe tother vij þat towch̄ mankende
herk now wełł man what I xal seyn̄ 65
and prent þise lawys wełł in þi mende.

1us Primum mandatum non habebis deos alienos.

¶ The fyrst comaundement of god as I ȝow say
of þe fyrst tabyl for sothe is this
þou xalt haue neythyr nyght nore day
noon other god but þe kyng of blysse 70
Vndyrstonde wele what menyth this
Euery man in hys degre
And sett neuyr ȝour hert a-mys
Vpon þis werdlys vanyte.

¶ Ffor if þ*ou* sett þ*i* loue so sor*e* 75
Vpon ryches and werdly good
þ*i* wurdly rycches þ*ou* takyst ev*er* mor*e*
evyn ffor þi god · as man ovyr wood
Amend þ*e* man *and* chaunge þ*i* mood
lese not þ*i* sowle for werdlys welth̄ 80
Only hym loue · which̄ bodyly ffood
doth ȝeve all̄ day *and* gostly helth̄.

Fo. 32v 2. Secundu*m* mandatu*m* · no*n* assumens nom*en* dei tui i*n* vanu*m*.

¶ The secund p*re*cept of þ*e* fyrst tabyll̄
þ*e* name of god take nevyr in vayne
Swer*e* none othis be noon fals fabyll̄ 85
þ*e* name of god þ*ou* nevyr dysteyn
Be whar*e* of othis for dowte of peyn̄
Among*es* ffelacheppe whan þ*ou* dost sytt
A lytyl othe þis is serteyn
may dampne thy sowle to helle pytt. 90

¶ Man whan þ*ou* art sett at þ*e* nale
and hast þ*i* langage as plesyth þe
loke þi*n* othis be non or smale
and ȝett alwey loke trewe thei be
But swer*e* not oftyn by rede of me 95
ffor yf þ*ou* vse oftyn tyme to swer*e*
it may gendyr custom̄ in the
by-war*e* of custom̄ ffor he wyl der*e*.

3. Terciu*m* mandatu*m* · memento vt sabbatu*m* *sanct*ificet.

¶ The iij^de^ comaundment of god as I rede
doth̄ bydde the · halwe well̄ þi*n* haly day 100
kepe þ*e* well̄ ffro synfull̄ dede
and care not gretly ffor rych̄ a-ray
A ryght por*e* man þis is non nay
of sympyl astat in clothis rent
may-be bettyr than rych̄ w*ith* garment*ys* gay 105
Oftyn tyme doth kepe þis comau*n*dment.

¶ Ffor rych̄ men do showe oftyn tyme pompe *and* p*r*ide
On halydays as oftyn is sene

Thyn halyday ye bryyst not clene
In glotony to lede ye lyff
In goddys hous zo yuldo bodono
honour zo god. both mayden & wyff

Quarta mandatu honora patrem tuum & matrem tuam.

Off ye secundo tabyll ye fyrst comaundment
And in ye ordyr ye mte ¶ soy in fay
ho byddyth ye eu mo lst hept bout
both fadyr & modyr to wyschop alway

Thols ye fadyr bo poor of aray
& yolde neu so yrch of goldo & good
zitt loko ye wyschop hy wyght & day
of whom ye hast both flesch & blood

In yis comaundmento includyd so
thy bodyly fadyr & modyr also
In cludyd also ¶ fynde in yis
thy gostly fadyr & modyr ye to

To ye gostly fadyr dyr yenoyeno do
ye gostly modyr so holy chorch
these thryn sans ye sowlo fro wo
eu thow to wyschop loko ye ye wyrch

Quinta mandatum non occides

The fyrst comaundement byddyth all no
Sole no man. no wyght ye ye wyll
vndyr fonde yis precept yno
Sole no wyght lst wynd not wyll

Wytbyd wyde wyghyght afte tymo gyett all
bo way ye fay of wytbyd langage
wyckyd spoch many on doth fyll
thy fay of spoch both not oblt yage

Whan por*e* men passe *and* go be-syde
At wurthy festys riche men wolł bene 110
Fo. 33 Thyn halyday þ*ou* kepyst not clene
In gloteny to lede þ*i* lyff
In goddys hous ȝe xulde be-dene
honour*e* ȝ*our* god · both mayden *and* wyff.

4. Quartu*m* mandatu*m* honora p*at*rem tuu*m* et m*at*rem tuam.

¶ Off þ*e* secunde tabyłł þ*e* fyrst comaundement 115
And in þ*e* ordyr þ*e* iiij^te^ I sey in fay
he byddyth þ*e* eu*er* mor*e* w*ith* hert bent
both ffadyr and modyr to wurchep Alway
Thow þ*at* þ*i* fadyr be por*e* · of array
and þow neu*er* so rych of golde *and* good 120
ȝitt loke þ*ou* wurchep hy*m* nyght *and* day
Of whom þ*ou* hast both fflesch and blood.

¶ In þis comaundmente includyd is
thi bodyli fadyr *and* modyr also
Includyd also I fynde in þis 125
thi gostly fadyr *and* modyr þ*er* to
To þ*i* gostly ffadyr evyr reuerens do
þ*i* gostly modyr is holy cherch
these tweyn saue þ*i* sowle fro woo
eu*er* them to wurchep loke þ*at* þ*ou* werch. 130

5. Quintum mandatum non occides.

¶ The ffyfft comaundement byddyth alł us
Scle no man · no whight þ*at* þ*ou* kyłł [1]
Vndyr stonde þis p*re*cept þus
Scle no wyght w*ith* wurd nor wyłł
Wykkyd worde werkyht oftyn tyme grett iłł 135
be war þ*er* fore of wykkyd langage
wyckyd spech many on doth spyłł
therfor*e* of spech beth not owt-rage.

Fo. 33^v^ 6. Sextu*m* mandatu*m* non makaberis.

¶ The sexte comaundement byddith every man
þ*at* no wyght lede no lecherous lay 140

[1] A final *e* has perhaps been erased here.

fforfet neu*er* be no woman
lesse þan þ*e* lawe a-lowe þ*i* play
Trespas nevyr w*ith* wyff ne may
w*ith* wedow nor w*ith* non othyr wyght
kepe þ*e* clene as I þ*e* say 145
to whom þ*ou* hast þ*i* trowth plyght.

7. Septimu*m* mandatu*m* non furtum facies.

¶ Do no thefte no thynge þ*ou* stele
þe vij^te^ precept byddyth þ*e* ful sor*e*
whyll þ*ou* arte in welth and wele
euyll gett good loke þ*ou* restor*e* 150
Off handys *and* dede be trewe euyr mor*e*
ffor yf þi*n* handys lymyd be
þ*ou* art but shent þ*i* name is lor*e*
in ffelde *and* town *and* in all cou*n*tr*e*.

8. Octauu*m* mandatu*m* · no*n* loquer*is* contra p*r*oximu*m* tuu*m* falsu*m* testi*m*oniu*m*.

¶ The viij^te^ p*r*ecept þus doth þ*e* bydde 155
Ffals wyttnes loke non þ*ou* ber*e*
þ*e* trowth nevyr mor*e* loke þ*at* þ*ou* hyde
w*ith* ffals wyttnes no ma*n* þ*ou* der*e*
Nowther ffor love ne dred ne fer*e*
Sey non other than trowth is 160
Ffals wytnes yf þ*at* þ*ou* rer*e*
Aȝens god þ*ou* dost grettly amys.

9. Nonu*m* mandatu*m* non desiderabis vxorem proximi tui et cetera.

Fo. 34 ¶ The ix^te^ precept of lawe of lyff
evyn þus doth bydde every man
desyre not þ*i* neybor*ys* wyff 165
þow she be fayr and whyte as swan
And þ*i* wyff brown ȝitt natt for than
þ*i* neybor*ys* wyff þ*ou* nevyr rejoyse
kepe þ*e* clene as evyr þ*ou* can
to þin owyn wyff *and* þi*n* owyn choyse. 170

10. Decimu*m* mandatu*m* no*n* concupisces domu*m* p*ro*ximi tui non seruu*m* no*n* ancillam̄ non bos no*n* asinu*m* nec o*mni*a que illi*us* su*n*t et cet*era*.

¶ The x^de^ comaundement of god and last is þis
thi neybor*ys* hous desyr*e* þ*ou* nowth
Maydon nor servaunt nor nowth of his
desyr*e* hem nevyr in wyłł nor thowth
Oxe ner*e* Asse þ*at* he hath bought 175
ner*e* no thynge þ*at* longyht hym to
Godys lawe must nedys be wrought
desyr*e* no thynge þi*n* neybor*e* ffro.

¶ The vj^te^ comaundement of lechory
doth exclude þ*e* synfułł dede 180
but theys tweyn last most streytly
both dede *and* thought þei do for-bede
In wyłł ner*e* thought no lechory þ*ou* lede
þ*i* thought *and* wyłł þ*ou* must refreyn̄
Ałł þ*i* desyr*e* as I þ*e* rede 185
in clennes of lyff þ*i* self restreyn̄.

¶ Ffrendys þese be þ*e* lawys þ*at* ȝe must kepe
therfor*e* every man sett wełł in mende
Fo. 34v Wethyr þ*at* þ*ou* do wake or slepe
these lawys to lerne þ*ou* herke ful hynde 190
And godys grace xal be þ*i* ffrende
he socowr*e* *and* saue ȝow in welth fro woo
Fare wełł gode frendys for hens wyłł I wende
my tale I haue taught ȝow my wey now I goo.

Explic*it* moyses.

Remainder of Fo. 34v—6½ inches left blank (except for a scribbled *wethyr* *þat* *þou* *do.* in another hand).

Fo. 35 I Am þ*e* p*ro*phete callyd Isaye Ysaias
Replett w*ith* godys grett influens
and sey pleynly be spyryte of p*ro*phecie **7**
þ*at* a clene mayde thourgh meke obedyens

Shall bere a childe which xal do resystens
Ageyn foule Zabulon þe devyl of helle
mannys soule ageyn hym to defens
Opyn in þe felde þe fend he xal felle.

¶ Where fore I seye || quod virgo concipiet
et pariet filium || nomen emanuel
Oure lyf for to saue he xal suffyr deth
and bye us to his blysse · in hevyn for to dwell
Of Sacerdotale lynage þe trewth I ȝow tell
Fflesch and blood to take god wyll be born
Joye to man in erth · and in hevyn Aungell
At þe chyldys byrth · joye xal make þat morn.

Radix Jesse

¶ Egredietur virga de radice jesse
Et flos de radice eius ascendet
A blyssyd braunch xal sprynge of me
That xal be swettere þan bawmys breth
Out of þat braunch in nazareth
A flowre xal blome of me jesse rote
The which by grace xal dystroye deth
and brynge mankende to blysse most sote.

Dauyd' Rex

¶ I am david of jesse rote
the fresch kyng by naturall successyon
and of my blood xal[1] sprynge oure bote
As god hym self hath mad promyssyon
Fo. 35v Of Regall lyff xal come suche foysoñ
þat a clene mayde modyr xal be
Ageyns þe devellys fals illusyoñ
with regall power to make man fre.

Jeremias propheta

¶ I am þe prophete Jeremye
And fullich a-corde in all sentence
with kyng dauid *and* with ysaie
Affermynge pleynly be-forn þis Audyens
That god of his high benyvolens
of prest *and* kynge wyll take lynage
And bye us all ffrom oure offens
in hevyn to haue his herytage.

[1] *spyr* miswritten before *sprynge* and crossed through.

Salamon rex

¶ I am Salamon þe secunde kynge
And þat wurthy temple for sothe made I
which þat is fygure of þat mayde ȝynge
þat xal be modyr of[1] grett messy.

Ezechieł propheta

A vysion̄ of þis fful veryly
I Ezechiel haue had also
Of a gate þat sperd was trewly
and no man but a prince myght þer-in go.

Roboas rex

¶ The iij^de^ kynge of þe jentyłł[2] jesse
my name is knowe kyng Roboas
of oure kynrede ȝitt men xul se
A clene mayde trede down̄ foule sathanas.

Micheas propheta

And I am a Prophete calde mycheas
I telle ȝow pleynly þat þus it is
Evyn lyke as Eve modyr of wo was
So xal a maydyn be modyr of blyss.

Abias Rex

Fo. 36 ¶ I that am calde kynge Abias
conferme for trewe þat ȝe han seyd
and sey also as in þis cas
þat alł oure myrth̄ comyth of a mayd.

Danyel propheta

I prophete Danyel am wełł apayed
in fygure of þis I saw a tre
alł þe fendys of hełł xałł ben affrayd
whan maydenys ffrute þer on þei se.

Asa Rex

¶ I kynge Asa be-leve alł þis
þat god wyłł of a maydyn be born̄
And vs to bryngyn to endles blys
Ruly on rode be rent *and* torn̄.

Jonas propheta

I jonas sey *þat* on *þe* iij^de^ morn̄
ffro deth̄ he xal ryse þis is a trew[3] tałł[4]

[1] *of* omitted and written above the line in darker ink.
[2] *jeng* first written and crossed through.
[3] Or *trow*—the vowel is blotted.
[4] There has been some subsequent attempt to correct *tałł* and *qwałł* to *tale*

fyguryd in me þe which longe beforn
lay iij days beryed w*ith* in þe qwall.[1] 70

Josophat rex

¶ And I josophat þe vj^te kynge serteyn
of jesse rote in þe lenyall successyon
All þat my progenitouris hath be-for me seyn
Ffeythfully be-leve w*ith*-owtyn all dubytac*ion*.

Abdias propheta

I abdias prophete make þis protestac*ion* 75
þat aftyr he is resyn to lyve onys a-ȝen
deth xal be drevyn to endles dampnac*ion*
And lyff xal be grawntyd of paradys ful pleyn.

Joras Rex

¶ And I Joras also in þe nu*m*bre of sefne
of jesse rote kynge · knowlych þ*at* he 80
aftyr his resurrecc*ion* returne xal to hefne
both god *and* verry ma*n* ther endles to be.

Abacuch propheta

I Abacuch prophete holde wele w*ith* the
whan he is resyn he xal up stye
in hevyn as juge sitt i*n* his se 85
Vs for to deme whan we xal dye.

Ozias Rex

Fo. 36^v ¶ And I Ozyas Kynge of hygh degre
Spronge of jesse rote dar[2] well sey this
whan he is gon to his dygnyte
he xal send þe sprytt to his discyplis. 90

Joell propheta

And I Joel knowe full trewe þat is
god bad me wryte in prophesye
he wolde sende down his sprytt i-wys
On ȝonge *and* olde ful sekyrlye.

Joathas rex

¶ My name is knowe kyng Joathan 95
the ix^e kynge spronge of jesse
Of my kynrede god[3] wol be man
mankend to saue and þ*at* joyth me.

and *qwale*. There are faint traces of an *e* in darker ink over the second *l* in each case.

[1] See note 4 on p. 59. [2] Or *dare*. [3] *gold* first written.

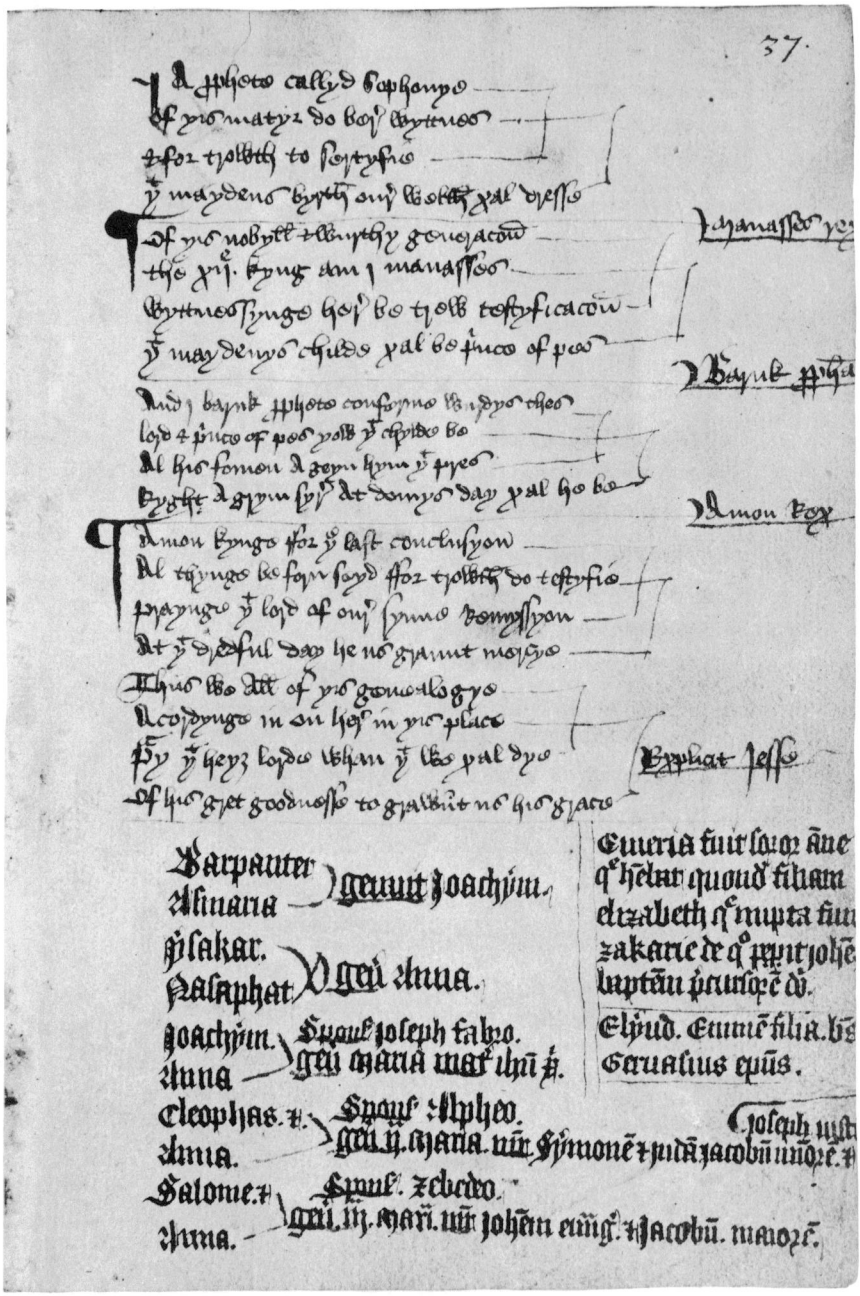

37.

A Photo called Sapheyro
of yis matyr do boy thytheos
tfor tolloth to soytyfie
ye maydens byth ouy lostth yal dysso
Of yis nobyll thrythy gonyacow
the xti. kyng am manasseos
thytheos kyngs hoy bo tolls tostyficacow
ye maydenys chudo yal bo fuco of yeo
And bayne Photo confoyne kindyo cheo
lord fuco of yos yolls ye chydo bo
Al his fomon a zeyn hym ye fyes
tyghte a zeyn sye at domys day yal ho bo
Amon kyngs for ye last conclusyon
Al thyngs be foyn soyd for tolloth do tostyfie
prayngs ye lord of ouy synne bomyssyon
At ye dedful day he no graunt mosyo
Thus be att of yis gomeologyo
Acordyngs in on hoy in yis place
Fy ye heyz lordes thym ye be yal dyo
Of his gyet goodnosse to graunt no his grace

Manasses yis

Bayne Photo

Amon kyng

Ezechias

Varpauter
Alinaria — genuit Joachim.
Sisakar.
Nasaphat — genuit Anna.
Joachim. — Syrach Joseph Sabyo.
Anna — genuit Maria mater Iesu.
Cleophas.
Anna. — Syrach Alpheo.
genuit Maria. uxor Simonis et Iudae Iacobi minoris.
Salome.
Anna. — Syrach Zebedeo.
genuit Maria uxor Iohannis evangelistae et Iacobi maioris.

Euuria fuit soror Anne
quae helisabet quoque filiam
elizabeth quae nupta fuit
zakarie de qua natus est Iohannes
baptista praecursor domini.

Elyud. Eumei filia. bis
Gerualius episcopus.

Joseph uxor

British Museum Cott. MS. Vespasian D. VIII, f. 37a
(Showing one of the marginal genealogies: cf. *Introduction*, p. xxxvii)

Aggeus *pro*pheta

W*ith* ȝow I do'holde þ*at* am prophete Aggee
Com of þe same hygh *and* holy stok 100
god of our*e* kynrede in dede born wyl be
From þ*e* wulf to saue al shepe of his flok.

Achas rex

¶ Off jesse kyng Achas is my name
þ*at* falsly wurchepyd ydolatrye
tyl Ysaie putt me in blame 105
And seyd a mayd xulde ber*e* messye.

Ozyas *pro*pheta

Off þ*at* byrthe wyttnes ber*e* I
A *pro*phete Osyas men me calle
And aftyr þ*at* tale of Isaye
þ*at* mayd xal ber*e* Emanuelle. 110

Ezechias rex

¶ My name is knowyn kyng Ezechyas
þe xj^te^ kyng of þis geneologye
And say ffor sothe as in þis cas
A mayde be mekenes xal brynge mercye.

Sophosas *pro*pheta

Fo. 37 I A *pro*phete callyd Sophonye 115
of þis matyr do ber*e* wyttnes
and for trowth to sertyfie
þ*at* maydens byrth̄ our*e* welth̄ xal dresse.

Manasses rex

¶ Of þis nobyl̄ *and* wurthy generac*i*on
the xij^e^ kyng am I manasses 120
wyttnessynge her*e* be trew testyfica*ci*on
þ*at* maydenys childe xal be p*ri*nce of pes.

Baruk *pro*pheta

And I baruk *pro*phete conferme wurdys thes
lord and p*ri*nce of pes þow þ*at* chylde be
Al his fomen ageyn hym þ*at* pres 125
Ryght a grym syr*e* at domys day xal he be.

Amon Rex

¶ Amon kynge ffor þe last conclusyon̄
Al thynge be-forn seyd ffor trowth̄ do testyfie
Praynge þ*at* lord of our*e* synne Remyssyon
At þ*at* dredful day he us graunt mercye 130

Thus we all of þis genealogye
Acordynge in on here in þis place
Pray þat heyȝ lorde whan þat we xal dye
Of his gret goodnesse to grawnt us his grace.

The remainder of Fo. 37 is filled with the following genealogical table and note written like those on Ff. 16, 17 and 21, 22:

Explicit Jesse.

Barpanter Asmaria	genuit Joachym	
Ysakar Nasaphat	gen. Anna	
Joachym Anna	gen.	sponsa Joseph fabro Maria mater ihesu Christi
Cleophas *et* Anna	gen. ij^{a}	sponsa Alpheo Maria mater Symonem et Judam Jacobum minorem et Joseph just[um
Salome *et* Anna	gen. iij^{a}	sponsa Zebedeo. Maria mater Johannem euangelistam et Jacobum majorem.

Emeria fuit soror Anne que habebat quondam filiam Elizabeth que nupta fui[t Zakarie de quo peperit Johanne[m baptistem precursorem domini Elyud. Eminē filia. beatus Geruasius episcopus

Fo. 37v Contemplacio

¶ Cryst conserve þis congregacion
fro perellys past · present *and* future
and þe personys here pleand · þat þe pronunciacion
of here sentens to be seyd · mote be sad *and* sure
And þat non oblocucyoñ · make þis matere obscure
but it may profite *and* plese eche persone present
ffrom þe gynnynge to þe endynge so to endure
þat cryst *and* every creature · with þe conceyte be content.

8

¶ This matere here mad · is of þe modyr of mercy
how be joachym And Anne · was here concepcion
Sythe offred into þe temple · compiled breffly
than maryed to joseph · and so folwyng þe salutacion
Metyng with Elyzabeth · *and* þer with a conclusyoñ
in fewe wordys talkyd þat it xulde nat be tedyous
to lernyd nyn to lewd · nyn to no man of reson
þis is þe processe[1] · Now preserve ȝow jhesus.

[1] *of oure* first written after *processe*, and crossed through.

¶ þ*er*fo*re* of pes I ȝow pray aꝗ þat ben here p*re*sent
and tak hed to oure talkyn · what we xal say
I be-teche ȝow þ*at* lorde þ*at* is evyr omnypotent
to governe ȝow in goodnes as he best may 20
In hevyn we may hym se
Now god þ*at* is hevyn kynge
sende us aꝗ hese dere blyssynge
and to his towre he mote vs brynge
Amen ffor Charyte. 25

The bottom of Fo. 37ᵛ is filled with the following two genealogical notes :

Qu*in*qu*e* sunt Anne { mater Samue, Vxor Rague } lis ; { Vxor Tob, Mat*er* be*a*te Mar, Anna[1] p*ro*phetissa } ie

Est Ysakar Anne pater : Melophat sic q*uoque* mat*er* vel Nasaphat[2]

¶ Ysakar[3]

Fo. 38

THe prestys of god offre sote Ensens
Vn-to here god *and* þ*er* fore they be holy
we þ*at* mynistere here in goddys p*re*sens
in vs xuld be fownd no maner of foly
Ysakar prynce of prestys am I 5
þ*at* þis holyest day here haue mynystrac*i*on
Certyfyenge aꝗ trib*us* in my cure specyaly
þ*at* þis is þ*e* hyest fest of oure solemnyzac*i*on.

¶ This we clepe festu*m* Encenniorum
þ*e* newe ffest · of which iij · in þ*e* ȝere we exercyse 10
now aꝗ þ*e* kynredys to jerusalem must cu*m*
In to þ*e* temple of god · here to do sacryfyse
Tho þ*at* be cursyd my dygnyte is to dysspyse
and þo þ*at* be blyssyd here holy sacrefyse to take
We be regal sacerdociu*m* · it p*er*teyneth vs to be wysse 15
be fastyng · be p*ra*yng · be almes · *and* at du tyme to wake.
Joachym

¶ Now aꝗ þis countre of Galyle

[1] This name appears to have been added, though possibly by the MS. scribe. It is not in more ornate book-hand like the others.

[2] This note is in red ink and in large liturgical script. The bottom half of the letters of the words *vel Nasaphat* has been cut away with the margin.

[3] A small reference sign in ink of the same colour as the text stands to the right of this name.

with þis cetye of Nazareth specyal
þis ffest to jerusalem must go we
to make sacrefyce to god eternal
My name is joachym a man in godys substancyall
Joachym is to say he *þat* to god is redy
so haue I be *and* evyr more xal
Ffor *þe* dredful domys of god sore dred I.

¶ I am clepyd Ryghtful why wole ȝe se
Ffor my godys in to thre partys I devyde
On to *þe* temple · *and* to hem *þat þer* servyng be
A nodyr to *þe* pylg*ri*mys *and* pore men · þe iij*de*[1] ffor he*m* *with* me abyde
Fo. 38v So xulde euery curat in þis werde wyde
ȝeve a part to his chauncel i-wys
A part to his parochoner*ys* *þat* to povert slyde
the thryd part to kepe for hy*m* *and* his.

¶ But blyssyd wyff anne sor*e* I drede
In *þe* temple þis tyme to make sacryfice
be-cawse *þat* no frute of vs doth *pro*cede
I fere me grettly *þe* prest wole me dysspice
Than grett slawndyr in *þe* tribus of vs xulde aryse
but þis I Avow to god *with* all *þe* mekenes I can
ȝyff of his mercy he wole a childe us devyse
we xal offre it up *in* to *þe* temple to be goddys man.

Anna

ȝ*our* swemful wurdys make terys trekyl down be my face
i-wys swete husbond *þe* fawte is in me
my name is Anne *þat* is to sey grace
we wete not how gracyous god wyl to us be
A woman xulde ber*e* cryst þese *pro*fecyes haue we
if[2] god send frute · *and* it be a mayd childe
with all reuerens I vow to his mageste
sche xal be her*e* foot mayd to mynyst*er* her*e* most mylde.

Joachym

¶ Now lete be it as god wole *þer* is no more
tweyn turtelys ffor my sacryfice *with* me I take

[1] This *de* is written in red ink.

[2] This is written in the margin, some miswritten letters standing at the beginning of the line.

and I be-seche wyff *and* evyr we mete more
þat hese grett mercy · vs meryer mut make.

Anna

For dred and ffor swem of ȝo*ur* wourdys I qwake
thryes I kysse ȝow w*ith* syghys ful sad
and to þe mercy of god · mekely[1] I ȝow be-take
and þo þat departe in sorwe god make þ*er* metyng glad.

Senior tribus

¶ Worchepful ser*e* joachym be ȝe redy now
aſt ȝo*ur* kynrede is come ȝow to exorte
Fo. 39 þ*at* þei may[2] do sacrifice at þ*e* temple w*ith* ȝow
Ffor ȝow be of grett wurchep as men ȝow report.

Joachym

Aſt synfuſt seke and sory · god mote comforte
I wolde I wer*e* as men me name
Thedyr in goddys name · now late us aſt resorte
A Anne · Anne · Anne · god scheeld us fro shame.

Anne

¶ Now am I left alone · sor*e* may I wepe
A husbond a-geyn god wel mote ȝow brynge
And fro shame *and* sorwe he mote ȝow kepe
tyl I se ȝow a-geyn · I kan not sees of wepynge.

Senior

Prynce of our*e* prestys · if it be ȝo*ur* plesynge
we be com mekely to make o*ur* sacrefice.

Ysakar

God do ȝow mede bothe elde *and* ȝynge
than devowtly we wyl begynne servyse.

℣ *There they xal synge þis sequens* · Benedicta sit b*eata* trinitas
And in þat *tyme Ysakar with his ministerys ensensyth* þe *Autere*
and *þan þei make her offryng* and *Isaker seyth*

¶ Comyth up serys and offeryth aſt now
ȝe þ*at* to do sacryfice worthy are
A-byde a qwyle ser*e* · Whedyr wytte[3] þ*ou*
þ*ou and* þ*i* Wyff arn barrany *and* bare
neyther of ȝow · ffruteful nevyr ȝett war*e*
Whow durste þ*ou* a-monge fruteful p*re*sume *and* Abuse

[1] Some letter (? I) erased before *mekely*.
[2] *may* omitted and written above the line.
[3] So in MS.

It is a tokyn þ*ou* art cursyd þar*e*
Where*ffore* w*ith* grett indygnac*ion* · þ*in* offeryng I refuse. 80
et *refudit sacrificium Joachim.*

¶ amonge all þis pepyl barreyn be no mo
there*fore* comyth up and offeryth here alle
þ*ou* joachym I charge þ*e* · fast out þ*e* temple þ*ou* go
et redit flendo.
than w*ith* goddys holy[1] wourde blysse ȝow I shalle
ministra[2] *cantando*

Fo. 39v (marked 40) Adiutori*um* n*ostrum* in n*omine* d*omini* 85
Qui fecit celum et terram — Ch*or*us
Sit nomen d*omini* benedictum — m*in*ister
Ex hoc nunc *et* usq*ue* in seculum — ch*or*us
Ep*iscop*us

Benedicat vos diuina[3] maiestas et vna deitas
✠Pater✠et filius✠et sp*iri*tus s*anc*tus — chor*us* · Amen. 90

Signando manu cum cruce solenniter[4] et *recedant tribus extra templum.*

Now of god *and* man blyssyd be ȝe alle
homward a-ȝen now returne ȝe
And in þis temple A-byde we xalle
to servyn god in trinyte.
Joachym

¶ A mercyfful lord what is this lyff 95
What haue I do lorde to haue þis blame
Ffor hevynes[5] I dar*e* not go hom to my wyff
And amonge my neybor*ys* · I dar*e* not abyde ffor shame
A Anne · Anne · Anne · Al o*ur* joye is turnyd to grame
Ffrom ȝo*ur* blyssyd ffelacheppe I am now exilyd 100
and ȝe here onys of þis ffowle fame
sorwe wyl sle ȝow · to se me thus revylyd.

¶ but son[6] god soferyth thys · vs must sofron[7] nede
now wyl I go to my sherherdys[8] · *and* w*ith* hem abyde

[1] *holy* omitted and written above the line.
[2] This word is rubbed—the end letters barely visible.
[3] Some word or words (? *et* vna) after *diuina* obliterated. [4] So in MS.
[5] *dar*e written before *hevynes* and crossed through.
[6] So in MS. for Sen?
[7] The *r* in *sofron* seems to be the correction of some other letter.
[8] So in MS.

and þer evyr more levyn in sorwe and in drede 105
shame makyth many man his hed for to hyde
ha how do ȝe felas in ȝow is lytel pryde [1]
how fare ȝe *and* my bestys þis wete wolde I veryly.

Primus pastor (a) [2]

A welcom̃ hedyr blyssyd mayster · we pasture hem ful wyde
they be lusty *and* fayr · *and* grettly multyply 110
how de ȝe mayster ȝe loke al hevyly
how doth̃ oure dame at hom sytt she *and* sowyht.

Joachym [3] (b)

to here þe speke of here it sleyth myn hert veryly
how I *and* sche doth god hym self knowyth̃ [4]
The meke god lyftyth up þe proude over throwyht (b)
Fo. 40 Go do what ȝe lyst se [5] ȝ*our* bestys not stray. 116

ii^us pastor (a)

Aftere grett sorwe mayster · evyr gret grace growyht
Sympyl as we kan · we xal for ȝow pray.

iii^us pastor (d)

¶ ȝa to pray ffor careful · it is grett nede
we all̃ wul prey ffor ȝow knelende 120
god of his goodnes send ȝow good spede
and of ȝ*our* sorwe ȝow sone amende.

Joachym [6]

¶ I am nott wurthy lord · to loke up to hefne
my synful steppys Anvemynyd þe grounde
I loth-folest [7] þ*at* levyth · þ*ou* lord hyest in þ*i* setys sefne 125
What art þ*ou* lord · What am I wrecche · werse þan an hownde
þ*ou* hast sent me shame · which̃ my*n* hert doth wounde
I thank þe more here fore · þan for all̃ my prosperite
þis is a tokyn þ*ou* lovyst me · now to the I am bounde
þ*ou* seyst þ*ou* art w*ith* hem · þ*at* in tribulac*i*on be. 130

[1] *Pastores* is scribbled underneath the words *for to hyde* and a rough line extending into the margin is drawn under this line on the other side.

[2] These letters seem to indicate some rearrangement, possibly to reduce the shepherds' parts to two.

[3] *ii^us pastor* first written and crossed through by the red loop of name of speaker; *Joachym* carelessly written, but probably by scribe of MS.

[4] The *y* seems to be written over an *h*. [5] Or *so*.

[6] Some indistinct marginal sign opposite Joachym's speech (? nota).

[7] MS. fo lest; cf. hy est in same line.

¶ And ho so haue þe · he nedyth not care thanne
my sorwe is feryng[1] I haue do sum offens
punchyth me lorde · *and* spare my blyssyd wyff anne
þat syttyth *and* sorwyth ful sore of myn Absens
Ther is not may profyte but prayour to ȝour presens 135
With prayorys prostrat · by-fore þi person I wepe
haue mende on oure a-vow · for ȝour mecħ magnyficens
and my lovyngest wyff Anne · lord for þi mercy kepe.

Anna

¶ A mercy lord mercy mercy mercy
we are synfolest it shewyth þat ȝe send us ałł þis sorwe 140
Why do ȝe thus to myn husbond lord why · why · why ·
for my barynes · he may amend þis · þi self *and* þou lyst to morwe
And it plese so þi mercy · þe my lord · I take to borwe
I xal kepe myn a-vow · qwhyl I leve and leste
I fere me · I haue offendyd þe · myn hert is ful of sorwe 145
Fo. 40v most mekely I pray þi pety · þat þis bale þou wyl breste.

Joachym

here þe Aungel descendith · þe hefne syngyng Exultet celum laudibus · resultet terra gaudijs Archangelorum gloria sacra canunt solemnia.

¶ Qwhat art þou in goddys name · þat makyst me a-drad
It is as lyth a-bowt me as al þe werd were fere.

Angelus

I am an Aungel of god · cōm to make þe glad
God is plesyd with þin helmes · *and* hath herd þi prayere 150
He seyth þi shame · þi repreff · *and* þi terys cler
God is a vengere of synne · *and* not nature doth lothe
Whos wombe þat he sparyth · *and* maketh barreyn her
he doth to shewe his myth · *and* his mercy bothe.

¶ Thu seest þat Sara was nynty ȝere bareyn 155
sche had a son Ysaac · to whom god ȝaff his blyssynge
Rachel also had þe same peyn
She had a son joseph · þat of Egypt was kynge
A strongere þan sampson · nevyr was be wrytynge
nor an holyere þan samuel it is seyd thus 160
ȝett here moderys were bareyn bothe in þe gynnynge
þe concepcion of ałł swycħ · it is ful mervelyous.

[1] The *r* seems to have been inserted—perhaps over another letter.

¶ And in þe lyke wyse · Anne þi blyssyd wyff
sche xal bere a childe · xal hygtħ mary
whicħ xal be blyssyd · in here body · *and* haue joys ffyff 165
and ful of þe holy goost · inspyred syngulyrly
Sche xal be offryd in to þe temple · solemply
þat of here[1] non evyl ffame · xuld sprynge thus
And as sche xal be bore · of a barrany body
So of here xal be bore · with-out nature Jhesus. 170
þat xal be savyour[2]

Fo. 41 (C quire) ¶ That xal be savyour · vnto al man-kende
In tokyn whan þou come · to jherusalem to þe gyldyn gate
þou xalt mete anne þi wyff · haue þis in þi mende
I xal sey here þe same here sorwys to rebate.

Joachym

Of þis imcomparabyl comfort · I xal nevyr for-gete þe date 175
my sorwe was nevyr so grett · but now my joy is more
I xal hom in hast be it nevyr so late
A Anne blyssyd be þat body · of þe xal be bore.

¶ Now fare wel myn shepherdys · governe ȝow now wysly.

1us pastor

Haue ȝe good tydyngys mayster · þan be we glad. 180

Joachym

Prayse god for me · for I am not wourthy.

ijus pastor

In feyth sere so we xal · with all oure sowlys sad.

iijus pastor

I holde it helpfful þat on of vs · with ȝow · be had.

Joachym

Nay abyde with ȝour bestys sone · in goddys blyssynge.

1us pastor

we xal make us so mery · now[3] þis is be-stad 185
þat a myle on ȝour wey · ȝe xal here us synge.[4]

Anne

¶ Alas ffor myn husbond · me is ful wo
I xal go seke hym · what so evyr be falle

[1] *of here* omitted and written above the line.

[2] This is given as the catchword at the bottom of the folio. There is some trace of other rubbed words—which may, however, be scribblings.

[3] *is* written before *þis* and crossed through.

[4] A rough line is drawn from *synge* to the edge of the page—? marking off the shepherds' part.

I wote not in erth which wey is he go
ffadyr of hefne ffor mercy · to ȝ*our* ffete I falle. 190

Angelus

Anne þ*i*n husbond ryght now I was w*ith*all
þ*e* aungel of god · þ*at* bar hym good tydynge
and as I seyd to hym so to þ*e* sey I xal
god hath herd þ*i* preyo*ur* · *and* þ*i* wepynge

¶ At þ*e* goldyn gate þ*o*u xalte mete hym ful mylde 195
and in grett gladnes returne to ȝ*our* hous
So be p*r*oces þ*ou* xalt conseyve *and* ber*e* A childe
Fo. 41v Whiche xal hyght mary · *and* mary xal ber*e* jh*e*sus
Which xal be savyo*ur* of all þ*e* werd *and* us
After*e* grett sorwe · evyr grett gladnes is had 200
now my*n* inbassett I haue seyd to ȝow thus
gooth in our*e* lordys name *and* in god beth glad.

Anne

¶ Now blyssyd be our*e* lorde *and* all his werkys ay
All heffne *and* erthe mut blysse ȝow for this
I am so joyful I not what I may say 205
þ*er* can no tounge telle what joye in me is
I to ber*e* a childe · þ*at* xal ber*e* all mannys blys
and haue my*n* hosbonde a-geyn · ho myth haue joys mor*e*
no creature in erth · is grauntyd mor*e* mercy i-wys
I xal hyȝe me to þ*e* ȝate · to be þ*er* befor*e*. 210

here goth þe Aungel A-ȝen to hefne.

¶ A blyssyd be o*ur* lord · my*n* husbond I se
I xalle on my*n* knes · *and* to hy*m*-ward crepe.

Joachym

A gracyous wyff Anne now fruteful xal ȝe be
Ffor joy of þis metyng[1] · in my sowle I wepe
Haue þis kusse · of clennesse · *and* w*ith* ȝow it kepe 215
in goddys name now go we wyff · hom to o*ur* hous.

Anne

þer was nevyr[2] joy sank in me so depe
now may we sey husbond · god is to us gracyous ¶ veryly[3].

[1] *wepy* first written and crossed through.

[2] *nevyr* omitted and written above the line in slightly darker ink, but probably by scribe of MS.

[3] The paragraph mark before *veryly* has been made over some black sign.

Joachym

ȝa *and* if we haue levyd wel here be-fore
I pray þe lord þin ore 220
so mote we levyn evyr more
And be þi grace more holyly.

Anne

Fo. 42 Now hom-Ward husbond I rede we gon
Ryth hom al to *our* place
to thank god þat sytt in tron̄ 225
þat þus hath sent us his grace.

Contemplacio

¶ Sovereynes ȝe han sen shewyd ȝow be-fore **9**
Of Joachym *and* Anne · here bother*ys* holy metynge
How *our* lady was conseyvid · *and* how she was bore
We passe ovyr þat · breffnes of tyme consyderynge
And how *our* lady in here tendyr age *and* ȝyng 5
In to þe temple was offryd *and* so forth̄ proced
Þis sentens sayd xal be hire be-gynnyng
now þe modyr of mercy · in þis be *our* sped.

¶ And as a childe of iij ȝere age · here she xal appere
to alle pepyl þat ben here present 10
and of here grett grace now xal ȝe here[1]
how she levyd evyr to godd*ys* entent
with grace
That holy matere we wole declare
tyl ffortene ȝere · how sche dyd fare 15
Now of ȝ*our* speche I pray ȝow spare
all̄ þat ben in þis place.

here joachym and Anne with oure lady betwen hem beyng al in whyte as a childe of iij ȝere age presente here in to þe temple thus seying joachym *.

[1] *here she xal* first written instead of *now xal ȝe here* and crossed through.

* *here jo hym and Anne with*—scribbled underneath this.

Fo. 42v

¶ Joachim

BLyssyd be oure lord ffayr ffrute haue we now
Anne wyff remembyr wole ȝe
Þat we made to god an holy a-vow
Þat oure fyrst childe · þe servaunt of god xulde be
The Age of mary oure dowtere is ȝer*ys* thre
Þer fore to thre personys *and* on god · lete us here present
þe ȝonger she be drawyn þe bettyr semyth me
and for teryeng of *our* a-vow · of god we myth be shent.

Anne

¶ it is as ȝe sey · husbond in dede
late us take mary *our* dowtere us be-twen
and to þe temple w*ith* here procede
Dowtere þe Aungel tolde us ȝe xulde be a qwen
Wole ȝe go se þat lord ȝ*our* husbond xal ben
and lerne for to love hym · *and* lede w*ith* hy*m* ȝ*our* lyff
telle ȝ*our* ffadyr *and* me her · ȝ*our* answere let sen
Wole ȝe be pure maydy*n* · *and* also goddys wyff.

Maria

¶ Ffadyr *and* modyr if it plesyng to ȝow be
ȝe han mad ȝ*our* a-vow · so sothly wole I
to be goddys chast seruaunt · whil lyff is in me
but to be goddys wyff · I was nevyr wurthy
I am þ*e* sympelest þ*at* evyr was born of body
I haue herd ȝow seyd · God xulde haue a modyr swete
þ*at* I may leve · to se hire · god graunt me for his[1] mercy
and Abyl me to ley my handys · vndyr hire fayr fete.

Et genuflectet Ad deum.

Joachym

¶ I-Wys dowtere · it is wel seyd
ȝe answere · *and* ȝe were twenty ȝere olde.

Anne

Fo. 43 whith ȝ*our* speche mary I am wel payd
can ȝe gon a-lone · lett se beth bolde.

Maria

To go to goddys hous · wole ȝe now be-holde
I am joyful thedyrward · as I may be.

[1] ȝ*our* first written, crossed out, and *his* written above the line in fainter ink and possibly by a later corrector. The crossing out by two crossed strokes is not the manner of the MS. scribe, and the *s* of *his* is slightly different.

Joachym

Wyff I ryght joyful oure dowtere to be-holde.

Anne

So am I wys husbond · now in goddys name go we.

Joachym

¶ Sere prince of prestes *and* it plese ȝow
we þat were barreyn · god hath sent a childe
to offre here to goddys service we mad oure avow
here is þe same mayde · mary most mylde.

Isakar

Joachym I haue good mende · how I ȝow revyled
I am ryght joyful þat god hath ȝove ȝow þis grace
to be amonge fruteful · now be ȝe reconsylid
Con̄ swete mary con̄ · ȝe haue a gracyous face.

Joachym flectendo Ad deum sic dicens

Joachym

¶ Now Ffadyr *and* sone *and* holy gost
on god *and* personys thre
we offre to þe lorde of myghtys most
Oure dowtere þi servaunt · evyr more to be.

Anna

Ther to most bounde evyr more be we
Mary in þis holy place leve ȝow we xall
In goddys name · now up go ȝe
oure fadyr · oure prest · lo doth ȝow call.

Maria

¶ Modyr *and* it plese ȝow · fyrst wole I take my leve
of my fadyr · *and* ȝow my modyr i-wys
I haue a fadyr in hefne · þis I be-leve
now good ffadyr · with þat fadyr ȝe me blysse.

Joachym

Fo. 43v[1] In nomine patris et filii et spiritus sancti.

Maria

Amen · Now ȝe good modyr.

Anne

In nomine patris et filij et spiritus sancti || Maria · Amen.

Maria

¶ Now oure lord thank ȝow for this
here is my fadyr *and* my modyr bothe
most mekely I beseche I may ȝow kys

[1] The writing on this folio is rather careless.

now for-ȝeve me yf evyr I made ȝow wrothe.
Et explexendo osculabit patrem et matrem.

Joachym

¶ Nay dowtere ȝe offendyd nevyr god nor man
lovyd be þat lord · ȝow so doth kepe.

Anne

Swete dowtyr thynk on ȝ*our* modyr An
ȝ*our* swemynge smytyht to my*n* hert depe.

[Maria][1]

Ffadyr *and* modyr I xal pray for ȝow *and* wepe
To god w*ith* al my*n* hert specyaly
blysse me day *and* nyght evyr her ȝe slepe
good ffadyr *and* modyr · and beth[2] mery.

Joachym

¶ A ho had evyr suche a chylde
nevyr creature ȝit þ*at* evyr was bor*e*
Sche is so gracyous she is so mylde
so xulde childyr to fadyr *and* modyr evyr mor*e*.

Anne

Than xulde thei be blyssyd · *and* plese god sor*e*[3]
husbond *and* it plese ȝow not hens go we xal
tyl mary be in þ*e* temple above thor*e*[3]
I wold not for al erthe se her*e* fal.

Ep*isco*pus

¶ Come gode mary · come babe I þ*e* call[4]
þ*i* pas pratyly to þis plas p*re*tende
þ*ou* xalt be þ*e* dowtere · of god Eternall
Fo. 44 If þ*e* fyftene grees · þ*ou* may Ascende
It is meracle if þ*ou* do · now god þ*e* dyffende
Ffrom babylony to hevy*n*ly jherusalem þis is þ*e* way
Every man þat thynk[5] his lyff to Amende
þ*e* fiftene psalmys · in memorye of þis mayde say.

Maria

Maria · et sic deinceps usque ad fine[6] *xv*^{cim} *psalmor*um.
The fyrst degre gostly applyed

[1] This name was omitted and is written in different ink and in another hand.
[2] The *th* of *beth* has been crossed through in different ink and with perpendicular strokes, not in the manner of the scribe.
[3] Or *sere, there*; the vowels are not clear.
[4] The writing of the first three lines of this stanza (bottom of folio) is smaller.
[5] Some other word has been altered to *thynk*. [6] So in MS.

It is holy desyre w*ith* god to be 85
In trobyl to god I haue cryed
And in sped · þat lord hath herde me.

Ad dominum cu*m* tribularer clamaui ⁝ et exaudiuit me.
The secunde is stody · w*ith* meke inquysissyon veryly
How I xal haue knowynge of godys wylle
To þe mownteynes of hefne I haue lyfte my*n* ey 90
Ffrom qwens xal comy*n* helpe me tylle.

Leuaui oculos meos in montes ⁝ vnde ue*n*iat auxilium m*ih*i.
The thrydde is gladnes in mende in hope to be
that we xall be savyd all thus
I am glad of these tydyng*ys* ben seyd to me
now xal we go · in to goddys hous. 95

Letatus sum in hijs que dicta su*nt* mihi: in domu*m* do*mi*ni ibim*us*.
The fourte is meke obedyence as is dette
to hym þat is a-bove þe planet*ys* sefne
to þe I haue myn eyn sette
þ*at* dwellys above þe skyes in hefne.

Ad te leuaui oculos meos ⁝ qui h*ab*itas in celis.
The ffyfte is p*ro*pyr confessyon 100
þat we be nought w*ith*-owth god thus
but god in vs haue habytac*i*on
Per aventure oure enemyes shulde swelle vs.

Fo. 44v Nisi quia d*omin*us er*at* i*n* nob*is* dicat n*un*c isr*ael* ⁝ nisi q*ui*a do*minus* erat i*n* nobis.
The sexte is confidens in goddys strenght A-lon
Ffor of all grace from hym comyth þe strem 105
they þ*at* trust in god as þe mownt syon
he xal not be steryd · endles þ*at* dwellyth in jherusalem.

Qui confidunt i*n* do*mi*no sicut mons Syon: no*n* *com*moue-b*itur* i*n* et*er*nu*m* q*ui* h*ab*itat i*n* hier*usal*em.
The sefte is vndowteful hope of immortalyte
In oure lorde is as gracy *and* mercy
Whan oure lord conu*er*tyth oure captiuite 110
Than Are we mad as joyful[1] mery.

In conu*er*tendo dom*inus* captiuitate*m* syon ⁝ *fac*ti sum*us* sicut co*n*solati.

[1] A letter has been erased between *joyful* and *mery*.

The eyted is contempt of veynglory in vs
Ffor hym þat Al mankende hath multyplyed
But yf oure lord make here oure hous
they An laboryd in veyn þat it han Edyfied. 115

Nisi d*omin*us edificau*er*it domum: in uanum laborau*eru*nt q*ui* edifica*n*t ea*m*.

The ny*n*te is a childely fer in dede
W*ith* A longyng love in oure lord þ*at* ay is
blyssyd Arn Aff they þ*at* god drede
Whiche þ*at* gon in his holy weys.

Beati o*mn*es q*ui* timent d*omin*um: qui ambulant i*n* vijs eius.

The tende is myghty soferauns of carnal temptac*ion* 120
ffor þ*e* fleschly syght*ys* be*n* fers and fel
ofte ȝough is ffowth with · with suech vexac*ion*
þ*ou*[1] seynge god say so · clepyd israel.

Sepe expugnauerunt me a iuuentute mea: dicat nu*nc* israel.

The Elefnte is accusatyff confessyon of iniquite
Of which ful noyous is þ*e* noyis 125
Fro depnes lord I haue cryed to the
Lord here in sped my sympyl voys.

Fo. 45[2] De profundis clamaui ad te d*omi*ne: d*omi*ne exaudi uocem mea*m*.

The twelfte is mekenes þ*at* is fayre *and* softe
In ma*n*nys sowle w*ith*-inne *and* w*ith*-owte
Lord my*n* herte is not heyued on lofte 130
nyn my*n* eyn be not lokynge a-bowte.

D*omi*ne no*n* est exaltatu*m* cor meu*m*: neq*ue* elati sunt oc*u*li mei.

The Threttene is ffeyth þ*er* w*ith*
w*ith* holy dedys don expresse

[1] Halliwell reads *þan*, but the MS. contraction *þ*ᵘ is that always used for *þou*, the contraction for *þan*, when it occurs, being *þ*ᵃn. The letter *u*(n) is ambiguous, and a difficulty on Fo. 97 ᵛ would be lessened by expanding *þ*ᵘ as *þan*.

[2] The writing on this folio is noticeably small.

haue mende lorde of davyth
And of All his swettnes. 135

Memento d*omi*ne d*auid* ⁝ et o*mn*is mansuetudinis eius.

The ffourtene is brothyrly concorde i-wys
þ*at* norchyth love of creatur*ys* echon
Se how good *and* how glad it is
bretheryn ffor to dwelle in on.

Ecce q*ua*m bonu*m* et q*ua*m jocundu*m* ⁝ h*ab*itare f*rat*res i*n* vnu*m*.

The ffyftene is gracyous · w*ith* on Acorde 140
whiche is syne of godly love semyth me
se now blysse our*e* lord
All þ*at* our*e* lordys servaunt*ys* be.

Ecce nunc b*ene*dicite d*omin*um ⁝ o*mn*es serui dominj.

Ep*iscop*us

A gracyous lord þis is A mervelyous thynge
þ*at* we se here all in syght 145
A babe of thre ȝer age so ȝynge
to come vp þese grecys[1] · so vp-ryght
It is An hey meracle *and* by goddys myght
no dowth of she xal be gracyous.

Maria

Holy ffadyr I be-seche ȝow forth ryght 150
Sey how I xal be rewlyd · in goddys hous.

Ep*iscop*us

Fo. 45v ¶ Dowter*e* god hath ȝovyn vs comaundement*ys* ten
Which shortely to say be comp*re*hendyd in tweyn
and þo must be kept of all crysten men
or ellys here jugement is p*er*petual peyn 155
ȝe muste love god severeynly[2] · *and* ȝo*ur* evyn crystyn pleyn
god fyrst ffor his hyȝ *and* sovereyn dygnyte
he lovyd ȝow fyrst · love hym a-geyn
ffor of love · to his owyn lyknes · he made the.

¶ Love ffadyr sone *and* holy gost 160
Love god þ*e* fadyr · ffor he gevyth myght
Love god þ*e* sone · ffor he gevyth wys*d*a*m* þ*ou* wost
Love god þe holy gost ffor he gevyth love *and* lyght

[1] So in MS. Halliwell gives *greeys*. [2] So in MS.

thre p*er*sonys *and* on god · þus love of ryght
w*ith* aꝗ þin hert · w*ith* aꝗ þ*i* sowle · w*ith* aꝗ þ*i* mende[1] 165
and w*ith* aꝗ þ*e* strenghthis in þ*e* be-dyght
þan love þ*in* evyn crystyn as þ*i* self w*ith*-owty*n* ende.

¶ Thu xalt hate no thynge but þe devyl *and* synne
god byddyth the lovyn þ*i* bodyly enmy
And as for ȝ*our* self her*e* · þus xal ȝe be-gynne 170
ȝe must serve · *and* wurchep god her*e* dayly
Ffor w*ith* prayȝer · w*ith* grace *and* mercy
Se the[2] haue · A resonable tyme to fede
thanne to haue a labo*ur* bodyly
þat þ*er* in be gostly *and* bodely mede. 175

¶ ȝ*our* A-bydynge · xal be w*ith* ȝour maydenys ffyve
Swyche tyme as ȝe wole haue consolac*i*on.

Maria

This lyff me lyketh as my lyve
of her*e* namys I be-seche ȝow to haue informac*i*on.

Ep*i*sc*o*pus

Fo. 46 There is þ*e* fyrst meditac*i*on 180
Contryssyon · compassyon · And clennes
And þ*at* holy mayde fruyssyon
W*ith* these blyssyd maydenes xal be ȝ*our* besynes.

Maria

¶ Her*e* is an holy ffelacheppe I fele
I am not wurthy Amonge he*m* to be 185
Swete syster*ys* to ȝow Aꝗ I knele
To receyve me I be-seche[3] ȝ*our* charyte.

Ep*i*sc*o*pus

They xal dowter*e* · And on þ*e* tother*e* syde se
ther ben sefne prestys in dede
to schryve · to teche · *and* to mynystryn to the 190
to lerne þ*e* goddys lawys · *and* scrypture to rede.

Maria

¶ Ffadyr knew I her*e* namys wele were I.

Ep*i*sc*o*pus

Ther is dyscressyon · devoc*i*on · dylexc*i*on · and deliberac*i*on

[1] *myght* first written and crossed through; *mende* written above the line.
[2] *Sethe* (in one word) in MS.
[3] *seke* first written and crossed through in red ink; *sethe* then apparently written and corrected by erasure to *seche*. A red dot stands under the *h*.

they xal tende upon ȝow besyly
With declaracion determynacion · dyvynacion 195
Now go ȝe maydenys · to ȝ*our* occupacion[1]
And loke ȝe tende þis childe tendyrly
and ȝe serys knelyth · *and* I xal gyve ȝow goddys benyson
in n*o*m*in*e p*a*tris *et* filij *et* sp*iri*tus s*anc*ti.

Et recedent cum ministris suis omnes virgines dicent· Amen.

¶ To ȝow ffadyr *and* modyr I me comende [Maria][2] 200
blyssyd be þ*e* tyme ȝe me hedyr brought.

Joachym

Dowtere þ*e* ffadere of our*e* feyth þ*e* mot defende
as he of his myght made alł thynge of nowth.

Anne

Mary to þ*i* sowle solas he sende
In whos wysdam alł þis werd was wrought. 205
go we now hens husbonde so hende *Hic joachim* et *anna recedent domum.*
for owth of car*e* now ar*e* we brought.

Maria

Fo. 46v Be þ*e* holy gost at hom be ȝe brought
Syster*ys* ȝe may go do what ȝe xałł *Ad virgines*
to serve god fyrst her*e* is al my thought 210
Be-forn þis holy awter*e* on my knes I fałł.

¶ Lord sefne petyc*i*ons I be-seche ȝow of her*e*
Ffyrst þ*at* I may kepe þ*i* love *and* þ*i* lawe
Þ*e* secunde to lovyn my*n* evyn crystyn as my self der*e*
Þ*e* thrydde from alł þ*at* þ*ou* hatyst me to w*ith*-drawe 215
The fourte Alł vertuys to þ*i* plesauns knawe
Þe fyfte to obey þ*e* ordenaryes of þ*e* temple ech̄on
Þ*e* sexte *and* þ*at* alł pepyl may serve þ*e* w*ith* Awe
þ*at* in þis holy temple fawte be non.

¶ The sefnte lord I haske w*ith* grett ffer*e* 220
þ*at* I may se onys in my lyve
þ*at* lady þ*at* xal goddys sone ber*e*
þ*at* I may serve her*e* w*ith* my wytt*ys* fyve
If it plese ȝow *and* ellys · it is not þ*er* w*ith* to stryve
w*ith* pr*a*yers pr*o*strat ffor þese gracys I wepe 225
O my god · devoc*i*on depe in me dryve

[1] The writing from this line to the bottom of the folio is smaller.
[2] No name of speaker given in MS.

þat myn hert may wake in *þe* · thow my body slepe.

here þe Aungel bryngyth manna in A cowpe of gold lyke to confeccions · þe hefne syngynge · þe Aungel seyth

¶ Merveyle not mekest maydon of my mynystrac*i*on
I am a good Aungel sent of god Aꝗ myght
W*ith* Au*n*gelys mete Ffor ȝ*our* Sustentac*i*on 230
ȝe to receyve it · Ffor Natural myght
Fo. 47 We Aungellys xul serve ȝow · day *and* nyght
now fede ȝow *þer-with* · in goddys name
we xal lerne ȝow *þe* lyberary[1] of our*e* lordys lawe lyght
ffor my sawys in ȝow shewyth sygnes of shame. 235

Maria

¶ To thank our*e* soveryen lord not sufficyth my mende
I xal fede me of þis fode my lord hath me sent
Aꝗ maner of savowr*ys* in þis mete I fynde
I felt nevyr non so swete ner so redolent.

Angel*us*

Eche day *þer* w*ith* ȝe xal be content 240
Aunge[2] alle howrys xal to ȝow aper*e*.

Maria

Mercy my maker*e* · how may þis be ment
I am þe sympelest creatur*e* · *þat* is levynge her*e*.

Angelus

¶ In ȝ*our* name Maria · ffyve letterys we han
M. Mayde most mercyfuꝗ *and* mekest *in* mende 245
A. Auerte of þe Anguysch *þat* Adam began
R. Regina of regyon Reyneng w*ith*-owty*n* ende
I. Innocent be Influens of Jesses kende
A. Aduocat most Autentyk ȝ*our* Antecer[3] Anna
hefne *and* helle her*e* kneys dowꝑ bende 250
Whan þis holy name of ȝow is seyd Maria.

Maria

¶ I qwake grettly for dred · to her*e* þis comendac*i*on
Good swete Aungel why wole ȝe sey thus.

Aungeꝗ

Ffor ȝe xal her*e* after*e*[4] hau*e* A salutac*i*on
þat xal þis excede it is seyd Amonge vs 255

[1] This word is underlined in pencil and a pencil cross has been made in the margin, probably by some modern reader.
[2] So in MS.; cf. Fo. 70ᵛ.
[3] Halliwell reads *autecer*.
[4] Some erasure after *aftere*.

The deyte þat dede xal determyn *and* dyscus
ȝe xal nevyr lady be lefte here A-lone.

Maria

Fo. 47v I Crye þe mercy lorde *and* þin erthe Cus
recomendynge me to þat godhyd · þat is tryne *in* trone.

hic osculet terram · here xal comyn Allwey An Aungel with dyvers presentys goynge and *comyng and in þe tyme þei xal synge in hefne þis hymꝑne* · Jhesu corona virginum · *And After* þer *comyth A minister fro þe busschop with A present* and *seyth*

Minister

¶ Prynce of oure prestes Ysakare be name
he hath sent ȝow hym self his servyce in dede
And bad ȝe xulde ffede ȝow spare for no shame
In þis tyme of mete · ne lenger ȝe rede.

Maria[1]

Recomende me to my fadyr sere · *and* god do hym mede
These vesselys A-ȝen sone I xal hym sende
I xal bere it my systerys I trowe þei haue more nede
goddys foyson is evyr to his servauntys hendyr þan we wende.

¶ Systerys oure holy ffadyr isakare
Hath sent us hese servyce here ryght now
Ffede ȝow þer of hertyly · I pray ȝow nat spare
and if owght be leve · specyaly I pray ȝow
That þe pore men þe relevys þer of haue now
Ffayn *and* I myth I wolde do þe dedys of mercy
Pore ffolk ffaryn god knowyth how
On hem evyr I haue grett pety.

¶ Contemplacio

lo sofreynes here ȝe haue seyn
in þe temple of oure ladyes presentacion
she was nevyr occapyed in thyngys veyn
Fo. 48 but Evyr besy in holy ocupacyon̄ **10**
And we be-seche ȝow of ȝoure pacyens
þat we pace þese materys so lythly Away
If þei xulde be do with good prevydens
Eche on wolde suffyce ffor An hool day
Now xal we procede to here dissponsacion
which Aftere þis was xiiij ȝere
tyme sufficyth not to make pawsacion

[1] The usual red loop round the name of speaker has been omitted here.

hath pacyens w*ith* vs we be-sech̄ ȝow her
And in short spas
The parlement of hefne sone xal ȝe se
and how goddys sone com̄ man xal he 15
And how þe salutac*i*on After*e* xal be
be goddys holy gras.

Remainder of Fo. 48—4¼ inches—left blank. Fo. 48ᵛ has been used for speeches to be inserted between the speeches on Fo. 50 and Fo. 50ᵛ in the next play.

Their place is made clear by red reference marks and also by rough capitals in black ink of a different shade from that of the MS.

Passage to be inserted between Fo. 50 and Fo. 50ᵛ.

Fo. 48ᵛ ¶ This Ansuer*e* grettly trobelyth me
to mak a vow to creatur*ys* it is lefful
Vovete *et* reddite · in scripture haue we
and to observe our*e* lawe also it is nedful 95
In þis to dyscerne to me it is dredful
þ*er*fore to cowcell[1] me in þis cas I calle
þ*e* holde *and* þe wyse *and* swiche as ben spedful
In þis sey ȝo*ur* a-vyse · I be-sech̄ ȝow Alle.

Mi*nister*

¶ To breke o*ur* lawe *and* custom · it wor*e* hard in dede 100
And on þ*at* other syde to do a-ȝen[2] scrypture
to ȝeve sentens in þis degre · ȝe must take goo[3] hede
ffor dowteles þis matere · is dyffuse *and* obscure
My*n* avyse her*e* in þis I ȝow ensure
þ*at* we prey all̄ god to haue relac*i*on 105
ffor be pr*ay*o*ur* · grett knowlech̄ me*n* recure
and to þis I counsell̄ ȝow to ȝeve assygnac*i*on.

Ep*iscop*us

[E] Trewly ȝo*ur* counsell̄ is ryght good *and* eylsum
And as ȝe han seyd so xal it be
I charge ȝow bretheryn *and* syst*er*ys hedyr ȝe com̄ 110
and to-gedyr to god now pr*ay* we
That it may plese his fynyte[4] deyte
knowleche in þis to sendyn vs Et *hic cantent veni creator*
Mekely eche man ffalle down on kne
And we xal be-gynne · Veni creator Sp*iri*tus 115

*and whan Veni creat*or *is don* þ*e buschop xal seyng*

Now lord god of lordys whysest of alle et cet*er*a.

[1] So in MS. [2] The *a* omitted and written above the line.
[3] So in MS. [4] So in MS.

Fo. 49 (D quire) [A] ¶ *Tunc venit Abysakar*[1] ep*iscopus*

10

LIstenyth lordyng*ys* bothe hye *and* lowe
And tendyrly takyth heyd on to my sawe
beth buxom *and* benyngne ȝ*our* busshopp to knowe
Ffor I am þ*at* lord þ*at* made þis lawe
W*ith* hertys so hende herkyn nowe 5
ȝoure damyselys to weddyng ȝa loke þ*at* ȝe drawe
þ*at* passyn xiiij ȝere for what þ*at* ȝe owe
þ*e* lawe of god byddyth þis sawe
þ*at* at xiiij ȝere of age
Euery damesel what so sche be 10
to þ*e* Encrese of mor*e* plente
xulde be browght in good degr*e*
On to her*e* spowsage.

Joachym

¶ Herke now Anne my jentyl spowse
how þ*at* þ*e* buschop his lawe hath tolde 15
þ*at* what man hath a dowtyr i*n* his house
þ*at* passyth xiiij ȝer*ys* olde
He muste her*e* brynge I herde hym Rowse
In to þ*e* tempyl a spowse to wedde
wher for our*e* dowtyr ryth good *and* dowse 20
In to þ*e* tempyl sche must be ledde
nd þ*at* anoon ryght sone.

Anne

Ser*e* I grawnt þ*at* it be so
A-ȝen þ*e* lawe may we not do
w*ith* her*e* to-gedyr lete us now go 25
I hold it ryght weyl done.

Joachym

¶ Ser*e* busshopp her*e* aftyr þi*n* owyn hest
we haue her*e* brought our*e* dowtyr der*e*
mary my swete childe she is ful prest
of Age she is ful xiiij ȝere. 30

[1] Halliwell reads *ab Ysakar* with the note 'Sic in MS pro Abysakar'; but the MS. has no capital *Y* and the initial *A* is of the form used indifferently for small or capital. The syllable *ab* is not joined to the *y*, but the *v* of the preceding *venit* is similarly detached.

Ep*iscopus*

Fo. 49v (marked 50) [B] Welcome joachym on to my*n* A-reste
bothe anne þ*i* wyff *and* mary cler*e*
now mary chylde to þ*e* lawe þ*ou* leste
and chese þ*e* a spowse to be þ*i* ffer*e*
þ*at* lawe þ*ou* must fful-ffylle.

Maria

A-ȝens þ*e* lawe wyl I nevyr be
but mannys ffelachep xal nevyr folwe me
I wyl levyn evyr in chastyte
be þ*e* grace of goddys wylle.

Ep*iscopus*

[C] ¶ A Ffayr*e* mayde why seyst þ*ou* so
what menyth the for to levyn chast
why wylt þ*ou* not to weddyng go
þ*e* cawse þ*ou* telle me *and* þ*at* in hast.

Maria

My ffadyr *and* my modyr sertys also
Er I was born ȝe may me trast
thei wer*e* bothe bareyn her*e* frute was do
they come to þ*e* tempyl at þ*e* last
to do her*e* sacryfice
By-cause they hadde nothyr frute ner*e* chylde
reprevyd þei wor*e* of wykkyd *and* wyllde
w*ith* grett shame þei wer*e* revylyd
Al men dede them dyspyce.

¶ My Ffadyr *and* my modyr thei wepte full sor*e*
fful hevy her*e* hertys wern of þis dede
w*ith* wepynge eyn þei preyd þ*er*for*e*
þ*at* god wolde socowr*e* hem *and* sende hem sede
Iff god wold graunt hem a childe be bor*e*
They be-hest þ*e* chylde her*e* lyff xulde lede
In goddys temple to serve evyr-mor*e*
and wurchep god in loue *and* drede
Than god fful of grace
Fo. 50 he herd her*e* longe p*ray*o*ur*
and þan sent hem both seed *and* flowr*e*[1]
whan I was born in her*e* bowr*e*
to þ*e* temple offryd I was.

[1] The *w* has been adapted from some other letter.

¶ Whan þ*at* I was to þ*e* temple brought
and offerde up to god Above
ther hestyd I as my*n* hert thought
to serve my god w*ith* hertyly love
Clennesse *and* chastyte my*n* hert owth 70
Erthely creatur*e* nevyr may shoue
Suc̄h clene lyff xuld ȝe nouht
In no ma*n*er wyse reprove
to þis clennesse I me take
this is þ*e* cawse as I ȝow tell̄ 75
þ*at* I w*ith* man wyll̄ nevyr mell̄
in þ*e* servyse of god wyl I evyr dwell̄
I wyl nevyr haue other make.

Ep*iscop*us

[D] ¶ A mercy god þese wordys wyse
of þis fayr mayde clene 80
thei trobyl my*n* hert in many wyse
her wytt is grett *and* þ*at* is sene
In clennes to levyn in godys servise
no man her*e* blame no*n* her*e* tene
And ȝit in lawe · þus it lyce 85
þ*at* suc̄h weddyd xulde bene
who xal expownd þis oute
þ*e* lawe doth after lyff of clennes
þ*e* lawe doth̄ bydde suc̄h maydenes expres
þ*at* to spowsyng they xulde hem dres 90
God help us in þis dowhte.

This Ansuere grettly trobelyth me ·/ et ce*tera* ut sup*ra*[1]

Now lord god of lordys wysest of All̄

Fo. 50v I pray þe lorde knelynge on kne
[F] w*ith* carefull̄ herte I crye *and* calle
þis dowteful dowte enforme þ*ou* me.

Angelus

¶[2] Thy prayo*ur* is herd to hyȝ hevyn halle 120

[1] The writing indicates that the scribe went on from *God help us in þis dowhte* to *Now lord god of lordys wysest of All̄* and inserted the reference to the interpolated speeches later.

[2] This paragraph sign should stand before the line *Now lord god*, &c.

god hath me sent her*e* dow̃ to the
to telle þ*e* what þ*at* þ*ou* do xalle
and how þ*ou* xalt be rewlyd in iche degr*e*
take tent *and* vndyrstond
This is goddys owyn byddyng 125
þ*at* all̃ kynsmen of dauyd þ*e* kyng
to þ*e* temple xul brynge her*e* du offryng
w*ith* whyte ȝardys in þ*er* honde.

¶ loke wele what tyme þei offer*e* ther*e*
all̃ her*e* ȝardys in þ*in* hand þ*ou* take 130
take heed whose ȝerde doth blome and ber*e*
and he xal be þ*e* maydenys make.

Ep*iscopus*

[G] I thank þ*e* lord w*ith* mylde cher*e*
thi wurde xal I werkyn w*ith*-owty*n* wrake
I xal send for hem bothyn fer *and* ner*e* 135
to werke þ*i* wyl I vndyr-take
A-non it xal be do
herk masangere þ*ou* wend þ*i* way
Dauyd · kynsmen as I þ*e* say
byd hem come offyr þis same day 140
and brynge white ȝardys Also.[1]

Nu*n*cius[2]

¶ Oy · al maner me*n* takyth to me tent[3]
that be owgth of kynrede to dauid þ*e* kyng
my lord þ*e* busshop hath for ȝow sent
to þ*e* temple þ*at* ȝe come w*ith* ȝo*ur* offryng̃.[4] 145

Fo. 51 Joseph
(E quire) In gret labore my lyff I lede 155
† myn*e* ocupasyou*n*[5] lyth in many place
ffor febylnesse of age my jorney I may not spede
I thank the gret god of thi grace.

[1] *Also* written twice and the first crossed out.

[2] Some faint letter written (? scribbled) above the *us* of *Nuncius*.

[3] These four lines are written smaller.

[4] This speech is continued on Ff. 53. Fo. 51 and 52 form a separate interpolated quire E. On Fo. 51 is written the dialogue between Joseph and the generaciones dauid' which follows, in a later cursive hand without rubrication. Ff. 51^{v} and 52 and 52^{v} are left blank. Roughly-made reference marks—as above—on Ff. 51 and 53 indicate the place of the interpolated dialogue.

[5] Or *ocupasyon̄*.

1us gen*er*acion*is* d*auid*ꝛ

what chere Josepꝯ what ys the case
That ye lye here on this grou*n*dꝛ [1].

Josepꝯ

age *and* febylnesse doth me enbrace
That I may nother weꝯ goo ne stou*n*dꝛ [2].

ij^us^ gen*er*acion[*is*

We be co*m*mandydꝛ be the beschoppys sondꝛ
That eu*er*y man of Dauydꝛ [3] Kynrede
In the tempyꝯ to offyr a wondꝛ
Therfor in this Jorney let vs p*ro*cede.

Josepꝯ [4]

Me to traveyꝯ yt is no nede
I prey you frend*ys* go forth yo*ur* wey.

iij^us^ gen*er*acion[*is*

yis comꝫ forth Joseph I you rede
andꝛ knowyth what the buschop woꝯ sey.

iiij^us^ gen*er*[ac*ion*is

Ther ys a maydꝛ whos name is clepydꝛ mary
Dought*er* to joachym̄ as it is toldꝛ [5]
here to mary thei woꝯ asay
To som̄ many [6] dowty *and* boldꝛ

Fo. 53 (F quire)

He chargight þ*at* ȝe hast ȝow · for he is redy bent
ȝow to receyve at ȝ*our* comyng
He byddyth ȝow ffer*th*ermore in handys þ*at* ȝe hent
A fayr*e* white ȝerde everych of ȝow ȝe bryng
In hyght

Tary not I pray ȝow
My lord as I say ȝow
now to receyve ȝow
is fuꝯ redy dyght.

Joseph [7]

¶ Benedicite I can not vndyr stande
What our*e* p*r*ince of prest*ys* doth menꝫ
þ*at* every man xuld come *and* brynge w*it*h hy*m* a whande

[1] Or *gronꝺ*; the stroke may be insignificant.
[2] Or *stonꝺ*.
[3] This may be a tailed *d* or *d* + *ys*.
[4] iii^us^ gen*er*acion first written and crossed through.
[5] *I haue herd* first written for *it is told*, and crossed through.
[6] So in MS. for *man*.
[7] *Pri*mus *generacionis* first written and crossed through.

Abyl to be maryed þ*at* is not I so mote I then
I haue be maydon evyr *and* evyr mor*e* wele ben
I chaungyd not ȝet of att my long lyff 180
and now to be maryed su*m* man wold wen
it is a straunge thynge An old man to take a ȝonge wyff.

¶ but nevyr-þ*e*-lesse no doute of we must forth to towne
now neybor*ys* *and* kynnysmen lete us forth go
I xal take a wand in my hand *and* cast of my gowne 185
yf I falle · þan I xalle · gronyn for wo
Ho so take A-way · my staff I say · he wer*e* my fo
ȝe be men · þ*at* may wele ren · go ȝe be-for*e*
I am old · *and* also colde · walkyng doth me wo
þ*er*for*e* now wole I · so my staff holde I · þis jurny to wor*e*. 190

Ep*iscopus*

[H] ¶ Ser*ys* ȝe xal vndyr-stande[1]
þ*at* þis is þ*e* cawse of o*ur* comynge
and why þ*at* ech of ȝow bryngyth a wande
ffor of god we haue knowynge
her*e* is to be maryde a mayde ȝynge 195
att ȝo*ur* roddys ȝe xal brynge vp to me
and on hese rodde þ*at* þ*e* holy gost is syttynge *hic portent virgas*
he xal þ*e* husbond of þis may be.

Joseph

Fo. 53v ¶ It xal not be I ley a grote
I xal a-byde be-hynde pr*e*uyly 200
now wolde god I wer*e* at hom in my cote
I am aschamyd to be seyn veryly.

P*rim*us gener*acion*is dauid

¶ To[2] wurchep my lord god hedyr am I come
her*e* ffor to offyr my dewe[3] offrynge
A fayr white ȝarde in hand haue I nome 205
* my lord ser*e* busshop at ȝo*ur* byddynge.

Se*cun*dus gener*acionis* d*auid*

Off dauythis kynred sert*ys* am I com

[1] From here to the bottom of the folio the writing is closer.
[2] *my* first written between *to* and *wurchep* and crossed through.
[3] *dw* written before *dewe* and crossed through.
* *To my woursyp* written (scribbled ?) here in different ink and different handwriting—not that of Fo. 51.

a ffayr white ȝarde in hand now I bryng
my lord þe busshop after ȝ*our* owyn[1] dom
þis ȝarde do I offre at ȝ*our* chargyng 210
Ryht here.

Terci*us* gene*ra*ci*onis* d*auid*ꝯ

And I a ȝarde haue both fayr *and* whyght[2]
here in my*n* hond it is redy dyght
And here I offre it forth w*ith*in syght
Ryght *in* good manere. 215

Quart*us* gene*ra*ci*onis* d*auid*ꝯ

¶ I am þe fourte of Dauid is kynn
and w*ith* my*n* offrynge my god I honoure
þis fayr whyte ȝarde is offryng my*n*
I trost in god of su*m* socoure
Comm on Joseph w*ith* offrynge þi*n* 220
and brynge up þi*n* as we han oure
þ*ou* taryst ryth longe be-hynde certeynn
why comyst not ffortht to goddys toure
Com on man for shame.

Joseph

Comm ȝa ȝa · god help fulll fayn I wolde 225
but I am so Agyd *and* so olde
þ*at* both my*n* leggys gyn to folde
I am ny Almost lame.

Ep*isco*pus

Fo. 54 A mercy lord I kan no sygne a-spy
It is best we go a-geyn to p*ra*yr. 230

Vox

he brought not up his rodde ȝet trewly
to whom þe mayd howyth to be maryed her.

Ep*isco*pus

¶ Whath joseph why stande ȝe there by-hynde
I-wys sere ȝe be to blame.

Joseph

Sere I kan not my rodde ffynde 235
to come þ*er* in trowth me thynkyht shame.

Ep*isco*pus Comyth thens[3]

[1] *owym* first written and corrected.

[2] The *gh* in this word may have been cancelled by the scribe. There appear to be two strokes through them.

[3] So written in the margin in MS.

Joseph

Sere he may Euyl go þat is ner lame
in soth̄ I com as fast as I may

Ep*iscop*us

offyr[1] up ȝ*our* rodde sere in goddys name 240
why do ȝe not as men ȝow pray.

Joseph

¶ Now in þe wurchep of god of hevyn
I offyr þis ȝerde as lely whyte
pr*ay*ng þat lord of gracyous stewyn̄
W*ith* hert · w*ith* wytt · w*ith* mayn w*ith* myght 245
And as he made þe sterr*ys* seven̄
þis sympyl offrynge þ*at* is so lyght
to his wurchep he weldygh̄ evyn
Ffor to his wurchep þis ȝerd is dygh̄t
lord god I þe pray 250
to my herte þ*ou* take good hede
and no thynge to my synful dede
Aftyr my wyl þ*ou* qwyte my mede
As plesyth to þ*i* pay.

¶ I may not lyfte my*n* handys heye 255
Lo · Lo · Lo · What se ȝe now.

Ep*iscop*us

A mercy mercy mercy lord we crye
þe blyssyd of god we se Art thou. *et clamant omnes* Mercy mercy.

Fo. 54v ¶ A gracyous god in hevyn trone
ryht wundyrful þ*i* werkys be 260
here may we se A merveyl one
A ded stok beryth flour*ys* ffre
Joseph in hert w*ith*-outyn mone
þ*ou* mayst be blyth̄ w*ith* game *and* gle
A mayd to wedde þ*ou* must gone 265
be þis meracle I do wel se
Mary is here name.

Joseph

What xuld I wedde god for-bede
I am an old man so god me spede

[1] *Helde* first written and crossed through; *offyr* written over the line.

and with a wyff now · to levyn in drede 270
It wore neyther sport nere game.

Ep*iscopus*

¶ A-ȝens god joseph þ*ou* mayst not stryve
god wyl þ*at* þ*ou* a wyff haue
þis fayr mayde xal be þ*i* wyve
she is buxum *and* whyte as laue. 275

Joseph

A shuld I haue here ȝe lese my lyff
Alas dere god xuld I now rave
An old man may nevyr thryff
W*ith* a ȝonge wyff so god me saue
nay nay sere lett bene 280
xuld I now in age begyn*ne* to dote
If I here chyde she wolde clowte my cote
blere myn ey *and* pyke out a mote
and þus oftyn tymes it is sene.

Ep*iscop*us

¶ Joseph now as I þ*e* saye 285
God hath assygnyd here to þ*e*
Fo. 55 þ*at* god wol haue do sey þ*ou* not nay
oure lord god wyl þ*at* it be so[1].

Joseph

A-ȝens my God not do I may
here wardeyn *and* kepere wyl I evyr be 290
But fayr maydon I þ*e* pr*ay*
Kepe þ*e* clene as I xal me
I am a man of age
therfore sere busshop I wyl þ*at* ȝe wete
þ*at* in bedde we xul nevyr mete 295
Ffor i-wys mayden suete
An Old man may not rage.

Ep*iscop*us

¶ This holyest virgyn xalt þ*ou* maryn now
ȝ*our* rodde floreschyth fayrest þ*at* man may se 299
þ*e* holy gost we se syttyht on a bow *Et hic cantent.* Bene-
now ȝelde we all preysyng to þ*e* trenyte. *dicta* sit be*ata* tr*in*itas.

¶ Joseph wole ȝe haue þis maydon to ȝ*our* wyff
And here honour *and* kepe as ȝe howe to do.

[1] *so* apparently by mistake as it destroys the rhyme.

Joseph

nay sere so mote I thryff
I haue ryght no nede þer-to. 305

Episcopus

Joseph it is goddys wyl it xuld be so
sey aftyr me as it is skyl.

Joseph

Sere *and* to performe his wyl I bow þer-to
Ffor all thynge owyght to ben at his wyl.

Episcopus *et* idem joseph

¶ Sey þan after me: here I take þe Mary to wyff 310
to hauyn to holdyn: as god his wyll with us wyl make[1]
and as longe as be-thwen us: lestyght oure lyff
to loue ȝow as my selff: my trewth I ȝow take.

nunc ad mariam sic dicens · Episcopus

Fo. 55v ¶ Mary wole ȝe haue þis man
And hym to kepyn as ȝour lyff. 315

Maria

In þe tenderest wyse fadyr as I kan
and with all my wyttys ffyff.

Episcopus

¶ Joseph with þis ryng now wedde þi wyff
and be here hand now þou here take.

Joseph

Sere with þis Rynge I wedde here Ryff 320
and take here now here ffor my make.

Episcopus

Mary mayd with-outyn more stryff
On to þi spowse þou hast hym take.

Maria

In chastyte to ledyn my lyff
I xal hym nevyr for-sake 325
but evyr with hym a-byde
And jentyll spowse as ȝe An seyd
lete me levyn as a clene mayd
I xal be trewe be not dysmayd
both terme tyme *and* tyde. 330

Episcopus

¶ Here is þe holyest matremony þat evyr was in þis werd
þe hyȝ names of oure lord we wole now syng hy

[1] *wylmake* (one word) in MS.

we aꝫ wole þis solempn dede record
devowtly · Alma chorus d*omi*ni · nu*nc* pangat no*mi*na su*m*mi.

¶ Now goth hom aꝫ in godys name
Where as ȝ*our* wonyng was be-fo*re*
Maydenys to lete her*e* go alone it wo*re* shame
It wold hevy ȝ*our* hert*ys* so*re*
ȝe xal blysse þ*e* tyme þat sche was bo*re*
now loke ȝe at hom her*e* brynge.

Maria

To haue ȝ*our* blyssyng ffadyr I falle ȝow be-fo*re*.

Ep*iscop*us

he blysse ȝow þ*at* hath non hendyng
In no*mi*ne p*at*ris *et* filij *et* sp*iri*tus *sancti*.

Ep*iscop*us

Fo. 56 ¶ Joseph þi selph art old of Age
And þi wyff of Age is ȝonge
and as we redyn in old sage
many man is sclepyr of tonge
þ*er* fo*re* Euyl langage for to swage
þ*at* ȝ*our* good fame may leste longe
iij damysellys xul dwelle w*ith* ȝow i*n* stage
W*ith* þ*i* wyff to be evyr mo*re* a-monge
I xal these iij her*e* take
Susanne þ*e* fyrst xal be
Rebecca þ*e* secunde xal go w*ith* the
Sephor*e* þ*e* thrydde · loke þ*at* ȝe thre
þis maydon nevyr ȝe for-sake.

Susanne

¶ Ser*e* I am redy Att ȝ*our* wyꝫ
w*ith* þis maydon for to wende.

Rebecca

ȝ*our* byddyng ser*e* xaꝫ fful-ffyl
and ffolwe þis maydon ffayr *and* hende.

Sephor

To ffolwe hyre it is good skyl
And to ȝ*our* byddynge wole I bende.

Joseph

Now ser*e* buschop hens go I wyl
for now comyth on to my mende
A mater*e* þ*at* nedful is.

Ep*iscop*us

Ffar*e* wel joseph *and* mary clere
I pray god kepe ȝow all̃ in fere
and sende ȝow grace in good manere
to serve þ*e* kynge of blysse.

Maria

¶ Ffadyr *and* modyr ȝe knowe þis cas
how[1] þ*a*t it now doth stonde w*ith* me
w*ith* my*n* spowse I must forth passe
and wott nevyr whan I xal ȝow se
Fo. 56v Therfor*e* I pray ȝow her*e* in þis plas
of ȝo*ur* blyssynge for charyte
and I xal spede þ*e* betyr *and* haue mor*e* gras
In what place þ*a*t evyr I be
On knes to ȝow I falle
I pray ȝow fadyr *and* modyr der*e*
to blysse ȝo*ur* owyn der*e* dowter*e*
and pray ffor me in all̃ manere
and I ffor ȝow all̃.

Joachym

¶ Almyghty god he mote þ*e* blysse
and my blyssynge þ*ou* haue Also
In all̃ godnesse god þ*e* wysse
on londe or on watyr wher evyr þ*ou* go.

Anna

Now god þ*e* kepe from every mysse
and saue þ*e* sownd in welth̃ from wo
I p*ray* þ*e* dowtyr þ*ou* onys me kys
or þ*a*t þ*i* modyr parte þ*e* fro
I pray to god þ*e* saue
I pray þ*e* mary my swete chylde
be lowe *and* buxhum meke *and* mylde
Sad *and* sobyr *and* no thyng wylde
and goddys blyssyng þ*ou* haue.

Joachym̃

¶ Ffor[2] wel joseph̃ *and* god ȝow spede
wher so ȝe be in halle or boure.

Joseph

Almyghty god ȝo*ur* weys lede
and saue ȝow sownd from all̃ doloure.

[1] *Whow* first written, the *W* crossed through. [2] So in MS.

Anna

Goddys grace on ȝow sprede[1] 400
Ffar*e* wel mary my swete fflowr*e*
Fare weyl joseph *and* god ȝow rede
Ffar*e* weyl my chylde *and* my tresowr*e*
Fo. 57 Ffar*e* wel my dowter*e* ȝyng.

Maria

Ffar*e* wel fadyr *and* modyr der*e* 405
at ȝow I take my leve ryght her*e*
god *þat* sytt in hevyn so cler*e*
haue ȝow in his kepyng.

Joseph

¶ Wyff it is ful necessary þis ȝe knowe
*þ*at I And my kyn-rede go hom be-for*e* 410
for in soth̄ we haue non hous of our*e* owe
þer fore I xal gon ordeyn · *and* thanne come ȝow for*e*
We Ar not ryche of werdly thynge
and ȝet of our*e* sustenauns we xal not mys
therfor*e* A-bydyth her*e* stylle to ȝ*our* plesynge 415
to worchep ȝ*our* god is all̄ ȝ*our* blysse.

¶ he *þ*at is *and* evyr xal be
Of hefne *and* helle Ryche kynge
in erth hath chosyn poverte
and all̄ Ryches *and* welthis refusynge. 420

Maria

Goth husbond in our*e* lordys blyssynge
he mote ȝow spede in all̄ ȝ*our* nede
and I xal her*e* a-byde ȝ*our* A-ȝen comynge
and on my sawter*e* book I xal rede
Now blyssyd be our*e* lord ffor this 425
of hefne *and* erthe *and* all̄ *þ*at beryth lyff
I am most bound to ȝow lord i-wys
ffor now I am bothe mayde *and* wyff.

¶ Now lord god dyspose me to p*r*ayo*ur*
*þ*at I may sey *þ*e holy psalmes of dauyth 430
wheche book is clepyd *þ*e sawter*e*
*þ*at I may preyse the · my god *þ*er w*ith*

[1] This and following three lines written smaller.

Fo. 57v Of þe vertuys þer of þis is þe pygth
It makyht sowles fayr þat doth it say
Angelys be steryd to help us þer with
it lytenyth therkeness *and* puttyth develys Away.

¶ þe song of psalmus is goddys dete
synne is put A-wey þer by
It lernyth A man vertuysful to be
It feryth mannys herte gostly
who þat it vsyth custommably
it claryfieth þe herte *and* charyte makyth cowthe
he may not faylen of goddys mercy
þat hath þe preysenge of god evyr in his mowthe.

¶ O holy psalmys · O holy book
Swetter to say than Any ony
þou lernyst hem love lord þat on þe look
and makyst hem desyre thyngys celestly
With these halwyd psalmys lord I pray the specyaly
* ffor all þe creatures qwyke *and* dede
þat þou wylt shewe to hem þi mercy
and to me specyaly þat do it rede.

¶ I haue seyd sum of my sawtere *and* here I am
at þis holy psalme in dede
Benedixisti domine terram tuam
In this holy labore · lord me spede.

Joseph

¶ Mary wyff *and* mayd most gracyous
displese ȝow not I pray ȝow so long I haue be
I haue hyryd for us a lytyl praty hous
and þer-in ryght hesely levyn wole we
Come forth mary *and* folwe me
to Nazareth now wele we go
Fo. 58 And all þe maydonys bothe ffayr *and* fre
with my wyff comyth forth also
Now lystenyth well wyff what I tell þe
I must gon owth hens fer þe fro
I wyll go laboryn in fer countre

* Here some words are scribbled in faint ink in the margin : of þe Vyn. R. Wych.

with trewth to maynteyn oure housholde so
þis ix monthis þ*ou* seyst me nowth
Kepe þ*e* clene my jentyl spowse 470
and alł þ*in* maydenys in þ*in* howse
þ*at* evyl langage I her*e* not rowse
Ffor hese love þ*at* alł hath wrought.

Maria

¶ I p*ray* to god he spede ȝ*our* way
and in sowle helth̄ he mote ȝow kepe 475
and sende ȝow helth̄ bothe nyth *and* day
he shylde *and* saue ȝow from al shenschepe
Now lord of grace to þ*e* I pray
with morny mood on kne I krepe
me saue from synne from tene *and* tray 480
with hert I mourne *with* eye I wepe
lord god of pete
Whan I sytt in my conclaue
alł my*n* hert on þ*e* I haue
Gracyous god my mayden-hed saue 485
Euyr clene in chastyte.

Fo. 58v

contemplacio

1*us* [1]

11

¶ Ffowr*e* thowsand · sex vndryd · four*e* ȝer*e* I telle
Man ffor his offens *and* ffowle foly
Hath loyn ȝer*ys* · in þ*e* peynes of helle
And were wurthy to ly þ*er*-in endlesly
But thanne xulde perysche ȝ*our* grete mercye 5
good lord haue on man pyte
haue mende of þ*e* prayo*ur* seyd by Ysaie
lete mercy meke þ*in* hyest mageste.

¶ wolde god þ*ou* woldyst breke þ*in* hefne myghtye
and com down her*e* in to [2] erth̄ 10
And levyn ȝer*ys* thre *and* threttye
thyn famyt ffolke *with* þ*i* fode to fede

[1] This figure is in red.
[2] *to* twice written, the first crossed through.

To staunche þ*i* thryste lete þ*i* syde blede
ffor erste wole not be mad redempc*i*on
Cu*m* vesyte vs in þis tyme of nede
of þ*i* careful creatur*ys* haue compassyon̄.

2[1] ¶ A woo to vs wrecchis of[2] wrecchis be
ffor god hath haddyd[3] ssorwe to sorwe
I prey þ*e* lord þ*i* sowlys com̄ se
How þei ly *and* sobbe ffor syknes *and* sorwe[4]
W*ith* þ*i* blyssyd blood ffrom balys hem borwe
thy careful creaturys cryenge in captyvyte
A tary not gracyous lord tyl it be to-morwe
The devyl hath dysceyved hem be hys iniquite.

¶ A q*uo*d Jeremye · who xal gyff wellys to my*n* eynes
þ*at* I may wepe bothe day *and* nyght
to se our*e* bretheryn in so longe peynes
her*e* myschevys Amende · may þ*i* mec̄h myght
Fo. 59 As gret as þ*e* se lord · was Adamys contryssyon ryght
Ffrom our*e* hed is falle þ*e* crowne
Man is comeryd in synne · I crye to þ*i* syght
Gracyous lord · Gracyous lord · Gracyous lord come downe.

Virtutes

¶ Lord plesyth it þin hyȝ domynac*i*on
On man þ*at* þ*ou* made to haue pyte
Patryarchys[5] *and* pr*o*phetys han made[6] supplycac*i*on
our*e* offyse is to presente · her*e*[7] pr*a*yer*ys* to the
Aungelys · Archaungelys we thre
þ*at* ben in þ*e* fyrst ierarchie
Ffor man to þi*n* hy mageste
Mercy · mercy · mercy we crye.

[1] This figure is in red.

[2] The original word *of* has been crossed out and *that* written in a different hand above.

[3] The initial *h* has been crossed through with a thick perpendicular stroke of different ink.

[4] The original phrase *ffor sykness* and *sorwe* has been altered in this different ink and presumably the same hand that wrote *that* for *of* above, to *bothe eve and morewe*. [Not impossibly the hand of the scribe of Ff. 95, 96.]

[5] Or *kys*. [6] Or *mad*.

[7] Some word was written between *her*e and *pr*a*yer*ys and crossed through.

¶ The Aungel lord þ*ou* made so gloryous
whos synne[1] hath mad hy*m* a devyl in helle
he mevyd man to be so contraryous
man repentyd · *and* he in his obstynacye doth dwelle
Hese grete males good lord repelle 45
And take man on to þ*i* grace
lete þ*i* mercy make hy*m* w*ith* Au*n*gelys dwelle
of locyfer*e* to restor*e* þ*e* place.

Pater

PRopter miseriam inopum
et gemitum pauperum } nu*n*c exurgam

¶ Ffor þ*e* wretchydnes of þ*e* nedy
And þ*e* porys lamentac*i*on 50
now xal I ryse þ*at* am Almyghty
tyme is come of reconsyliac*i*on
Fo. 59v My p*r*ophetys w*ith* prayers haue made supplicac*i*on
my contryte creaturys crye all for comforte
All my*n* Aungellys in hefne · w*ith*-owte cessac*i*on 55
they crye þ*at* grace to man myght exorte.

Veritas

¶ Lord I am þ*i* dowter*e* trewth
þ*ou* wylt se I be not lor*e*
thyn vnkynde creatur*ys* to saue wer*e* rewthe
the offens of man hath grevyd þ*e* sor*e* 60
Whan Adam had synnyd þ*ou* seydest þor*e*
þ*at* he xulde deye *and* go to helle
And now to blysse hym to resstor*e*
twey contraryes mow not to-gedyr dwelle.

¶ Thy trewthe lord xal leste w*ith*-owtyn ende 65
I may in no wyse ffro þ*e* go
þ*at* wretche þ*at* was to þ*e* so vnkende
he may not haue to meche wo
He dyspysyd þ*e* *and* plesyd þ*i* ffo
þ*ou* art his creato*ur* · *and* he is þ*i* creature 70
þ*ou* hast lovyd trewthe · it is seyd evyr mo
þ*er*fore in peynes · lete hym evyr mor*e* endur*e*.

[1] A blotted *hath* between *synne* and *hath* crossed through.

M*isericordia*

¶ O Ffadyr of mercy · *and* god of comforte
þ*at* counselł us in eche trybulac*i*on
lete ȝ*our* dowtere mercy to ȝow resorte 75
And on man þ*at* is myschevyd haue compassyon
hym grevyth fful gretly his transgressyon̄
Ał hefne *and* erthe crye ffor mercy
Fo. 60 Me semyth þ*er* xuld be non excepc*i*on
ther prayers ben offeryd so specyally. 80

¶ Threwth Sseyth she hath evyr be than
I graunt it wel she hath be so
and þ*ou* seyst endlesly · þ*at* mercy þ*ou* hast kept ffor man̄
than mercyabyl lorde kepe us bothe to
Thu seyst · Veritas mea *et* m*isericordia* mea cu*m* ip*so* 85
Suffyr not þ*i* sowlys than in sorwe to slepe
þ*at* helle hownde þ*at* hatyth þ*e* byddyth hy*m* ho
þ*i* love man no lengere lete hy*m* kepe.

Justicia

¶ Mercy me merveylyth what ȝow movyth
ȝe know wel I am ȝ*our* systere ryghtwysnes 90
God is ryghtful *and* ryghtffulnes lovyth
man offendyd hym þ*at* is endles
Ther fore his endles punchement may nevyr sees
Also he forsoke his makere þ*at* made hym of clay
And þ*e* devyl to his mayster he ches 95
xulde he be savyd · nay nay nay.

¶ As wyse as is god he wolde A be
this was þ*e* Abhomynabyl p*resumpc*ion
it is seyd ȝe know wel þis of me
þ*at* þ*e* ryghtwysnes of god hath no diffynic*i*on 100
Therffore late þis be oure conclusyon̄
he þ*at* sore synnyd ly stylle in[1] sorwe
he may nevyr make A seyth be reson
whoo myght thanne thens hym borwe.

M*isericordia*

Fo. 60v ¶ Systyr Ryghtwysnes ȝe Are to vengeabyl 105
Endles synne god endles may restore

[1] A thin stroke in darker ink above this *n* noted by Manly appears to be an accidental mark.

Above all hese werkys god is mercyabyl
þow he for-sook god be synne · be feyth he for-sook hy*m* nev*er*
þ*e* mor*e*
And þow he p*re*sumyd nevyr so sor*e*
ȝe must consyder þ*e* frelnes of mankende 110
lerne[1] *and* ȝe lyst þis is goddys lor*e*
þ*e* mercy of god is w*ith*-owtyn ende.

Pax

¶ To spar*e* ȝ*our* speches syster*ys* it syt
It is not onest in vertuys to ben dyscenc*i*on
the pes of god ovyr comyth all wytt 115
þow trewth *and* ryght sey grett reson
ȝett mercy seyth best to my pleson
ffor yf mannys sowle xulde abyde in helle
be-twen God *and* man evyr xulde be dyvysyon
And than myght not I pes dwelle. 120

¶ Ther*e*for*e* me semyth best ȝe thus Acorde
than hefne *and* erthe ȝe xul qweme
putt bothe ȝ*our* sentens in our*e* lorde
And in his hyȝ wysdam lete hym deme
This is most syttynge me xulde seme 125
And lete se how we ffowr*e* may all A-byde
þ*at* mannys sowle it xulde p*er*ysche it wor*e* sweme
or þ*at* ony of vs ffro other*e* xulde dyvyde.

Veritas

¶ In trowthe her*e*-to I consente
I wole prey our*e* lord it may so be. 130

Justicia

I Ryghtwysnes am wele contente
Ffor in hym is very equyte.

M*isericord*ia

Fo. 61 ¶ And I mercy ffro þis counsel wole not fle
tyl wysdam hath seyd I xal ses.

Pax

Her*e* is god now · her*e* is vnyte 135
hefne *and* erth is plesyd w*ith* pes.

ffilius

¶ I thynke þ*e* thoughtys of pes · *and* nowth of wykkydnes
this I deme to ses ȝ*our* contraversy

[1] The original first word of this line—probably a miswritten *lerne*—has been obliterated.

HOMERTON COLLEGE LIBRARY

If Adam had not deyd · peryschyd had ryghtwysnes
And Also trewth had be lost *þer*-by 140
T*er*rewth[1] *and* ryght wolde chastyse ffoly
ȝiff a-nother deth come not · mercy xulde perysch
þan pes were exyled ffynyaly
So tweyn dethis must be ȝow fowr*e* to cherysch.

¶ But he þ*a*t xal deye ȝe must knawe 145
þ*a*t in hym may ben non iniquyte
þ*a*t helle may holde hym be no lawe
But þ*a*t he may pas at hese lyberte
Qwer*e* swyche on his p*re*vyde *and* se
And hese deth · for mannys deth xal be redempc*i*on 150
Att hefne *and* erth seke now ȝe
Plesyth it ȝow þis conclusyon.

Veritas

¶ I trowthe haue sowte þe erthe · w*ith*-owt *and* w*ith*-inne[2]
and In sothe þ*er* kan non be fownde
þ*a*t is of o day byrth · w*ith*-owte synne 155
nor to þ*a*t deth wole be bownde.

M*isericord*ia

I mercy haue ronne · þ*e* hevynly Regyon rownde
and þ*er* is non of þ*at* charyte
þat ffor man wole suffre A deddly wounde
I I[3] can nott wete how þis xal be. 160

Justicia

Fo. 61v[4] ¶ Sur*e* I can fynde non sufficyent
Ffor servauntys vn-profytable we be ech on
He[5] love nedyth to be ful Ardent
that for man to helle wolde gon.

Pax

That god may do is non but on 165
þ*er* for*e* þis is Pesys[6] A-vyse

[1] Or *Trewth*; the stroke marking the contraction may be a flourish.

[2] *with inne and with owt* first written. [3] So repeated in MS.

[4] The writing becomes more spaced and remains so till Fo. 66. It is the hand of the same scribe.

[5] The original *he* has been corrected by a different hand to *hes* with a very faint *s*.

[6] The original *Pesys* has been corrected by a different hand to *be hys*. [Not the hand of the scribe of Ff. 95, 96.]

he þ*at* ȝaff þis cou*n*seltt · lete hy*m* ȝeve þ*e* comforte A-lon
Ffor þ*e* conclusyon · in hym · of altt þese lyse.

Ffilius

¶ It peynetħ me · þ*at* man I mad
þ*at* is to seyn peyne I must suffre fo*re* 170
A counsel of þ*e* trinite must be had
Whiche of vs xal man restor*e*.

Pater

In ȝo*ur* wysdam son · man was mad thor*e*
And in wysdam[1] was his temptac*i*on
þ*er*for sone sapyens ȝe must ordeyn her*e*-for*e* 175
and se how of man may be salvac*i*on.

Filius

¶ Ffadyr he þ*at* xal do þis must be bothħ god *and* man
lete me se how I may were þ*at* wede
And sythħ in my wysdam he be-gan
I am redy to do þis dede. 180

Sp*irit*us S*anct*us

I the holy gost · of ȝow tweyn do p*r*ocede
this charge I wole take on me
I love to ȝo*ur* lover xal ȝow lede[2]
þis is þ*e* Assent of our*e* vnyte.

M*isericord*ia

Fo. 62 Now is þ*e* loveday mad of us fowre fynialy 185
now may we leve in pes · as we were wonte
Misericordia *et* veritas obviauerunt sibi
Justicia *et* pax · osculate sunt
et *hic osculabunt pariter omnes.*

Pater

¶ Ffrom vs god Au*n*gel Gabryel þ*ou* xalt be sende
In to þ*e* countre of Galyle 190
the name of þ*e* cyte Nazareth is kende
to A mayd · w[e]ddyd[3] to A man is she
Of whom þ*e* name is joseph se
of þ*e* hous of davyd bore
The name of þ*e* mayd ffre 195
Is Mary þ*at* xal Al Restore.

[1] A miswritten letter before *was*.
[2] *procede* first written for *ȝow lede*.
[3] The *e* has disappeared through a flaw in the MS.

Ffilius

¶ Say þat she is w*ith*-owte wo and ful of grace
And þat I þe son of þe godhed of here xal be bore
Hyȝe þe þou were there A pace
ellys we xal be there the be-ffore 200
I haue so grett hast to be man thore
In þat mekest *and* purest virgyne
Sey here she xal restore
Of ȝow Aungellys þe grett Ruyne.

Sp*iri*tus S*anc*tus

Fo. 62v ¶ And if she Aske þe how it myth be 205
telle here I þe holy gost xal werke al this
Sche xal be savyd thorwe oure vnyte
In tokyn here bareyn cosyn Elyzabeth is
Qwyk w*ith* childe · in here grett Age i-wys
Sey here to vs is no thynge impossyble [1] 210
Here body xal be so ful-fylt w*ith* blys
Þat she xal sone thynke · þis sownde credyble.

Gabriel

¶ In thyn hey inbassett lord I xal go
It xal be do w*ith* a thought
be-holde now lord I go here to 215
I take my fflyth *and* byde nowth

Ave maria [2] gratia plena Dominus tecum.
Heyl fful of [3] grace god is w*ith* the
Amonge Aꝏ women blyssyd art thu'
here þis name Eva · is turnyd Aue
þat is to say w*ith*-owte sorwe ar ȝe now. 220

¶ Thow sorwe in ȝow hath no place
ȝett of joy lady ȝe nede more
Fo. 63 Therfore I Adde And sey Fful of grace
Ffor so Ful of grace was nevyr non bore
ȝett who hath grace he nedyth kepyng sore 225
therfore I sey god is w*ith* the

[1] The words *no thynge impossyble* are written in larger form and, to a less degree, ll. 208–16.

[2] *Maria* is crossed through in a different ink, possibly to make the line correspond exactly with the translation in the line following.

[3] The word *of* written twice, the second crossed through.

Whiche xal kepe ȝow endlesly thore
So amonge Alł women blyssyd[1] are ȝe.

Maria

¶ A mercy god þis is a mervelyous herynge[2]
In þe Aungelys wordys I am trobelyd her
I thynk how may be þis gretynge
Aungelys dayly to me doth Aper
But not in þe lyknes of man þat is my fer[3]
And Also thus hyȝly to comendyd be
and am most vn-wurthy I can-not Answere
grett shamfastnes and grett dred is in me.[3]

Gabryel

¶ Mary in þis take ȝe no drede
Ffor At God · grace ffownde haue ȝe
ȝe xal conceyve in ȝour wombe in dede
A childe þe sone of þe trynyte
Fo. 63v His name of ȝow · jhesu · clepyd xal be
He xal be grett · þe son of þe hyest · clepyd of kende
and of his ffadyr davyd · þe lord xal ȝeve hym þe se
Reynyng in þe hous of jacob · of which regne xal be no[4] ende.

Maria

¶ Aungel I sey to ȝow
In what manere of wyse xal þis be
Ffor knowyng of man I haue non now
I haue evyr more kept *and* xal my virginyte
I dowte not þe wordys ȝe han seyd to me
But I Aske how it xal be do.

Gabryel

The holy gost xal come fro A-bove to the[5]
and þe vertu of hym hyest xal schadu þe so.

¶ Ther fore þat holy gost of þe xal be bore
he xal be clepyd þe son of god sage
And se Elyzabeth ȝour cosyn thore

[1] The double *s* is a correction in different ink—*blylled* first written.
[2] *thynge* first written and corrected by scribe.
[3] From *But not in þe* to *is in me* the writing is still larger.
[4] MS. n̂.
[5] This speech *The holy gost . . . wyl seyn* and especially the words *impossyble to goddys vsage* are written in larger form.

She hath conseyvid A son in hyre Age
This is þe sexte monyth of here passage ·
Fo. 64 Of here þat clepyd was bareyn
no thynge is impossyble to goddys vsage
they thynkyth longe to here what ȝe wyl seyn 260

here þe Aungel makyth a lytyl restynge and *mary be-holdyth hym* and *þe Aungel seyth*

¶ Mary come of *and* haste the
And take hede in thyn entent
Whow þe holy gost · blyssyd he be
A-bydyth þin answere *and* þin assent
Thorwe wyse werke of dyvinyte 265
the secunde persone verament
is mad man by fraternyte
With-inne þi self in place present.

¶ Fferther more take hede þis space
Whow alle þe blyssyd spyrytys of vertu 270
þat are in hefne by-ffore goddys face
And alle þe gode levers *and* trew
That Are here · in þis erthely place
thyn owyn kynrede · þe sothe ho knew
And þe chosyn sowlys · þis tyme of grace 275
þat Are in helle *and* byde rescu[1].

Fo. 64v ¶ As Adam · Abraham · *and* davyd in fere
And many othere of good reputacion
þat þin Answere desyre to here
and þin Assent to þe incarnacion 280
In which þou standyst · As persevere[2]
of All man-kende savacion
Gyff me myn Answere · now lady dere
to All these creaturys comfortacion.

Maria

¶ With All mekenes I clyne to þis A-corde 285
Bowynge dowm my face with All benyngnyte

[1] *rescu* roughly scratched through in different ink and three dots above and three below in this ink; *þer* (?) *rescu* roughly written in yet another ink and not the scribe's hand—possibly that of writer of Ff. 95–6.

[2] *persyvere* first written, the *e* thickly written over the *y*.

Se here þe hand-mayden of oure lorde
Aftyr þi worde · be it don to me.

Gabryel

Gramercy my[1] lady ffre
Gramercy · of ȝour Answere on hyght
Gramercy · of ȝour grett humylyte
Gramercy · ȝe lanterne off lyght.

here þe holy gost discendit with iij bemys to our *lady · the sone of be godhed nest with iij bemys · to þe holy gost · the fadyr godly with iij bemys to þe sone · And so entre Att thre to here bosom* · and *Mary seyth*

Maria

Fo. 65 A now I ffele in my body be
parfyte god *and* parfyte man
havyng Al schappe · of chyldly carnalyte
Evyn Al at onys · þus god be-gan.

¶ Nott takynge ffyrst o membyr *and* sythe A-nother
but parfyte childhod ȝe haue A-non
of ȝour hand-mayden · now ȝe haue mad ȝour modyr
With-owte peyne in Fflesche *and* bon
Thus conceyved nevyr woman non
þat evyr was beynge in þis lyff
O myn hyest ffadyr in ȝour trone
It is worthy ȝour son · now my son · haue A prerogatyff.

¶ I can not telle what joy what blysse
now I fele in my body
Aungel Gabryel I thank ȝow for thys
most mekely recomende me · to my faderys mercy
To haue be þe modyr of god fful lytyl wend I
Now myn cosyn Elyzabeth ffayn wold I se
how sche hath conseyvid as ȝe dede specyfy
Now blyssyd be þe hyȝ trynyte.

Gabryel

Fo. 65v ¶ Ffare weyl turtyl · goddys dowtere dere
Ffare wel goddys modyr · I þe honowre
Ffare wel goddys sustyr · *and* his pleynge fere
Ffare wel goddys chawmere *and* his bowre.

[1] *my* omitted and written above the line.

Maria

¶ Ffar*e* wel Gabryel specyalye
Ffare wel goddys masanger*e* expresse
I thank ȝow for ȝo*ur* traveyl hye
Gramercy of ȝo*ur* grett goodnes. 320

¶ And namely of ȝo*ur* comfortabyl massage
Ffor I vndyrstande by inspyrac*i*on
þ*a*t ȝe knowe by syngular*e* p*re*uylage
most of my sonys incarnac*i*on
I p*ra*y ȝow take it in to vsage 325
be A custom ocupac*i*on
to vesyte me ofte be mene passage
ȝo*ur* p*re*sence is my comfortac*i*on.

Gabriel

¶ At ȝo*ur* wyl lady so xal it be
ȝe gentyllest of blood · *and* hyest of kynrede 330
þ*a*t reynyth in erth̄ in ony degre
be pryncypal incheson of þ*e* god-hede.

¶ I comende me on to ȝow · þ*ou* trone of þ*e* trinyte[1]
O mekest mayde now þ*e* modyr of jh*e*su
Fo. 66 qwen of hefne · lady of erth̄ · *and* empres of helle be ȝe 335
soco*ur* to Alt̄ synful · þ*a*t wole to ȝow sew
Tho*ur*[2] ȝo*ur* body beryth þ*e* babe · our*e* blysse xal renew
to ȝow modyr of mercy · most mekely I recomende
and as I began I ende · w*ith* An Ave new
Enjonyd hefne *and* erth̄ · w*ith* þ*a*t I Ascende. 340

Ave maria gr*ati*a plena
Dominus tecum · uirgo sesena[3] } *Aṅgeli cantando istam sequenciam.*[4]

Remainder of Fo. 66—5½ inches—and Fo. 66ᵛ left blank.

[1] The more cramped writing is resumed with this line. [2] MS. Thd.
[3] *sesena* first written ; trace of correction to *serena*.
[4] The words *And þan mary seyth* follow on as part of this direction, but have been crossed through in red ink.

Fo. 67
(G quire)

¶ Joseph

HOw dame how · vn-do ȝoure dore vn-do
Are ȝe at hom why speke ȝe notht [1]

12

Susanna

Who is ther why cry ȝe so
telle us ȝo*ur* herand wyl ȝe ought.

Joseph

Vn-do ȝo*ur* dore I sey ȝow to
Ffor to com in is all my thought.

Maria

it is my spowse þ*at* spekyth us to
On-do þe dore his wyl were wrought.

¶ well-come hom my*n* husbond dere
how haue ȝe ferd in fer cou*n*tre.

Joseph

To gete oure levynge w*ith*-owtyn dwere
I haue sore laboryd ffor þe *and* me.

Maria

husbond ryght gracyously now come be ȝe
it solacyth me sore · sothly to se ȝow [2] in syth.

Joseph

Me merveylyth wyff surely · ȝo*ur* face I can not se
but as þe sonne w*ith* his bemys · qu*an* he is most bryth.

Maria

¶ Husbond it is as it plesyth oure lord þ*at* grace of hy*m* grew
Who þ*at* evyr be-holdyth me veryly
they xal be grettly steryed to vertu
ffor þis ȝyfte *and* many moo good lord gramercy.

Joseph

¶ How hast þ*ou* ferde jentyl mayde
whyl I haue be out of londe.

Maria

Sekyr sere beth nowth dysmayde
Ryth aftyr þe wyl of goddys sonde.

joseph

That semyth evyl I am afrayd
þ*i* wombe to hyȝe doth stonde
I drede me sore I am be-trayd

[1] The words *how hast* follow as the beginning of a third line, but have been crossed out.

[2] ȝ*w* written before ȝ*ow* and crossed through.

Su*m* other man þ*e* had in honde
Fo. 67v Hens sythe þ*at* I went
Thy Wombe is gret it gynnyth to ryse
than hast þ*ou* be-gownne a synfull gyse
telle me now in what wyse
thy self þ*ou* Ast þus schent.

¶ Ow dame what þinge menyth this
w*ith* childe þ*ou* gynnyst ryth gret to gon
Sey me mary þis childys fadyr ho is
I p*ray* þ*e* telle me *and* þ*at* anon

Maria

The fadyr of hevyn *and* ȝe it is
other fadyr hath he non
I dede nevyr forfete w*ith* man i-wys
Wher-fore I p*ray* ȝow amende ȝ*our* mon
this childe is goddys *and* ȝ*our*.

Joseph

Goddys childe þ*ou* lyist in fay
God dede nevyr jape so w*ith* may
And I cam nevyr ther I dar*e* wel say
ȝitt so nyh þ*i* bour*e*
But ȝit I sey mary whoos childe is this.

Maria

Goddys *and* ȝoure I sey i-wys.

Joseph

¶ ȝa ȝa all Olde men to me take tent
and weddyth no wyff in no kynnys wyse
þ*at* is a ȝonge wench be my*n* a-sent
ffor doute *and* drede *and* swych servyse
Alas Alas my name is shent
all men may me now dyspyse
and seyn olde cokwold þ*i* bow is bent
newly now aft*er* þ*e* frensche gyse
Alas *and* welaway
Fo. 68 Alas dame why dedyst þ*ou* so
Ffor þis synne þ*at* þ*ou* hast do
I the for-sake *and* from þ*e* go
Ffor onys evyr *and* Ay.

Maria

¶ Alas gode spowse why sey ȝe thus[1]
Alas der*e* hosbund a-mende ȝ*our* mod
It is no man but swete jh*e*sus
he wyll be clad in flesch *and* blood
and of ȝ*our* wyff be born̄.

Sephor

Ffor sothe þ*e* Au*n*gel þus seyd he
þ*at* goddys sone in trynite
Ffor mannys sake a man wolde be
to save þ*at* is for-lorn.

Joseph

¶ An Au*n*gel allas alas fy for schame
ȝe syn now in þ*at* ȝe to[2] say
to puttyn an Au*n*gel in so gret blame
Alas alas let be do way
It was su*m* boy be-gan þis game
þ*at* clothyd was clene *and* gay
and ȝe ȝeve hym now an Au*n*gel name
Alas alas *and* wel away
þ*at* evyr this game be-tydde
A dame what thought haddyst þ*ou*
Her*e* may all men þis proverbe trow
þ*at* many a man doth bete þ*e* bow
Another man hath þ*e* brydde.

Maria

Fo. 68v A gracyous god in hefne trone
comforte my spowse in þis hard cas
mercyful god A-mend his mone
as I dede nevyr so gret trespas.

Joseph

¶ Lo · Lo Ser*ys* · What told I ȝow
þ*at* it was not for my prow
A wyff to take me to
An þ*at* is wel sene now
Ffor mary I make god A-vow
is grett w*ith* childe lo
Alas why is it so
to þ*e* busshop I wole it telle

[1] *So* first written instead of *thus*. [2] *to* miswritten before *to* and crossed through.

þat he *þe* law may here do
W*ith* stonys here to qwelle.

¶ Nay nay ȝet god ffor-bede
þat I xuld do *þat* vegeabyl[1] dede
but if I wyst wel qwy 100
I knew never w*ith* here so god me spede
tokyn[2] of thynge i*n* word nor dede
þat towchyd velany
nevyr *þe* les what for thy
þow she be meke *and* mylde 105
W*ith*-owth mannys company
she myght not be w*ith* childe.

¶ But I ensure my*n* was it nevyr
thow *þat* she hath not don here devyr
rather than I xuld[3] pleyny*n* opynly 110
Serteynly ȝitt had I levyr
Ffor sake *þe* cou*n*tre ffor evyr
and nevyr come i*n* here co*m*pany
Ffor *and* men knew þis velany
In repreff þei wolde me holde 115
and ȝett many bettyr than I
ȝa · hath ben made cokolde

Fo. 69 * Now alas whedyr xal I gone
I wot nevyr whedyr nor to what place
ffor oftyn tyme sorwe comyth sone 120
and longe it is or it pace
no comforte may I haue here
I-wys wyff *þou* dedyst me wronge
Alas I taryed from *þe* to longe
Aff men haue pety on me[4] amonge 125
Ffor to my sorwe is no chere.

Maria

¶ God *þat* in my body Art sesyd
þou knowist my*n* husbond is dysplesyd
to se me i*n* þis plight

[1] So in MS.
[2] *nevyr þe les* first written as beginning of this line, and crossed through.
[3] A contraction mark stands above this word by a scribal slip.
* No paragraph sign in MS. [4] MS. *onime* corrected by dot under *i*.

Ffor vnknowlage he is desesyd 130
and þ*er*fore help þ*at* he were esyd
þ*at* he myght knowe þ*e* ful p*er*fyght
Ffor I haue levyr abyde respyt
to kepe þ*i* sone in p*ri*uite
grauntyd · by þ*e* holy spyryt 135
þan þ*at* it xulde be opynd by me.

Deus

¶ Descende I sey my*n* Au*n*gelle
on to joseph for to telle
such as my wyl is
byd hym w*ith* mary A-byde *and* dwelle 140
Ffor it is my sone fful snelle
þ*at* she is w*ith* i-wys.

Angelus

Almyghty god of blys
I am redy ffor to wende
wedyr as þ*i* wyl is 145
to go · bothe fer *and* hynde

* Joseph Joseph þ*ou* wepyst shyrle
ffro þ*i* wyff why comyst þ*ou* owte.

Joseph

Fo. 69v Good sere lete me wepe my ffylle
Go forthe þ*i* wey *and* lett me nowght. 150

Angelus

In þ*i* wepynge þ*ou* dost ryght ylle
A-ȝens god þ*ou* hast mys-wrought
Go chere þ*i* wyff w*ith* herty wylle
and chawnge þ*i* chere Amende þ*i* thought
Sche is a ful clene may 155
I telle þ*e* god wyl of here be born
And sche clene mayd as she was be-forn
to saue mankynd þ*at* is for-lorn
Go chere hyre þ*er*fore I say.

Joseph

¶ A lord god benedicite 160
of þ*i* gret comforte I thank the
þ*at* þ*ou* sent me þis space
I myght wel A wyst parde

* No paragraph sign in MS.

So good a creature as she
wold nevyr A done trespace 165
For sche is ful of grace
I know wel I haue myswrought
I walk to my pore place
and Aske ffor-gyfnes I haue mys-thought.

¶ Now is þe tyme sen At eye 170
þat þe childe is now to veryfye
which xal saue mankende
As it was spoke be prophesye
I thank þe god þat syttys on hye
with hert wyl *and* mende 175
þat evyr þou woldyst me bynde
to wedde · mary to my wyff
þi blysful sone · so nere to fynde
In his presens · to lede my lyff.

¶ Alas ffor joy I qwedyr *and* qwake 180
Alas what hap now was this
Fo. 70 A mercy mercy my jentyl make
mercy I haue seyd al Amys
Att þat I haue seyd here I for-sake
ȝour swete fete now lete me kys. 185

Mary

Nay lett be my fete not þo ȝe take
my mowthe ȝe may kys i-wys
and welcom̃ on to me.

Joseph

Gramercy myn owyn swete wyff
gramercy myn hert my love my lyff 190
xal I nevyr more make suche stryff
be-twyx me *and* þe.

¶ A mary mary wel þou be
and blyssyd be þe frewte in the
goddys sone of myght 195
now good wyff fful of pyte
as be not evyl payd with me
þow þat þou haue good ryght
As for my wronge in syght

to wyte þe *with* ony synne 200
had þou not be A vertuous wythe
god wold not A be þe *with*-inne.

¶ I knowlage I haue don A-mys
I was nev*er* wurthy i-wys
ffor to be þi*n* husbonde 205
I xal amende After*e* thys
ryght as þin owyn wyl is
to serve þe at foot *and* honde
and þi chylde bothe to vndyr-stonde
to wurchep hym w*ith* good Affecc*i*on 210
and þ*er*fore telle me *and* nothynge whonde
the holy mater*e* of ȝo*ur* concepc*i*on.

Maria

Fo. 70v ¶ At ȝowr*e* owyn wyll as ȝe bydde me
ther cam An Aunge[1] hyght Gabryell
and gret me ffayr *and* seyd Aue 215
And ferther mor*e* to me gan tell
God xulde[2] be borne of my bode
þ*e* ffendys pouste ffor to ffelle
þorwe þe holy gost as I wel se
þus god in me · wyl byde *and* dwelle. 220

Joseph

¶ Now I thank god w*ith* spech *and* spelle
þ*at* Euyr mary I was weddyd to the.

Mary

it was þe werk of[3] god as I ȝow telle
now blyssyd be þ*at* lord · so purveyd for me.

Remainder of Fo. 70v—4¾ inches—left blank.

Maria

Fo. 71[4] BVtt husbond of oo thyng I pray ȝow most mekely 13
I haue knowyng þat our*e* Cosyn Elizabeth w*ith* childe is
þ*at* it plese ȝow to go to her*e* hastyly
If owught we myth comforte her*e* it wor*e* to me blys.

Joseph

A godys sake · is she w*ith* childe sche 5
than wole her*e* husbond zakarye be mery

[1] So in MS.; cf. Fo. 47. [2] *xulde* is written twice in the MS.
[3] *of* omitted and written above the line.
[4] The writing on Ff. 71, 71v, and 72 is slightly less cramped.

In Montana they dwelle · fer hens so moty the
In *þe* cety of juda[1] I knowe it veryly
It is hens I trowe myles two *and* ffyfty
We are lyke to be wery or we come at *þat* same 10
I wole[2] w*ith* a good wyl · blyssyd wyff mary
now go we forthe than in goddys name.

Maria

¶ Goth husbond þow it be to ȝow peyne
this jurny I pray ȝow · lete us go fast
ffor I am schamfast of *þe* pepyl to be seyne 15
and namely of men *þe*r of I am A-gast
Pylgrymag*ys and* helpyng*ys* wolde be go in hast
þe mor*e þe* body is peynyd *þe* mor*e* is *þe* mede
Say ȝe ȝo*ur* devoc*i*onys *and* I xal my*n* I cast[3] (?)
now in þis jurny god mote us spede. 20

Joseph

Amen Amen · *and* evyr mor*e*
lo wyff lo · how starkly I go be-for*e*. *et sic transient circa placea*m.

Comte*m*plac*i*o[4]

Fo. 71v ¶ Sovereynes vndyrstondyth *þat* kynge davyd here
Ordeyned ffour*e and* twenty prestys of grett devoc*i*on
In *þe* temple of god · After*e* her*e* let[5] aper*e* 25
þei weryd clepyd su*m*mi sacerdotes · ffor her*e* mynistrac*i*on
And on was prynce of prestys · havynge d*omi*nacyon̄
A-monge whiche was An old prest clepyd zakarye
and he had An old woman[6] to his wyff of holy co*n*versac*i*on
whiche hyth Elizabeth *þat* nevyr had childe verylye. 30

¶ In hese mynistrac*i*on the howr*e* of incense
the Aungel Gabryel · Apperyd hym to
þat hese wyff xulde conseyve he ȝaff hy*m* intelligence
hese juge · hese vnwurthynes *and* Age not be-levyd so
The plage of dompnesse · hise lippis[7] lappyd lo[8] 35

[1] A miswritten letter after *juda* crossed through.
[2] *wyl* first written and crossed through.
[3] Halliwell reads *reast* (?). There is a small stroke between the *j* and *cast* as if the scribe had begun to write *jn*. Cf. l. 148.
[4] So in MS. [5] So in MS. for *lot*?
[6] *wyff* first written and crossed through.
[7] *lippis* omitted and written above the line.
[8] *to* first written for *lo* and crossed through.

thei wenten hom *and* his wyff was conseyvenge
this concepc*i*on gabryel[1] tolde our*e* lady to
and in soth sone After*e* · þ*a*t sage sche was sekynge
And of her*e* tweyners metyng
her*e* gynnyth þ*e* proces
now god be our*e* be-gynnynge
and of my tonge I wole ses.

Joseph

¶ A · A · Wyff Infeyth I am wery
therfor*e* I wole sytt downe *and* rest me ryght her*e*
lo Wyff · her*e* is þ*e* hous of zakary
Wole ȝe I clepe Elyzabeth to ȝow to A-per*e*.

Maria

Fo. 72 Nay husbond And it plese ȝow I xal go ner
now þ*e* blyssyd trynite · be in þis hous
A cosyn Elizabeth · swete modyr what cher
ȝe grow grett · A my god how ȝe be gracyous.

Elizabeth

¶ A-non as I herd of ȝow þis holy gretynge
mekest mayden *and* þe modyr of god mary
be ȝ*our* breth þ*e* holy gost vs was inspyrynge
þ*a*t þ*e* childe in my body enjoyd gretly
And turnyd down on his knes · to our*e* god reverently
whom ȝe bere in ȝ*our* body þis veryly I ken
ffulfyllyd w*ith* þ*e* holy gost þus lowde I cry
blyssyd be þ*ou* A-monge All women.

¶ And blyssyd be þ*e* frute of þ*i* wombe also
þ*ou* wurthyest virgyne *and* wyff þ*a*t evyr was wrought
how is it þ*a*t þ*e* modyr of god me xulde come to
þ*a*t wrecche of all wrecchis · A whyght wers þ*a*n nought
And þ*ou* art blyssyd þ*a*t be-levyd veryly in þ*i* thought
þ*a*t þ*e* wurde of god xulde profyte in the
but how þis blyssydnes A-bought was brought
I can not thynk nyn say how it myght be.

Maria

¶ To þ*e* preysynge of god cosyn this seyd mut be
whan I sat in my lytyl hous · on to god praynge
Gabryel come *and* seyde to me Ave

[1] The first syllable of *Gabryel* repeated before *tolde* and crossed through.

ther I conceyvyd god · At my consentynge 70
Fo. 72v Parfyte god · *and* parfyte man · At onys beynge
than þe Aungel seyd on to me
þat it was sex monethys syn ȝ*our* conseyvynge
Þis cawsyth my comynge cosyn[1] ȝow to co*m*forte *and* se.

Elizabeth

¶ Blyssyd be ȝe cosyn · ffor ȝ*our* hedyr comynge 75
How I conseyvyd I xal to ȝow say
þe Aungel Apperyd · þe howr*e* of[2] incensynge
Seynge I xulde conseyve · *and* hym thought nay
Sethe ffor his mystrost · he hath be dowm̃ Al-way
and þus of my concepc*i*on I haue tolde ȝow su*m*. 80

Maria

ffor þis holy psalme I be-gynne her*e* þ*is* day.
MAgnificat,' anima mea dominum
Et exultauit spiritus meus ⁄ in deo salutari meo.

Elizabeth

Be þe holy gost w*ith* joye goddys son is in þe cu*m*
þ*at* þ*i* spyryte so injouyid þe helth of þ*i* god so.

Maria

Quia respexit humilitatem ancille sue
ecce eni*m* ex hoc beatam me dicent o*m*n*e*s gen*e*rac*i*on*e*s.

Elizabeth

Ffor he be-held þe lownes of hese hand-maydeȝe[3]
so ferforthe ffor þ*at* · Alł gen*e*rac*i*onys blysse ȝow *in* pes. 85

Maria

Quia fecit m*ih*i magna qui potens est
et sanctum nomen eius.

Elizabeth

Ffor grett thyng*ys* he made *and* also myghtyest
And ryght holy is þe name of hym in vs.*

Maria

Fo. 73 Et m*isericordi*a eius a progenie in progenies
timentib*us* eum.

Elizabeth

ȝa þe mercy of hym · fro þ*at* kynde in to þe kynde of pes
Ffor alł þ*at* hym drede · now is he cum.

[1] Some miswritten letter obliterated before *cosyn*.
[2] *of* omitted and written above the line.
[3] The *de* is in darker ink. It looks as if *may ȝe* was originally written.
* *Et* written at the bottom of the folio.

Maria

Fecit potenciam in brachio suo ⁝
dissperſit superbos mente cordis sui.

Elizabeth

The por*e* in his ryght Arme · he hath mad so
þ*e* prowde to dyspeyr*e* · And þ*e* thought of her*e* hertys only.

Maria

Deposuit potentes de sede
et exaltauit humiles.

Elizabeth

The prowde men · fro hey setys put he
And þ*e* lowly vpon heyth · in þ*e* sete of pes.

Maria

Esurientes impleuit bonis ⁝
et diuites dimisit i*n*anes.

Elizabeth

Alle þ*e* por*e* *and* þ*e* nedy · he fulfyllyth w*ith* his goodys
And þ*e* Ryche · he fellyth to voydnes.

Maria

Suscepit israel puerum suum ⁝
recordatus est misericordie sue.

Elizabeth

Israel ffor his childe · vp-toke he to cu*m*
On his mercy to thynk · ffor hese þ*at* be.

Maria

Sicut locutus est ad p*at*res n*ost*ros
abrah*am* et semini eius in secula.

Elizabeth

As he spak her*e* to our*e* forfaderys in clos
Abrah*am* *and* to Alł hese sed of hym i*n* þis werd sa.

Maria

Gloria patri et filio ⁝
et spiritui sancto.

Elizabeth

Preysyng be to þ*e* fadyr i*n* hevyn lo
þ*e* same to þ*e* son · her*e* be so
þ*e* holy gost Also to ken ⁝

Maria

Fo. 73v Sicut erat i*n* principio et nunc *et* semper ⁝
et in secula seculor*um* amen.

Elizabeth

As it was in þe begynnynge · *and* now is · *and* xal be forevyr
and in this werd · in alł good werkys to abydyn then.

Maria

¶ This psalme of p*r*ophesye seyd be-twen vs tweyn
In hefne it is wretyn w*ith* Au*n*gellys hond
evyr to be songe · *and* Also to be seyn
Euery day Amonge us at our*e* eve song.

¶ but cosyn Elyzabeth I xal ȝow her*e* kepe
and þis thre monethis Abyde her*e* now
tyl ȝe han childe to wasche skor*e* *and* swepe
and in alł þ*at* I may to comforte ȝow.

Elizabeth

A ȝe modyr of god · ȝe shewe us her*e* how
we xulde be meke þ*at* wrecchis her*e* be
Alł hefne *and* herthe wurcheppe ȝow mow
þ*at* ar*e* trone *and* tabernakyl of þ*e* hyȝ trinite.

Joseph

¶ A how do ȝe · how do ȝe · ffadyr zacharye
we ffalle ffast in Age w*ith*-owte oth̄
why shake ȝe so ȝo*ur* hed · haue ȝe þ*e* palsye
Why[1] speke ȝe not ser*e* I trowe ȝe ar*e* not wroth.

Elizabeth

Nay wys ffadyr joseph · þ*er* to he wer*e* ful loth
it is þ*e* vesytac*i*on of god · he may not speke veryly
lete us thank god · þ*er*ffor both̄
he xal remedy it · whan it plesyth his mercy.[2] *

Joseph

¶ Of ȝo*ur* dissese thynkys no greff[3]
thank god of al adu*er*syte
Ffor he wyl chastyse *and* repreff

[1] Some miswritten letter at the beginning of the line obliterated.

[2] The asterisk here indicates the following alternative continuation with *Contemplacio* as name of next speaker written in the bottom margin :

* his mercy

Si placet. { come I pray ȝow specialy
I-wys ȝe Ar*e* welcome mary — Elizabeth
ffor þis comfortabelest comynge good god gramercy.

Co*n*tem*p*lac*i*o.

[3] From here to the end of the play, with the exception of the first eight lines of Contemplacio's speech, two lines are written as one divided by two parallel black strokes or the stop : and a red stroke.

þo þat he lovyth most hertyle
Mary I hold best *þat* we go hens
we haue fer hom *with*-owt fayl. 130

Maria

Al redy husbond *with*-out defens
I wyl werke be *ȝour* coun*s*ayl
Cosyn be *ȝour* leve · *and* *ȝour* lycens
for homward now us must travayl
of þis refreschynge i*n* *ȝour* pre*s*ens 135
god ȝeld ȝow *þat* most may Avayl.

Elizabeth

Fo. 74[1] ¶ Now cosynes bothe god ȝow spede
and wete ȝow wele *with*-owty*n* mo
ȝour pre*s*ens *com*fortyth me in dede
and *þer* fore now am I ryght wo 140
That ȝe my ffrendys *and* my kynrede
þus sone now xul parte me fro
but I pr*a*y god he mote ȝow lede[2]
in every place wher so ȝe go.

here mary and *elizabet partyn* and *elizabeth goth to zakarie and seyth*

¶ Good husbond ryse up I be-seke ȝow 145
and go we to *þe* te*m*ple now fast
to wurchep god *with* *þat* we mow
and thank hy*m* bothe this is my cast
Of *þe* tyme *þat* is comynge now
ffor now is cu*m* mercy *and* venjauns is past 150
God wyl be born for ma*n*nys prow
to brynge us to blysse *þat* eu*er* xal last.

Con*tem*placio

¶ lystenyth sovereynys here is conclusyon
how *þe* Aue was mad · here is lernyd vs
þe Aungel seyd · Ave *grat*ia plena d*omin*us tecum ·
benedicta tu in mulieribus ·
Elizabeth seyd · et benedictus · 5
fructus uentris tui· thus *þe* chirch addyd Maria And Jhesus· her

[1] The writing becomes more cramped and the text extends into the right-hand margin on this folio.

[2] *spede* first written.

who seyth oure ladyes sawtere dayly ⁄ ffor A ȝer þus
he hath pardon · ten thousand And eyte hundryd ȝer.

¶ Than ferther to oure matere for to procede
Mary with elizabeth abod þer stylle
iij monthys fully as we rede
thankynge god with hertly wylle
α [1] A lord god what hous was þis on
þat þese childeryn *and* here moderys to
as mary *and* elizabeth jhesus *and* john
and joseph *and* zakarye Also.

¶ And evyr oure lady a-bod stylle þus
tyl johan was of his modyr born
and þan zakarye spak i-wus
þat had be dowm *and* his spech lorn
he *and* Elizabeth prophesyed as þus
they mad Benedictus · them be-forn
and so Magnificat · And · Benedictus
ffyrst in þat place þer made worn.

¶ Whan all was don oure lady fre
tokę here leve than aftere this
At Elizabeth *and* at zakarie
And kyssyd johan *and* gan hym blys
Now most mekely we thank ȝou of ȝour pacyens
and beseke ȝou of ȝour good supportacion
If here hath be seyd ore don Any inconuenyens
we Asygne it to ȝour good deliberacion
Be-sekynge to crystys precious passyon
conserve *and* rewarde ȝour hedyr comynge
with Aue we be-gunne · *and* Aue is oure conclusyon
Ave regina celorum · to oure lady we synge.

[1] This red ink sign is otherwise only found in this MS. (in larger form) in the second Passion Play and in the Assumption Play, where it is used to mark couplet rhymes. In black it is used in both Passion Plays and occasionally elsewhere to mark stage directions.

Fo. 74v ¶ Den

¶ A-voyd Ser*ys* · And lete my lorde þe buschop come
And syt in þe courte · þe lawes ffor to doo
And I xal gon in þis place · them for to somowne
tho þat ben in my book · þe court ȝe must com too
I Warne ȝow here att abowte 5
þat I somown ȝow att þe rowte
loke ȝe fayl for no dowte
at þe court to pere
both Johan Jurdon *and* Geffrey Gyle
Malkyn mylkedoke *and* fayr mabyle 10
Stevyn sturdy *and* Jak at þe style
and sawdyr sadelere.

¶ Thom tynkere *and* betrys belle
peyrs pottere *and* whatt at þe welle
Symme Smalfeyth *and* kate kelle 15
and bertylmew þe bochere
kytt cakelere *and* colett crane
gylle fetyse *and* fayr jane
powle pewterere *and* pernel prane
and phelypp þe good flecchere. 20

¶ Cok crane *and* davy drydust
Luce lyere *and* letyce lytyl trust
Miles þe myllere *and* colle Crake crust
bothe bette þe bakere *and* Robyn rede[1]
And loke ȝe rynge wele in ȝo*ur* purs 25
Ffor ellys ȝo*ur* cawse may spede þe wurs
þow þat ȝe slynge goddys curs
Evy*n* at my*n* hede Ffast co*m* A-way[2]
Bothe boutyng þe browstere *and* sybyly slynge
Megge mery wedyr *and* sabyn[3] sprynge 30
Tyffany Twynkelere ffayle ffor no thynge
The courte xal be þis day.[4]

[1] The words '*and* Robyn rede' are enclosed in a red loop as if the name of a speaker.

[2] The words *Ffast* com *A-way* are similarly enclosed in a red loop.

[3] *spy* miswritten before *sprynge* and crossed through.

[4] The following note is written at the foot of the page in ink of the same

Fo. 75 *hic intrabit pagetum de purgacione Marie et joseph · hic dicit primus detractor*

AA · serys god saue ȝow alle
here is a fayr pepyl in good ffay
Good ser*ys* telle me what me*n* me calle
I trowe ȝe kan not be þis day
ȝitt I walke wyde *and* many way
but ȝet þ*er* I come I do no good
to reyse slawdyr is al my lay
bakbyter*e* is my brother of blood.

¶ Dede he ought come hedyr in al þis day
now wolde god þ*at* he wor*e* her*e*
and be my trewthe I dar*e* wel say
þ*at* Yf we tweyn to-gedyr a-per*e*
Mor*e* slawndyr we to xal a-rer*e*
w*ith*-in an howr*e* thorwe-outh this town
than evyr þ*er* was þis thowsand ȝer*e*
and ellys I[1] shrewe ȝow bothe vp *and* downe.

¶ Now be my trewthe I haue a syght
Evyn of my brother lo wher*e* he is
Welcom der*e* brother my trowthe I plyght
ȝowr*e* jentyl mowth let me now kys.

Se*cun*dus det*rac*tor

Gramercy brother so haue I blys
I am ful glad we met þis day.

1*us* detractor

Ryght so am I brothyr i-wys
mech gladder*e* than I kan say.

¶ but ȝitt good brother I ȝow pray
telle alle þese pepyl what is ȝ*our* name.
Ffor yf þei Knew it my lyf I lay
they wole ȝow wurchep *and* speke gret fame.

colour and by the same or a contemporary hand, in the same script as the note on the five Annes, Fo. 37v, but neater and not rubricated:

a. 14. kal. ap*ri*lis · Translat*io* San*cte* Marie Magdalene · Et s*anc*ti Joseph' sponsi d*ei* ge*nitr*icis Marie.

e. x. kal. ap*ri*lis Adam creatus est.

[1] *s* converted to *I*.

Fo. 75v ij^us detractor

I am bakbytere þat spyllyth all game
bothe kyd *and* knowyn in many a place. 30

1us detractor

be my trowth I seyd þe same
and ȝet sum seyden þou xulde haue evyl grace.

ij^us detract[or

¶ Herk reyse sclaundyr canst þou owth telle
of Any newe thynge þat wrought was late.

1us detract[or

With-in a short whyle a thynge be-felle 35
I trowe þou wylt lawhȝ ryght wel þer Ate
Ffor be trowth ryght mekyl hate
If it be wyst þer of wyl growe.

ij^us detractor

If I may reyse þer with de-bate
I xal not spare þe seyd to sowe. 40

1us detractor

¶ Syr in þe tempyl a mayd þer was
calde mayd mary þe[1] trewth to tell
Sche semyd so holy withinne þat plas
men seyd sche was ffedde with holy Aungell
Sche made A vow with man nevyr to melle 45
but to leve chast *and* clene virgine
How evyr it be · here wombe doth swelle
and is as gret as þinne or myne.

ij^us detractor

¶ ȝa þat old shrewe joseph my trowth I plyght
was so Anameryd upon þat mayd 50
þat of hyre bewte whan he had syght
He sesyd nat tyll had here a-sayd.

1us detractor

Fo. 76 A nay nay wel wers she hath hym payd
Sum fresch ȝonge galaunt she loveth wel more
þat his leggys to here hath leyd 55
and þat doth greve þe old man sore.

ij^us detractor

¶ be my trewth al may wel be
ffor fresch *and* fayr she is to syght

[1] Some miswritten letter before *þe* crossed through.

And such a mursel as semyth me
Wolde cause A ȝonge man to haue delyght. 60

1us detractor

Such a ȝonge damesel of bewte bryght
And of schap so comely Also
Of hire tayle ofte tyme be lyght
and rygh[1] tekyl vndyr þe too.

ijus detractor

¶ that olde cokolde was evyl be-gylyd 65
to þat fresche wench whan he was wedde
now muste he faderyn A-nothyr ma*n*nys chylde
and w*ith* his swynke he xal be fedde.

1us detractor

* A ȝonge man may do mor*e* cher*e* in bedde
to A ȝonge wench þan may An olde 70
þ*at* is þ*e* cawse such lawe is ledde
þat many a man is a kokewolde.

hic sedet episcopus Abiȝachar inter duos legis doctores et audientes hanc de-famocionem vocat ad se detractores dicens

Ep*iscop*us

¶ Herke ȝe felawys why speke ȝe such schame
of þ*at* good virgyn ffayr mayd mary
ȝe be a-cursyd · so hir*e* for to defame 75
She þ*at* is of lyff so good *and* holy
Of hir*e* to speke suche velany
ȝe make my*n* hert ful hevy of mood
Fo. 76v I charge ȝow sese of ȝour*e* fals cry
Ffor sche is sybbe of my*n* owyn blood. 80

ijus detractor

¶ Syb of þi kyn þow þ*at* she be
all gret w*ith* chylde hir*e* wombe doth swelle
Do calle her*e* hedyr þ*i*-self xal se
þ*at* it is trewthe þ*at* I þ*e* telle.

1us detractor

Ser*e* ffor ȝo*ur* sake I xal kepe cowncelle 85
ȝow for to greve I am ryght loth
But lest syr*ys* lyst what seyth þ*e* belle
Our*e* fayr mayd now gret w*ith* childe goth.

[1] So in MS. * A paragraph mark has been erased here.

primus doctor legis

¶ Take good heed serys what ȝe doth say
A-vyse ȝow wele what ȝe present
ȝyf þis be fownd fals a-nothyr day
Ful sore ȝe xal ȝour tale repent.

ij^us detractor

Sere þe mayd for sothe is good *and* gent
bothe comely *and* gay *and* a fayr wench
And feetly with help sche can consent
to set A cokewolde on þe hye benche.

ij^us doctor legis

¶ ȝe be to besy of ȝour langage
I hope to god ȝow fals to preve
It were gret rewthe she xuldė so outrage
or with such synne to myscheve.

Episcopus

This evy talys my hert doth greve
of hire to here such fowle dalyawnce
If she be fowndyn in suche repreve
she xal sore rew here governawns.

¶ Symon somnore in hast wend þou þi way
byd Joseph *and* his wyff be name
Fo. 77 At þe coorte to Appere þis day
here hem to pourge of here defame
Sey þat I here of hem grett schame
and þat doth me gret hevynes
if þei be clene with-owtyn blame
byd hem come hedyr *and* shew wyttnes.

Deo

¶ All redy sere I xal hem calle
here at ȝour courte for to appere
And yf I may hem mete with all
I hope ryght sone þei xal ben here
A-wey serys lete me com nere
A man of wurchep here comyth to place
of curtesy me semyth ȝe be to lere
Do of ȝour hodys with an evyl grace.

¶ Do me sum wurchep be-for my face
or be my trowth I xal ȝow make
If þat I rolle ȝow up in my race
Ffor fere I xal do ȝour ars qwake

But ȝit su*m* mede *and* ȝe me take
I wyl w*ith*-drawe my gret rough toth
gold or sylvyr I wol not for-sake
but evyn as alł somnor*ys* doth.

¶ A Joseph good day w*ith* þ*i* ffayr spowse
my lorde þ*e* buschop hath for ȝow sent
it is hym tolde þ*at* in þi*n* house
A cuckolde is bowe is ech̄ nyght bent
he þ*at* shett þ*e* bolt · is lyke to be schent
Ffayr*e* mayde þ*at* tale ȝe kan best telłe
now be ȝour*e* trowth̄ telle ȝo*ur* entent
dede not þ*e* Archer*e* plese ȝow ryght welł.

Maria

Fo. 77v ¶ Of god in hevyn I take wyttnes
þ*at* synful werk was nevyr my thought[1]
I am a mayd ȝit of pur*e* clennes
lyke as I was in to þis werd brought.

Deñ

Othyr wyttnes xal non be sought
þ*ou* art w*ith* childe eche man may se
I charge ȝow bothe ȝe tary[2] nought
but to þ*e* buschop com̄ forth w*ith* me.

Joseph

¶ To þ*e* buschop w*ith* ȝow we wende
of our*e* purgac*i*on hawe we no dowth.

Maria

Almyghty god xal be our*e* frende
Whan þ*e* trewthe is tryed owth.

Deñ

ȝa on þis wyse excusyth her*e* every scowte
Whan her*e* owyn synne hem doth defame
but lowly þ*an* þei gyn to lowth
Whan þei be gylty *and* fowndyn in blame.

¶ Ther-fo*re* com forth̄ cokewolde be name
þ*e* busschop xal ȝo*ur* lyff appose
Com forth̄ Also ȝe goodly dame
A clene huswyff as I suppose
I xal ȝow tellyn w*ith*-owtyn glose.

[1] *werk* first written and crossed through.
[2] *t* written over some other letter.

and ȝe were myn *with* owtyn lak
I wolde ech day be-schrewe ȝ*our* nose
and ȝe dede brynge me such a pak. 160

¶ My lord þe buschop here haue I brought
þis goodly copyl at ȝ*our* byddyng
and as me semyth as be here fraught
Ffayr chylde lullay sone must she syng.

1us detractor

Fo. 78 To here a credyl *and* ȝe wolde brynge 165
ȝe myght saue mony in here purse
be-cawse she is ȝ*our* cosyn ȝynge
I pray ȝow sere lete here nevyr fare þe wers.

Ep*iscop*us

¶ Alas mary what hast þ*ou* wrought
I am a schamyd evyn for þ*i* sake 170
how hast þ*ou* chaungyd þ*in* holy thought
dude old joseph w*ith* strenght þe take
Or hast þ*ou* chosyn a-nother make
by whom þ*ou* art þus brought in schame
telle me who hath wrought þis wrake 175
how hast þ*ou* lost þ*in* holy name.

Maria

¶ My name I hope is saff *and* sownde
god to wyttnes I am a mayd
of ffleschly lust *and* gostly wownde
In dede · nere thought · I nevyr a-sayd. 180

1us doctor leg*is*

how xulde þ*i* wombe þus be arayd
So grettly swollyn as þ*at* it is
but if su*m* man þe had ovyr-layd
þ*i* wombe xulde never be so gret i-wys.

ij^{us} doctor leg*is*

¶ Herke þ*ou* joseph I am afrayd 185
þ*at* þ*ou* hast wrought þis opyn synne
þis woman þ*ou* hast þus be-trayd
w*ith* gret flaterynge or su*m* fals gynne.

ij^{us} detractor

Now be my*n* trowth ȝe hytte þe pynne
w*ith* þ*at* purpose in feyth I holde 190

telle now how þ*ou* þus hir*e* dudyst wynne
Or*e* knowlych̄ þ*i* self ffor a cockewold̄.

Joseph

Fo. 78v[1] ¶ Sche is for me a trewe clene mayde
And I for hir*e* am clene Also
of ffleschly synne I nevyr a-sayde 195
Sythyn þ*at* sch̄ was weddyd me to.

Ep*iscop*us

Thu xalt not schape from vs ȝitt so
Ffyrst þ*ou* xalte tellyn us a-nother lay
Streyt to þ*e* Awter þ*ou* xalt go
þ*e* drynge of vengeawns þ*er* to a-say. 200

¶ her*e* is þ*e* botel of goddys vengeauns
this drynk xal be now þ*i* purgac*i*on
þis [hath][2] suche vertu by goddys ordenauns
þ*at* what man drynk of þis potac*i*on
And goth̄[3] s*er*teyn in p*ro*cessyon̄ 205
her*e* in þis place þis Awter*e* abowth
If he be gylty sum maculacion
Pleyn in his face xal shewe it owth.

¶ Iff þ*ou* be gylty telle us lete se
Ou*er* godys myght be not to bolde 210
If þ*ou* p*re*sume[4] *and* gylty be
god þ*ou* dost greve many afolde.

Joseph

I am not gylty as I fyrst tolde
All̄ myghty god I take wytnes.

Ep*iscop*us

than þis drynke in hast þ*ou* holde 215
and on p*ro*c̄essyon anon þ*e* dresse.

hic joseph bibit et *sepcies circuiuit altar*e *dicens* Joseph

¶ This drynk I take w*ith* meke entent
as I am gyltles to god I pray
Fo. 79 Lord as þ*ou* art omnypotente
on me þ*ou* shewe þ*e* trowth þis day. *modo bibit.* 220
About þis Awter*e* I take þ*e* way

[1] The writing on this page is freer and less neat.

[2] The word *hath* is written above the line in another ink—but probably by the scribe.

[3] MS. *Angoth* with *d* written above the line.

[4] *presue* first written and crossed through.

O gracyous god help þ*i* servaunt
As I am gyltles A-ȝen ȝon may
þ*in* hand of mercy þis tyme me graunt.

Deñ

¶ This olde shrewe may not wele gon
longe he taryeth to go A-bowth
lyfte up þ*i* feet sett forth þ*i* toñ
or be my trewth þ*ou* getyst a clowte.

ij[48] detractor

now ser*e* evyl Thedom com to þ*i* snowte
What heylyght þ*i* leggys now to be lame
þ*ou* dedyst hem put ryght freschly owte
Whan þ*ou* dedyst pley w*ith* ȝon ȝonge dame.

1 detractor

¶ I p*ra*y to god gyf hym myschawns
hese leggys[1] her*e* do folde for Age
but w*ith* þis damysel whan he dede dawns
þ*e* olde charle had ryght gret corage.

Deñ

The shrewe was þan sett in a dotage
and had good lust þ*at* tyme to pleyn
ȝaff sche not ȝow cawdel to potage
whan ȝe had don to comforte ȝ*our* brayn.

Joseph

¶ A gracyous god help me þis tyde
ageyn þis pepyl þ*at* me doth fame
as I nevyr mor*e* dede towch her*e* syde
þis day help me fro werdly schame
A-bowte þis awter*e* to kepe my fame
vij tymes I haue gon rownd abowte
Fo. 79v If I be wurthy to suffyr blame
O ryghtful god my synne shewe[2] owughte.

Ep*iscop*us

¶ Joseph w*ith* hert thank god þ*i* lorde
Whos heyȝ mercy doth the excuse
Ffor þ*i* purgac*i*on we xal recorde
W*ith* hyr*e* of synne þ*ou* dedyst nev*er* muse
But Mary þ*i*-self mayst not refuse
Att grett w*ith* chylde we se þ*e* stonde

[1] The *y* is written over an original *e* (in the same ink).
[2] *sw* miswritten before *shewe*.

What mystyr man dede þe mys-vse 255
Why hast þou synned Ageyn þin husbonde.

Maria

¶ I trespacyd nevyr with erthely wyght
Þer of I hope þurowe goddys sonde
Here to be purgyd be-fore ȝour syght
Ffrom all synne clene · lyke as myn husbonde 260
Take me þe botel out of ȝour honde
Here xal I drynke be-forn ȝour face
A-bowth þis Awtere than xal I fonde
Vij tymes to go by godys grace.

1us doctor legis

¶ Se þis bolde bysmare wolde presume 265
Ageyn god to preve his myght
þow goddys vengeauns hyre xuld consume[1]
Sche wyl not telle hyre fals delyght
þou art with chylde we se in syght
to us þi wombe þe doth accuse 270
þer was nevyr woman ȝitt in such plyght
þat ffrom mankynde hyre kowde excuse.

1us detractor

¶ in Ffeyth I suppose þat þis woman slepte
Fo. 80 With-owtyn all coverte whyll þat it dede snowe
And a flake þer of in to hyre mowthe crepte 275
and þer of þe chylde in hyre wombe doth growe.

ijus detractor

Than be-ware dame for this is wel i-knowe
whan it is born yf þat þe sunne shyne
it wyl turne to watyr ageyn as I trowe
ffor snow on to watyr doth evyr more reclyne[2]. 280

ijus doctor legis

¶ with goddys hyȝ myght loke þou not jape
of þi purgacion wel þe Avyse
yf þou be gylty þou mayst not schape
be-ware evyr of god þat ryghtful justyce
if god with vengeauns set on þe his syse 285
not only þou but all þi kyn is schamyd
bettyr it is to telle þe trewth devyse

[1] Some word—possibly *persume*—first written and crossed through.
[2] *reu* first written and crossed through.

than god for to greve *and* of hym be gramyd.

Maria

¶ I trostyn in his grace I xal hym nevyr greve
his servaunt I am in worde dede *and* thought 290
A mayd yndefyled I hope he xal me *pre*ve
I pray ȝow lett me nought.

Ep*isco*pus

Now be þat good lord þat alł þis werd hath wrought
If god on þe shewe ony manyr tokyn
purgac*io*n I trowe was nevyr so der*e* bowth 295
If I may on the in Any wyse be wrokyn.

¶ holde her*e* þe botel *and* take a large draught
and abowth the Awter*e* go þ*i* pr*o*cessyon.

Maria

to god in þ*is* case my cawse I haue be-taught
lorde thorwe þi*n* helpe I drynke of þis potacyon. 300

hic beata uirgo bibit de potacione et postea circuiuit altare dicens

Maria

Fo. 80v (marked 80 twice, one figure faded and perhaps crossed through)

God as I nevyr knew of ma*n*nys maculac*io*n
but evyr haue lyued in trewe virginite
send me þis day þi*n* holy consolac*io*n
þat alł þis fayr peple my clennes may se.

¶ O gracyous god as þ*ou* hast chose me 305
ffor to be þ*i* modyr of me to be born
saue þ*i* tabernacle þ*at* clene is kepte for þ*e*
which now am put at repref *and* skorn
Gabryel me tolde w*ith* wordys he be-forn
þ*at* ȝe of ȝ*our* goodnes wold be-come my chylde 310
help now of ȝ*our* hyȝness my wurchep be not lorn
A der*e* sone I p*ray* ȝow help ȝ*our* modyr mylde.

Ep*isco*pus

¶ Almyghty god what may þis mene
Ffor alł þ*e* drynke of goddys potacyon
þis woman w*ith* chylde is fayr *and* clene 315
w*ith*-owtyn fowle spotte or maculac*io*n
I can nat be non ymagynac*io*n
P*re*ve hyr*e* gylty *and* synful of lyff
it shew*ith* opynly by her*e* purgac*io*n
Sche is clene mayde bothe modyr *and* wyff. 320

1[us] detracto[r

¶ Be my fadyr sowle here is gret gyle
be-cawse sche is syb of ȝ*our* kynreed
þ*e* drynk is chaungyd by su*m* fals wyle
þ*at* sche no shame xuld haue þis steed.

Ep*iscopus*

Be-cawse þ*ou* demyst þ*at* we do falshede
and for þ*ou* dedyst hem fyrst defame
þ*ou* xalt ryght her*e* magre þ*in* heed
beforn alł þis pepyl drynk of þ*e* same.

1[us] detractor

Fo. 81 ¶ Syr in good ffeyth oo draught I pulle
If these to drynker*ys* haue not Alł spent.

hic bibit et scenciens dolorem in capite cadit et *dicit*

Out out Alas what heylith my sculle
A my*n* heed w*ith* ffyr*e* me thynkyht is brent
Mercy good mary I do me repent
of my cursyd *and* ffals langage[1].

Maria

Now god lord in hevyn omnypotent
of his gret mercy ȝ*our* seknes aswage.

Ep*iscop*us

¶ We alł on knes fałł her*e* on grownd
þ*ou* goddys hande-mayd prayng for grace
alł cursyd langage *and* schame on sownd[2]
good mary ffor-ȝeve us her*e* in þis place.

Maria

Now god for-ȝeve ȝow alł ȝowr*e* trespace
and also for-ȝeve ȝow alł defamac*i*on
þ*at* ȝe haue sayd both more and lesse
to my*n* hynderawnce *and* maculac*i*on.

Ep*iscop*us

¶ Now blyssyd virgyne we thank ȝow alłe
of ȝour*e* good hert *and* gret pacyens
we wyl go w*ith* ȝow hom to ȝ*our* hałłe
to do ȝow servys w*ith* hyȝ reverens
I thank ȝow hertyly of ȝour*e* benevolens

Maria

On to ȝ*our* owyn hous I pray ȝow ȝe goo

[1] A thick line has been drawn over the original line dividing this speech and the next.

[2] The *w* has been written by the scribe over another letter—possibly a *u*.

and take þis pepyl hom w*ith* ȝow hens
I am not dysposyd to passyn hens froo.

Ep*iscop*us

* Than ffar*e* wel mayden *and* pur*e* virgyne
far*e* wel trewe hand-mayd of god in blys
we all to ȝow lowly inclyne 355
and take our*e* leve of ȝow as wurthy is.

Maria

Fo. 81v All myghty god ȝ*our* weys wys se[1]
Ffor þ*at* hyȝ lord is most of myght
he mote ȝow spede þ*at* ȝe not mys
In hevyn of hym to haue A syght. 360

Joseph

¶ Honouryd in hevyn be þ*at* hyȝ lord
whos endles grace is so habundaunt
þ*at* he doth shewe þ*e* trewe recorde
of iche wyhgt*e* þ*at* is his trewe servaunt
That lord to wurchepe w*ith* hert plesaunt 365
we bothe be bownd ryght on þis place
which our*e* purgacyon us dyde graunt
and prevyd us[2] pur*e* by hieȝ grace.

Maria

¶ Ffor sothe good spowse I thank hy*m* hyȝly
Of his good grace for our*e* purgac*i*on 370
our*e* clennesse is knowyn ful opynly
be vertu of his grett consolac*i*on

Remainder of Fo. 81v—4⅞ inches—left blank.

exxplicit cum gaudio amen.[3]

Fo. 82 ¶ Joseph

LOrd What travayl to man is Wrought
Rest in þis werd be-hovyth hy*m* non̄ 15
Octauyan our*e* Emp*erour* sadly hath be-sought
Our*e* trybute hy*m* to bere · ffolk must forth ichon
It is cryed in every bourgh *and* cety be name 5

* No paragraph sign here in MS.

[1] This *se* stands after *wys* (apparently in the same ink and the same hand) as a separate word.

[2] Some miswritten word or a second *pure* between *us* and *pure* crossed through.

[3] Scribbled in a later hand enclosed in rudely scribbled flourishes.

I þ*a*t am a por*e* tymbre wryth · born of þ*e* blood of dauyd
þ*e* Emper*or*y*s* comawndement I must holde w*ith*
And ellys I wer*e* to blame.

¶ Now my wyff mary · what sey ȝe to this
For sekyr nedys I must fforth̃ wende
On to þ*e* cyte of bedleem ffer hens i-wys
Þus to labor*e* I must my body bende.

Maria

My*n* husbond *and* my spowse w*ith* ȝow wyl I wende
A syght of þ*a*t cyte ffayn wolde I se
If I myght of my*n* Alye ony þ*er* ffynde
It wolde be grett joye on to me.

Joseph

¶ My Spowse ȝe be w*ith* childe . I fere ȝow to kary
Ffor me semyth it wer*e* werkys wylde
but ȝow to plese ryght ffayn wold I
ȝitt women̄ ben ethe to greve · whan þei be w*ith* childe
now latt us Fforth wende as ffast as we may
and al-myghty god spede us in our*e* jurnay.

Maria

¶ A my swete husbond · wolde ȝe telle to me
What tre is ȝon̄ standynge vpon ȝon hylle.

Joseph

Ffor sothe mary it is clepyd A chery tre
In tyme of ȝer*e* · ȝe myght ffede ȝow þ*er*on ȝ*our* Ffylle.

Maria

¶ Turne A-geyn husbond *and* beholde ȝon tre
how þ*a*t it blomyght now so swetly.

Joseph

Fo. 82v Cum on Mary þ*a*t we worn At ȝon Cyte
or ellys we may be blamyd I telle ȝow lythly.

Maria

¶ Now my spowse I pray ȝow to be-hold
How þ*e* cheryes growyn vpon ȝon tre
Ffor to haue þ*er*-of ryght ffayn I wold
and it plesyd ȝow to labor*e* so mech̄ for me.

Joseph

¶ ȝ*our* desyr*e* to ffulfylle I xal Assay sekyrly
Ow · to plucke ȝow of these cheries · it is a werk wylde
Ffor þ*e* tre is sō hyȝ · it wol not be lyghtly
þ*er*fore lete hy*m* pluk ȝow cheryes · be-gatt ȝow w*ith* childe.

Maria

¶ Now good lord I pray þe · graunt me þis boun
to haue of þese cheries *and* it be ȝ*our* wylle
now I thank it god · þis tre bowyth to·me dowñ
I may now gaderyn A-nowe · *and* etyn my ffylle.

joseph

¶ Ow · I know weyl I haue offendyd my god i*n* trinyte
Spekyng to my spowse these vnkynde wurdys
Ffor now I beleve wel it may non other be
but þ*a*t my spowse beryght þ*e* kyngys son of blys
He help us now at our*e* nede
of þ*e* kynrede of jesse · worthely were ȝe bor*e*
Kyng*ys* *and* patryarkys ȝow be-ffor*e*
alł þese wurthy · of ȝ*our* kynred · wor*e*
as clerkys in story rede.

Maria

¶ Now gramercy husbond for ȝ*our* report
in our*e* weys wysely late us forth wende
þ*e* fadyr alł myghty he be our*e* comfort
þ*e* holy·gost gloryous · he be our*e* frende.

Joseph

Fo. 83 ¶ Heyl wurchepful ser*e* *and* good day
A ceteceyn of þis cyte · ȝe seme to be
of herborwe ffor spowse *and* me I ȝow p*ra*y
Ffor trewly þis woman is fful were
And fayn At reste ser*e* wold she be
We wolde ffulffylle þ*e* byddynge of our*e* emper*our*
Ffor to pay trybute As ryght is our*e*
and to kepe our*e* self ffrom dolowre
We a*r*e come to þis cyte.

Ciues

¶ Ser*e* ostage in þis town know I non
þin Wyff *and* þ*ou* in for to slepe
this cete · is be-sett · w*ith* pepyl every won
And ȝett þei ly w*ith*-owte fful every strete.

¶ W*ith*-inne no wałł man comyst þ*ou* nowth
Be þ*ou* onys w*ith*-inne þ*e* cyte gate
on-ethys in þ*e* strete A place may be sowth
þ*er* on to rest w*ith*-owte debate.

Joseph

¶ Nay sere debate · þat wyl I nowth
aꝗ such thyngys passyn my powere
but ȝitt my care *and* aꝗ my thought 75
is for Mary my derlynge dere.

¶ A Swete Wyff · what xal we do
wher xal we logge þis nyght
on to þe ffadyr of heffne pray we so
vs to kepe ffrom every wykkyd whyt. 80

Ciues

¶ Good man O word I wyl þe sey
If þou wylt do by þe counsel of me
ȝondyr is An hous of haras þat stant be þe wey
Amonge þe bestys herboryd may ȝe be.

Maria

Fo. 83v ¶ Now þe fadyr of hefne he mut ȝow ȝelde 85
His sone in my wombe forsothe he is
He kepe þe *and* þi good be fryth *and* ffelde
go we hens husbond for now tyme it is.

* But herk now good husbond a new relacyon̄ [1]
which in my self I know ryght weꝗ 90
Cryst in me hath take in-carnacion
Sone wele be borne þe trowth̄ I fele.

¶ In þis pore logge my chawmere I take
here for to A-byde þe blyssyd byrth̄
of hym þat aꝗ þis werd dude make 95
be-twyn myn sydys I fele he styrth̄.

Joseph

¶ god be þin help spowse it swemyth me sore
þus febyly loggyd *and* in so pore degre
goddys sone amonge bes*tys* to be bore
his wou*n*dyr werkys ffulfyllyd must be. 100

¶ In An hous þat is desolat w*ith*-owty Any waꝗ
Ffyer nor wood non here is.

Maria

Joseph myn husbond a-bydyn here I xal
ffor here wyl be born þe Kyng*ys* sone of blys.

* No paragraph sign in the MS. here.
[1] *ral* first written (for *rel*) and crossed through.

Joseph

¶ Now jentyll wyff be of good myrth 105
and if ȝe wyl owght haue telle me what ȝe thynk
I xal not spar*e* for schep nor derth [1]
now telle me ȝo*ur* lust of mete *and* drynk.

Maria

¶ Ffor mete *and* drynk lust I ryght nowth
all-myghty god my fode xal be 110
now þ*at* I am in chawmer*e* brought
I hope ryght well my chylde to se
Fo. 84 Therfor*e* husbond of ȝo*ur* honeste
A-voyd ȝow hens out of þis place
And I a-lone w*ith* humylite 115
her*e* xal abyde goddys hyȝ grace.

Joseph

¶ All redy wyff ȝow for to plese
I wyl go hens out of ȝo*ur* way
and seke su*m* mydwyuys ȝow for to ese
Whan þ*at* ȝe trauayle of childe þis day 120
Ffar*e* well trewe wyff *and* also clene may
God be ȝo*ur* comforte in trinyte.

Maria

to god in hevyn for ȝow I pray
He ȝow p*re*serve wher so ȝe be.

hic dum joseph est Absens parit Maria filium vnigenitum.

Joseph

¶ Now god of whom comyth all releffe 125
And as all grace in þ*e* is grownde
So saue my wyff from hurt *and* greffe
tyl I sum mydwyuys for her*e* haue fownde
Travelynge women in car*e* be bownde
w*ith* grete throwys whan þei do grone 130
god helpe my wyff þ*at* sche not swownde
I am ful sory sche is a-lone.

¶ It is not co*n*uenyent a man to be
þ*er* women gon in travalynge
wher for*e* su*m* mydwyff fayn wold I se 135
my wyff to helpe þ*at* is so ȝenge.

[1] *derke* first written and crossed through.

Fo. 84v

zelomy

¶ Why makyst þ*ou* man suche mornyng
tell̃ me su*m* dele of ȝ*our* gret mone.

Joseph

My wyf is now in gret longynge
trauelyng of chylde *and* is a-lone
Ffor godys loue þ*at* sytt in trone
as ȝe myd-wyuys þ*at* kan ȝ*our* good
help my ȝonge spowse in hast a-none
I drede me sor*e* of þ*at* fayr food.

Salome

be of good cher*e* *and* of glad mood
we ij mydwyuys w*it*h þ*e* wyll̃ go
þ*er* was nevyr woman in such plyght stood
but we were redy her*e* help to do.

¶ My name is Salomee all̃ men me knowe
Ffor a mydwyff of wurthy fame
Whan women travayl grace doth growe
þ*er* as I come I had nevyr shame.

zelomye

And I am zelomye men knowe my name
We tweyn w*it*h the wyl go togedyr
and help þ*i* wyff fro hurt *and* grame
Com̃ forth̃ joseph go we streyth̃ thedyr.

Joseph

¶ I thank ȝow damys · ȝe comforte my lyff
streyte to my spowse walke we þ*e* way
In þis por*e* logge lyght mary my wyff
hÿr*e* for to comforte gode frendys a-say.

Salome

We dar*e* not entre þis logge in fay
þ*er* is þ*er*-in so gret bryghtnes
mone be nyght nor su*n*ne be day
Fo. 85 Shone nevyr so cler*e* in þ*er* lyghtness*e*.

zelomye

¶ In to þis hous dare I not gon̄
þ*e* wou*n*dyrffull̃ lyght doth me affray.

Joseph

than wyl my-self gon in Alon
and cher*e* my wyff if þ*at* I may

Att heyl maydon *and* wyff I say
how dost þ*ou* far*e* telle me þ*i* cher*e* 170
the for to comforte in gesyn*e* þis day
tweyn gode mydwyuis I haue brought her*e*.

¶ The for to helpe þ*at* art in harde bonde
zelomye · And salomee be com̄ w*ith* me
Ffor dowte of drede w*ith*-owte þei do stond 175
and dar*e* not come in for lyght þ*at* they se.

Maria

hic maria subridendo dicat · Maria.

The myght of þ*e* godhede in his mageste
Wyl not be hyd now at þis whyle
the chylde þ*at* is born wyl p*reve* his modyr fre
A very clene mayde *and* þ*er*fore I smyle. 180

Joseph

¶ Why do ȝe lawghe[1] wyff ȝe be to blame
I pray ȝow spowse do no mor*e* so
In happ þ*e* mydwyuys wyl take it to grame
and at ȝo*ur* nede helpe wele non do *
Iff ȝe haue nede of mydwyuys lo 185
p*er* auenture thei wyl gon hens
þ*er*for be sad *and* ȝe may so
And wynnyth att þ*e* mydwyuis good diligens.

Maria

Fo. 85v ¶ husbond I p*ray* ȝow dysplese ȝow nowth
þow þ*at* I lawghe · *and* gret joye haue 190
her*e* is þ*e* chylde þis werd hath wrought
born now of me þ*at* att thynge xal saue.

Joseph

I Aske ȝow grace for I dyde raue
O gracyous childe I aske mercy
As þ*ou* art lord *and* I but knaue 195
Ffor-ȝeue me now my gret foly.

¶ Alas mydwyuis what haue I seyd
I pray ȝow com to us mor*e* ner*e*
Ffor her*e* I fynde my wyff a mayd
and in her*e* Arme a chyld hath her*e* 200

[1] A miswritten final letter (? *te* or *e*) deleted and *e* written above.
* *ȝellony* roughly scribbled here in faint ink.

bothe mayd *and* modyr sch̄ is in ffere
þat god wole haue · may nevyr more fayle
modyr on erth̄ was nevyr non clere
With-owth sche had in byrth̄ travayle.

zelomy

¶ In byrth trauayle muste sche nedys haue
or ellys no chylde of here is born.

Joseph

I pray ȝow dame *and* ȝe vowch̄-saue
com se þe chylde my wyff be-forn.

Salome

Grete god be in þis place
Swete systyr how fare ȝe.

Maria

I thank þe fadyr of his hyȝ grace
his owyn son *and* my chylde here ȝe may se.

zelomye

¶ Aſt heyl mary *and* ryght good morn̄
Who was mydwyfe of þis ffayr chyld.

Maria

he þat no thynge wyl haue for-lorn̄
Sent me þis babe *and* I mayd mylde.

zelomye

Fo. 86 (H quire) ¶ With honde lete me now towch̄ *and* fele
yf ȝe haue nede of medycyn̄
I xal ȝow comforte *and* helpe ryght wele
As other women yf ȝe haue pyn̄.

Maria

Of þis fayr byrth̄ þat here is myn̄
Peyne nere grevynge fele I ryght non̄
I am clene mayde *and* pure virgyn̄
tast with ȝour hand ȝour-self a-lon

hic palpat zelomye beatam mariam virginem dicens

zelomy

¶ O myghtfull god haue mercy on me
A merveyle þat nevyr was herd be-forn
Here opynly I fele *and* se
A fayr chylde of a maydon is born̄
And nedyth no waschynge as other don̄
Fful clene *and* pure for soth̄ is he
with-outyn spot or ony polucyon̄
his modyr nott hurte of virgynite.

¶ Coom ner*e* gode systyr Salome
be-holde þ*e* brestys of þis clene mayd
Fful of fayr mylke how þ*at* þei be 235
And hyr*e* chylde clene as I fyrst sayd
As other ben · nowth fowle arayd
but clene *and* pur*e* bothe modyr *and* chylde
Of þis matyr I am dysmayd[1]
to se them both̄ thus vndefyled. 240

Salome

¶ It is not trewe it may nevyr be
þ*at* bothe be clene I can not be-leve
a mayde mylke haue[2] · nev*er* man dyde se
ne woman ber*e* chylde w*ith*-owte grett greve.

Fo. 86v ¶*I xal nevyr trowe it but I it p*re*ve 245
W*ith* hand towchynge but I Assay
in my conscience it may nevyr cleue
þat sche hath chylde *and* is a may.

Maria

¶ ȝow for to putt clene out of dowth
towch̄ w*ith* ȝ*our* hand *and* wele a-say 250
Wysely ransake *and* trye þ*e* trewthe owth
Whethyr I be fowlyd or a clene may.[3]

hic tangit salomee marie et cum arescerit manus eius vlulando[4] *et quasi flendo dicit*

Salomee

¶†Alas Alas *and* Wele a-Waye
Ffor my grett dowth *and* fals beleve
myne hand is ded *and* drye as claye 255
my fals vntrost hath wrought myscheve.

¶ Alas þ*e* tyme þ*at* I was born̄
thus to offende A-ȝens goddys myght
my*n* handys power is now all̄ lorn

[1] *dyf* first written for *dys* and crossed through.

[2] The word *haue* is crossed through in darker ink and the contraction for *ys* has been added, also in darker ink and another hand, to *mayd*.

* This paragraph sign seems inserted by mistake.

[3] *ransak* roughly scribbled in the margin here.

[4] There is an erasure in this word; the second *l* is adapted from some other letter.

† This paragraph sign again seems inserted by mistake.

Styff as a stykke *and* may nowth plyght
Ffor I dede tempte þis mayde so bryght
And helde A-ȝens here pure clennes
In grett myscheff now am I pyght
Alas alas ffor my lewdnes.

¶ O lord of myght þou knowyst þe trowth
þat I haue evyr had dred of þe
on every power whyght evyr I haue rowthe
and ȝove hem almes for loue of þe
Bothe wyff *and* wedowe þat Askyght for the
And frendles chylderyn þat haddyn grett nede
Fo. 87 I dude them cure *and* all for the
and toke no rewarde of them nor mede.

¶ Now as a wrecch ffor fals be-leve
þat I shewyd in temptynge þis mayde
my hand is ded *and* doth me greve
Alas þat evyr I here assayde.

Angelus[1]

Woman þi sorwe to haue de-layde[2]
wurchep þat childe þat þer is born
towch þe clothis þer he is layde
ffor he xal saue all þat is lorn.

Salomee

¶ O gloryous chylde *and* kynge of blysse
I aske ȝow mercy for my trespace
I knowlege my synne · I demyd a-mys
O blyssyd babe grawnt me sum grace
Of ȝow mayde Also here in þis place
I aske mercy knelynge on kne
moste holy mayde grawnt me solace
Sum wurde of comforte sey now to me.

Maria

¶ As goddys Aungel to ȝow dede telle
my chyld is medycyn ffor every sor
towch his clothis be my cowncelle
ȝowre hand ful sone he wyl restor.

[1] From here to the end of this play the initial of the name of speaker is marked with a red stroke.

[2] *leyde* originally written, the *e* altered in slightly darker ink to *a*.

hic salomee tangit fimbriam Christi dicens Salomee

A now blyssyd be þis chylde eu*er* mor*e*
þ*e* sone of god for sothe he is
Hath helyd my*n* hand þ*at* was for-lor*e* 295
thorwe ffals be-leve *and* demynge a-mys.

Fo. 87v ¶ In every place I xal telle þis
Of a clene mayd þ*at* god is born
And in our*e* lyknes god now clad is
Mankend to saue þ*at* was for-lorn 300
His modyr a mayde as sche was be-forn̄
natt fowle polutyd as other women be
but fayr *and* fresch̄ as rose on thorn̄
Lely wyte · clene w*ith* pur*e* virginyte.

¶ Of þis blyssyd babe my leve now do I take 305
and also of ȝow · hyȝ modyr of blysse
Of þis grett meracle mor*e* knowlege to make
I xal go telle it in iche place i-wys.

Maria

Ffar*e* wel good dame *and* god ȝ*our* wey wysse
In alł ȝ*our* jurnay god be ȝ*our* spede 310
and of his hyȝ mercy þ*at* lord so ȝow blysse
þ*at* ȝe nevyr offende mor*e* in word thought nor*e* dede.

zelomy

¶ And I Also do take my leve her*e*
Of alł þis blyssyd good company
praynge ȝ*our* grace bothe fer*e* *and* ner*e* 315
On us to spede ȝ*our* endles mercy.

Joseph

¶ The blyssyng of þ*at* lord þ*at* is most myghty
mote sprede on ȝow in every place
of alł ȝ*our* enmyes to haue þ*e* victory
God þ*at* best may grawnt ȝow his grace. Amen. 320

Fo. 88 left blank.

Fo. 88v *Angelus ad pastores dicit* [Gloria in excelsis deo.[1]]

Angelus

JOye to god þat sytt in hevyn̄
And pes to man on erthe grownde
A chylde is born be-nethe þe levyn
thurwe hym many ffolke xul be vn-bownde
Sacramentys þer xul be vij
Wonnyn þurowe[2] þat childys wounde
Therfore I synge A joyful stevene
þe flowre of frenchep now is founde
God þat wonyght on hyȝ[3]
he is gloryed mannys gost to wynne
he hath sent salue to mannys synne
Pes is comyn to mannys kynne
thorwe goddys sleytys slyȝ[4].

1us pastor

¶ Maunfras maunfras felawe myne
I saw a grett lyght with shene[5] shyne
ȝit saw I nevyr so selkowth[6] syne
Shapyn vpon þe skyes
It is bryghtere þan þe sunne bem
It comyth ryght ouer alł þis rem
Evyn above bedleem̄
I saw it brenne thryes.

ijus pastor

¶ Thu art my brother boosras
I haue beholdyn þe same pas
I trowe it is[7] tokenynge of gras

[1] The words *Gloria in excelsis deo* have been crossed through in a different ink.

[2] This word has been crossed out in darker ink, and *thorough* (?) or *therogh* (?)—there has been some double correction—written above in another hand.

[3] An *e* has been added to *hyȝ* in different ink.

[4] *sleytys* is partly erased, and over *slyȝ* is written *hye*, and the words *wysdam̄ I saye* are added in the writing of the scribe of the interpolated Ff. 95, 96, and 112.

[5] *shene* crossed out and *bryght* written above in the hand of the same later scribe.

[6] *mervelus a* written by the same hand over *selkowth*, which is barely decipherable.

[7] Some miswritten letter after *is* crossed through by scribe.

Fo. 89 þ*at* shynynge shewyght be-forn̄ 25
Balaam spak in p*ro*phesye
A lyght xuld shyne vpon þe skye
Whan A sone[1] of a mayd marye
In bedleem were i-born.

iij^us pastor

¶ thow I make lyty noyse 30
I am an herde þ*at* hattyht moyse
I herde carpynge of a croyse
of Moyses in his lawe
Of a mayd a barne born̄ *
On a tre he xulde be torn̄ 35
delyver folkes þ*at* arn forlorn
The chylde xulde be slawe.

1us pastor

¶ Balaam spak in p*ro*phecie
out of jacob xuld shyne a skye
many ffolke he xulde bye 40
w*ith* his bryght blood
Be þ*at* bryght blod þat he xulde blede
he xal us brynge fro þ*e* develys drede
as a duke most dowty in dede
thorwe his deth̄ on rode. 45

ij^us pastor

¶ Amos spak w*ith* mylde meth̄
A frute swettere than bawmys breth̄
His deth̄ · xulde slen oure sowlys deth̄
And drawe us all̄ from helle
Ther fore such̄ lyght goth be-forn̄ 50
Fo. 89v In tokyn þ*at* þ*e* childe is born̄
Whiche xal saue þ*at* is for-lorn̄
As p*ro*phetys gonne spelle.

[1] *sone* crossed out and *chyld̄* written above in the later hand.
* These lines have been altered by the second hand thus :

thow I make lytyl noyse of this
I am an herde mañ þ*at* hattyht sayyng Amys
I herde spekyng of a chyld̄ of blys
of Moyses in his lawe
Of a mayd a child̄ xuld̄ be borne.

iij^{us} pastor

¶ Danyel þe *pro*phete þus gan speke
wyse god from woo us wreke
þi bryght hevyn *þou* to-breke
and medele *þe* w*ith* a mayde
This p*ro*phecye is now spad
Cryst in our*e* kend is clad
þerfore mankend may be glad
As p*ro*phet*ys* be-forn han seyd.
Gloria in excelsis deo *cantent.*

1us pastor

B[1] ¶ The p*ro*phecye of boosdras is spedly sped
now leyke we hens as *þat* lyght us lede
myght we se onys *þat* bryght on bed
our*e* bale[2] it wolde vnbynde
We xulde shadyr[3] for no shour*e*
buske[4] we us hens to bedleem bour*e*
to se *þat* fayr fresch flowr*e*
the mayde mylde in mynde.

ij^{us} pastor

¶ Lete us ffolwe w*ith* all our*e* myght
W*ith* songe *and* myrth we xul us dyght
and wurchep w*ith* joye *þat* wurthy wyght
þat lord is of mankynne
Lete us go fforthe[5] fast on hye

Fo. 90 And honowr*e* *þat* babe wurthylye
w*ith* merthe songe *and* melodye
haue do þis songe be-gynne.

tunc pastores cantabunt · stella celi extirpauit · quo facto ibunt ad querendum christum.

1us pastor

A Ey · Ey · þis was a wondyr note

[1] The marginal letters B, A, C indicate the order of the passages. They are slightly rubricated. It will be noted that a third shepherd's speech is missing here. The disturbance of the text is connected with the cancelling of the angels' song at the beginning of the play. It is not irrelevant that the humorous passage marked A closely resembles a passage in the Chester Shepherds' play. It would appear to be a borrowed interpolation replacing a speech from the third shepherd.

[2] *bale* crossed out and *sorow* written above by the later hand.

[3] *shadyr* crossed out and *not let* written above by the later hand.

[4] *buske* crossed out and *go* written against it in the margin by the later hand.

[5] This *ff* is miswritten with three strokes.

þat was now songyn above *þe* sky
I haue *þat* voys fful wele I wote 80
þei songe gle glo glory.

ij^{us} pastor

¶ Nay so moty the so was it nowth
I haue *þat* songe fful wele i-nvm̄ [1]
In my wytt weyl it is wrought
It was gle glo glas glum. 85

iij^{us} pastor

¶ The songe me thought it was glory
and aftyr-warde he seyd us to
þer is a chylde born xal [2] be a prynce myghty
Ffor to seke *þat* chylde I rede we go.

1us pastor

C ¶ Heyle floure of flour*ys* fayrest i-fownde 90
Heyle perle peerles p*r*ime rose of p*r*ise
heyl blome on bedde we xul be vn-bownde
w*ith* *þi* blody woundys *and* werkys full wyse
heyl god grettest I grete *þe* on grownde
þe gredy devyl xal grone grysly as a gryse 95
whan *þou* wynnyst *þis* worlde w*ith* *þi* wyde wounde
and puttyst man to paradys w*ith* plenty of prys
to loue *þe* is my delyte
Fo. 90v [marked 90] Heyl floure fayr *and* fre
Lyght from *þe* trynyte 100
Heyl blyssyd mote *þou* be
heyl mayde*n* fayrest in syght.

ij^{us} pastor

¶ Heyl floure ovyr fflowr*ys* fowndyn in fryght [3]
Heyl Cryst kynde in oure kyth
Heyl werker of wele to wony*n* us wyth 105
Heyl wynnere i-wys
Heyl fformere *and* ffrende
Heyl ffellere of *þe* fende
Heyl clad in oure kende
heyl p*r*ince of paradys. 110

[1] The *v* has been written by the scribe (?) over some other letter—?*o* or a half-formed *u*.

[2] *xal* written twice in MS. crossed through in different ink.

[3] So in MS. for *fryth*.

iij^{us} pastor

¶ Heyl lord ou*er* lordys *þat* lyggyst ful lowe
Heyl kynge ovyr kyng*ys* *þi* kynrede to knowe
Heyl comely knyth *þe* deuyl to ou*er* throwe
Heyl flowr*e* of alle
Heyl werker*e* to wynne
bodyes bowndyn in synne
Heyl in a bestys bynne
Be-stad[1] in a stalle.

joseph

¶ Herdys on hylle[2]
beth not stylle
but seyth ȝ*our* wylle
to many A man
How god is born
þis mery morn
þat is for-lorn
fyndyn he can.

1us pastor

¶ We xulle telle
be dale *and* hylle
How harwer*e* of helle
was born *þis* nyght
myrthis to melle
and fendys to quelle
þat wer*e* so felle
Aȝens his ryght.

ij^{us} pastor

Fo. 91 ¶ Ffar*e* wel babe *and* barne of blys
Ffar*e* wel lord *þat* lovely is
þe to wurchep *þi* feet I kys
on knes to *þe* I falle
The to wurchepe I falle on kne
alle *þis* werd may joye of *þe*
now far*e* wel lorde of grett pouste
ȝa far*e* wel kynge of alle.

[1] This *d* is adapted from some miswritten letter.

[2] From line 119 to line 134 two lines are written as one divided by two black strokes and one red.

iij^us^ pastor

¶ Thow I be þe last þat take my leve
ȝit fayre mullynge take it nat at no greve
now fayre babe wele mut þou cheve 145
Ffayr chylde now haue good day
Ffare weyl myn owyn dere derlyng
I-wys þou art a ryght fayr thyng
Ffare wel my lorde *and* my swetyng
Ffare wel born in pore Aray. 150

Maria

¶ Now ȝe herd-men wel mote ȝe be
Ffor ȝoure omage *and* ȝour syngynge
my sone xal a-qwyte ȝow in hefne se
and ȝeue ȝow alł ryght good hendynge.

Amen.

Remainder of Fo. 91—3⅜ inches—left blank.

Fo. 91ᵛ blank except for the following scribblings in different hands of a later date than the MS.:

1. W. Will*iam* Dere
   Polerd.
2. Wyll*iam* Dere.
3. John Hasycham.
4. John Taylphott of Parish Bedonson
   wee that will not when we paie
   when we would we shall saie* nay.

¶ Herode[

Fo. 92[1]

AS a lord in ryalte in non Regyon̄[2] so ryche
And rulere of alł remys I ryde *in* ryal a-ray
Ther is no lord of lond in lordchep to me lyche
non lofflyere[3] non lofsu*m*mere evyr lastyng is my lay.
*[1] Of bewte *and* of boldnes I bere ever-more þe belle
Of mayn *and* of myght I mast*er* every man 18
I dy*n*ge w*ith* my dowtynes þe devyl down to helle
Ffor bothe of hevyn *and* of herth̄ I am kyng sertayn̄.

* Or *find*.

[1] Ff. 92 and 93 are narrower than the rest, more margin having been cut away. Half the number 18 has been thus clipped.

[2] *Regyon̄* omitted and written above the line.

[3] The *ff* is written over some other letter.

*[1] No paragraph sign in MS. here.

¶ I am þe comelyeste kynge clad in gleterynge golde
ȝa·*and* þe semelyeste syre þat may be-stryde a stede
I welde att my wyłł ałł wyght*ys* upon molde.

* ¶ ȝa and wurthely I am wrappyd in a wurthy wede
ȝe knyght*ys* so comely bothe curteys *and* kene
to my paleys wyl I passe · fułł prest I ȝow plyth
ȝe dukys so dowty ffolwe me be-dene
On to my ryal paleys þe wey lyth ful ryght.

¶ Wyghtly fro my stede I skyppe dowñ in hast
to my*n* heyȝ hallys I haste me in my way
ȝe mynstrełł of myrth blowe up a good blast
Whyłł I go to chawmere *and* chaunge my*n* array.

1*us* rex

¶ Heyl be ȝe kyng*ys* tweyne
Fferre rydyng out of ȝ*our* regne
me thynkyth be ȝ*our* present*ys* seyne
ȝe sekyn oure sauyo*ur*
Ffro Saba haue I folwyd fferre
the glemynge of ȝon gay sterre
a chyldys blood xal bye vs dere
þ*at* þ*er* is born i*n* best*ys* boure.

¶ My name is kynge Baltazare
Fo. 92v Of prophetys speche I am ware
Therfore a ferre wey I fare
A maydenys childe to seche
Ffor he made man of þe moolde
and is kynge of hevyn holde
I wyl hym offere þe rede golde
as reson wyl me teche.

Se*cun*dus rex

¶ Melchizar þ*at* my name is kydde
in hote loue my*n* hert is hydde
to þe blosme upon his bedde
born by best*ys* bynne
in tarys I am kynge w*ith* crowne
by bankys *and* brymmys browne
I haue trauaylid by many a towne
my lordys love to wynne.

* Paragraph sign inserted a line too high.

¶ I seke hym w*ith* ensens sote 45
of all prestys he xal be rote
his bryght blood xal be our*e* bote
to brynge vs out of bende
The childe xal be chosyn a preste
in all vertuys ffowndyn meste 50
be-forn his fader*ys* fayr breste
Ensens he xal up sende.

Terci*us* Rex

¶ In ypotan And Archage
I am kynge knowyn in kage
to seke a childe of Semlant sage 55
I haue faryn ryght fferre
Jasper is my name knowyn
in many countres þ*at* ar*e* my*n* owyn
thorwe byttyr blastys þ*at* gyn blowy*n*
I stryke after*e* þe sterre. 60

Fo. 93 ¶ I brynge myrre to my p*re*sent
A byttyr lyco*ur* verament
Ffor he xal th[o]lyn[1] byttyr dent
In a maydonys flesch is clad
on byttyr tre he xal be bent 65
Man *and* god omnypotent
w*ith* byttyr betynge his fflesch be rent
tyl all his blood be bledde.

Her[ode

¶ Now I regne . . . þ*e*[2] kynge A-rayd ful Rych
Rollyd in ry*nggys and* robys of array 70
Dukys w*ith* dentys I dryve in to þ*e* dych
my dedys be ful dowty demyd be day
I xall marryn þ*o* men þ*at* r . . . n on a mych*e*[3]
and þ*er*-inne sette here sacrement*ys* sottys . . . say[4]

[1] *tholyn* has been in great part erased and the word *suffyr* written over it by the second (Ff. 95, 96) hand.

[2] Over an illegible word and *þe, lyk a* has been written by the later hand.

[3] The latter part of this line has been altered in the second hand to (*þo*) *heretyk*ys (*þ*at) *beleuen a-mysse.*

[4] *þ*er-*inne* has been changed to *þ*er*in* ; *here* to *there*; and from *sottys* to the end the line has been altered by erasure and adaptation to: *falsse þ*ey *are*

þer is no lorde in þis werde[1] *þat* lokygh me lyche[2] 75
Ffor to lame l . . . r*ys*[3] of *þe* lesse lay
I am jolyer*e* than *þe* jay
Stronge thevys to steke
þat wele our*e* lawys breke
on *þo* wrecchis I wyłł be wreke 80
and hont hem vndyr ha[y.

¶ In kyrtyl of ca*m*maka kynge am I cladde
Cruel *and* curryd in my*n* crowne knowe
I sytt in vndyr[4] sesar in my se . . e[5] sadde
Sorwyn to sottys such seed wyłł I sowe 85
Boys now blaberyn bostynge of a baron bad
In bedde (?)[6] is born be bestys suche bost is blowe
Fo. 93v I xal prune *þat* paphawk *and* pr*e*vyn hy*m* as a pad[7]
Scheldys and shaftys sh . . . lh . . . y sowe[8]
my knyght*ys* xaln rydyn on rowe[9] 90
knave[10] chylderyn ffor to qwelle
be mahound dyngne duke of helle
Sowr*e* deth his lyff xałł selle
Such threttt wolde me throw[11].

¶ Styward[12] bolde 95*
walke *þou* on wolde[13]

I say. The words cannot be deciphered with certainty as the adaptation is confused.

[1] An *l* has been inserted in this word.

[2] *lyche* has been altered to *lyke i-wysse.*

[3] This word, altered by the reviser to *heretykkys*, may be lo[ve]r*ys* or le[ve]r*ys*.

[4] *in* has been altered to *here*, and the *v* of *vndyr* to *o*.

[5] The later writer has *sette*, the *tt* being written over an erasure.

[6] This word is uncertain; the reviser has altered it to *bedlem.*

[7] *prune* altered to *prycke*, and *paphawk* to *paddoke.*

[8] Two or three words of some thirteen letters indecipherable in this line. The reviser alters *shaftys* to *sperys* and continues: *schałł I ther* (*sowe*).

[9] *xaln* (or *xuln*) altered to *xalle*, and an *a* inserted between *on* and *rowe.*

[10] A *k* between *knave* and *chylderyn* crossed through by original scribe.

[11] *threttt* altered to *thrett*ys and *throw* to *ouerthrow.*

[12] *Styward* crossed through. Only the last letters *or* of the substituted word remain, the rest having been cut away with the margin.

* From line 95 to line 150 two short lines are printed in one, divided as before.

[13] *ou*er *mowlde* substituted by second hand for *on wolde.*

and wisely beholde
Aꝑꝑ Aboute
Iff Any thynge
shuld greve þ*e* kynge 100
brynge me tydynge
if þ*er* be ony dowte.

Senescall*us*

[¶][1] Lord kynge in crowne
I go fro towne
bẏ bankys browne 105
I wyꝑꝑ a-byde
And w*ith* erys lyste
est *and* west
iff any geste
on grownde gynnyth glyde. 110

[*H*]*erodes nup*[2]. . .

Tunc ibit senescallus et obuiabit tribus regibus et *dicit eis*

[¶][1] Kyng*ys* iij
vndyr þis tre
In þis countre
why wyꝑꝑ ȝe abyde
Herowde is kynge 115
of þis wonynge
On to his dwellynge
now xul ȝe glyde.

1[us] Rex

[¶][1] Now lede us alle
to þ*e* kyng*ys* halle 120
how it befalle
we pray to the
wyttys to wete
he may us pete
In flesshe be glete 125
godys frute fre.

[1] These paragraph marks have been cut away with the margin.

[2] This, which appears to be a stage direction, is faintly written in the later hand in the margin, and has been partly cut away. The second word (or words) is not certainly decipherable—? nup*er*. . . .

Senescall*us*

[¶][1] Ffolw*ith* in stownde
vpon þis grownde
to þ*e* castel rownde
I xal ȝow tech̄
Where[2] kynge gynny[3] wyde
vp in þis tyde
in pompe *and* pryde
his myght gynnyth reche.

[¶][1] Ser*e* kyng in trone
her*e* comyth a-none
Fo. 94 by strete *and* stone
kyng*ys* thre
They ber*e* present
what thei haue ment
ne whedyr thay arn bent
I can not se.

Herodes Rex

¶ I xal hem craue
what they haue
Iff they raue
or waxyn wood
I xal hem reve
her*e* wyttys deve
her*e* hedys cleve
And schedyn her*e* blood.

1*us* Rex

¶ Heyl be þ*ou* kynge in kage ful hye
Heyl we nyghe þin halle ryght nye
Knowyst þ*ou* ought þ*at* chylde slye
He is born her*e* a-bowth
He is born of a mayd ȝynge
he xal be kynge ou*er* every kynge
We go to seke þ*at* louely thynge
to hym ffayn wolde I lowth.

[1] Paragraph marks cut away with the margin.
[2] The *h* omitted and written small above the line by the scribe.
[3] *gynny* altered by reviser to *wonyt*.

ij^{us} Rex

¶ Balaam spak in p*ro*phecy
A sterr*e* xulde ful louelye 160
lythtyn vpon mayd marye
comy*n* of jacobys kynne
þ*e* childe is born *and* lyth̄ here by
Blomyd in a madenys body
A sterr*e* hath strekyn upon þ*e* sky 165
and ledde us fayr be fenne.

iij^{us} Rex

¶ The sterre hath ledde us out of þ*e* est
to seke a baron̄ born best
he xal be kynge of myght*ys* mest
as p*ro*phecy gy*n*neth spelle 170
we be kyng*ys* in wey wery
Syr kynge ffor þ*i* curtesy
telle us to þ*at* childe so louely
in what town̄ gy*n*nyth he dwelle.

herodes Rex

Fo. 94v ¶ ȝe thre kyng*ys* rekenyd be rowe 175
ley now downe ȝ*our* wurdys lowe
Such̄ a carpynge is vnknowe
on-rekenyd i*n* my regne
I am a kynge of hyȝ degre
þ*er* xal non ben above me 180
I haue flore*ns* *and* fryhthis fre
parkys *and* pownd*ys* pleyne.

¶ But goth to fynde þ*at* ȝe sech̄
and yf ȝe knowe such̄ a lech̄
and ȝe hy*m* fynde I ȝow be-sech̄ 185
comyth a-ȝen be me
And I xal be both̄ blyth̄ *and* bowne[1]
þ*at* all̄ worchep to hy*m* be done
w*ith* reuerens I xal seke hy*m* sone
and hono*ur* hy*m* on kne. 190

¶ *and* þ*er*fore kyng*ys* I ȝow pray
whan ȝe haue don ȝ*our* jurnay

[1] *do* first written for *bo* and crossed through.

come aȝen þis same way
þe trewth̄ to me to telle
come *and* telle me as ȝe spede 195
and I xal qwyte ryght wel ȝ*our* mede
w*ith* gold *and* tres*our* *and* rych̄ wede
w*ith* furr*ys* rych̄ *and* wurth pelle.

1us Rex

¶ Kynge haue good day
I go my way 200[1]
to seche
lord of myght
he xal be ryght
our*e* leche.

ij^us^ Rex

¶ Kynge fful sterne 205
be felde *and* ferne
I goo
to sekyn a kynge
he takyth wonynge
in woo. 210

iij^us^ Rex

¶ If we hy*m* fynde
our*e* kynge ful kynde
be a may
Ffrom kynge *and* qwen
we comy*n* aȝen 215
þis day.

transient.[2] Herodes Rex

¶ A Fy Fy on talys þ*at* I haue ben tolde
her*e* be-forn my cruel kne
how xulde a barn̄ wax so bolde
be bestys yf he born be 220
Fo. 95[3] he is yong *and* I am̄ old̄
An hardy kyng of hye degre

[1] From line 199 to line 216 three lines are written in one, divided as before.
[2] This word is in a later hand, but not apparently that of the Ff. 95–6 scribe.
[3] Ff. 95 and 96 are interpolated and the writing is different, being the same as that of Fo. 112 and as that in which alterations are made in several plays, notably in the Shepherds and Magi plays and the Resurrection play. Ff. 95, 96, and 112 are not rubricated.

This daye the kynggys xal be[1] kold̄
If þey cum ageyne be me
My goddys I xall̄ vp-reyse 225
A derke devyll̄ with falsnese I saye
Shall cast a myst in þe kynggys eye
Be bankys *and* be[2] dalys drey·
þat be derk þei xall̄ cum this weyys.

primus [Rex

Go we to sek owr lord̄ *and* our lech 230
yon stere will̄ us tech þe weyis full̄ sone
To saue vs from̄ myschyff god̄ I here be-sech
on̄ to his joyis þat we may rech
I pray hem of this bone.

α *Tunc ibunt reges cum muneribus ad ihesum* et *primus rex dicit*

Heyle be þou kyng Cold̄ clade 235
heyll̄ with maydynnys mylk fade
heyll̄ I cum to þe with gold̄ glade
As wese wrytyng bere it record̄
gold̄ is þe[3] rycheste metall̄
And̄ to weryng most ryall̄ 240
gold̄ I gyff þe in this hall̄
And know þe for my lorde.

ij^us r[ex

Lorde I knele vpon my kne[4]
Sote encence I offere to the
Thow xalte be þe fyrst of hyȝ degre[5] 245
non so mekell̄ of myght
In goddys howse as men xall̄ se
Thow xalt honour þe trynite
iij personys in oon gode free
And all̄ oo lord̄ of myght. 250

Fo. 95v Lord̄ I knele downe be thy bede iij^us r[ex
In maydyns fleshe þou arte hede
Thy name xal be[6] wyde rede
And kyng ouer all̄ kynggys

[1] *xalbe* (one word) in M.S.
[2] A miswritten letter before *be* crossed through.
[3] Omitted and written above the line.
[4] *Knele* first written and crossed through.
[5] *dr* first written and crossed through.
[6] *xalbe* as above in MS.

Byttyr myre to þ*e* I brynge
ffor bytt*er* dent*ys* oñ þ*e* þ*ei* xall dyng
and byttyr deth xall be þ*y* endyng
And þ*er*for I make mornyng [1].

Maria

Kynggy*s* kynde
ffrome þ*e* fende
god yow defende
homwarde ȝe wende
and to yo*ur* plac*ys* ȝe lende
þ*at* ȝe xulde te*n*de.

p*ri*m*us* rex

Now haue we þ*e* place fownde
To Herode go we this stownde
w*ith* ow*r* word*ys* we were bownde
That we xulde cu*m* a-geyne
go we a pace *and* sey ow*r* spech
ffor we haue fownde ou*r* lord *and* lech
All þ*e* truth we wyll hem tech
how þ*e* kyng is borne of a quene.

ij^{us} rex

Myn hede is [2] hevy as lympe of leede
But yf I slepe I am a-drede
My witt xall fare þ*e* wurse
I wax hevy in lyme *and* flamke [3]
Downe I ley me vpoñ this banke
Vnder this bryght sterre i-wys.

iij^{us} rex

Broþ*er* I must lye þ*e* bye
I will go neu*er* ou*er* þ*is* stye
Tyll I haue a slepe
The yong kyng *and* h*is* moþ*er* mary
Saue vs all [4] frome eu*er*y velany
Now Cryst vs save *and* kepe.

P*ri*m*us* rex

Fo. 96 Such hevynese haue vs cawght [5]
I must drynk w*ith* yow a drawght

[1] First written *mornyngys* as the rhyme demands. The *ys* contraction has, however, been erased.

[2] *is* omitted and written above the line. [3] So in MS.

[4] *all* omitted and written above the line.

[5] This line, written too far to the left, is crossed out and rewritten.

To slepe a lytyll whyle [1]
I am hevy heed *and* footte
I xulde stumbyll at resch *and* root
And I xuld goo a myle. 290

x [2] *Hic dormiunt reges* et *venit Angelus* et *dicit eis*

Ange[lus

ȝe kynggys on this hill
Werk ȝe not aftyr herodys wyll
for [3] yf ȝe do he wyll yow kyll
þis day or nyght
My lorde yow sent this tydyng 295
to [4] Rest yow kynggys in rych clothyng
And when ȝe rysyn *and* goo to your dwellyng
Tak home þe wey full ryght

whether þat ȝe be [5] wakyn or slepe [6]
My lorde god xall yow keppe 300
In goode tyme ȝe dede down drepe
To take yowr rest
herode to þe devyll he tryst.

loke ferþer

Fo. 96v—left blank.

Fo. 97 [7] herowdys to þe devyl he tryste
to marre ȝow in a myrke [8] myste
my lord god is ful of lyste 305
to glathe ȝow for his geste.

¶ And þerfore kyngys whan ȝe ryse
Wendyth ferth be weys wyse
þer ȝour hall be [9] sett in syse
in dyverse londe 310

[1] *wyll* first written and crossed through.
[2] This and the similar sign on Fo. 95 are very roughly made.
[3] *for* omitted and written in the margin against *yf*.
[4] *to* omitted and written in the margin against *Rest*.
[5] *be* omitted and written above the line.
[6] *slepyne* first written and crossed through.
[7] With Fo. 97 the original hand is resumed.
[8] *myrke* altered by reviser to *thyrke*.
[9] Some erasure between *hall* and *be*.

ffadyr[1] of god in all thynge
hath ȝow grawntyd his swete blyssynge
he xal ȝow saue ffrom all shendynge
with his ryght honde.

Tunc surgant reges et dicat

Primus Rex

¶ A bryght sterre ledde us in to bedleem
A bryghtere thynge I saw in drem
bryghtere than þe sunne beem̃
An Aungell I saw ryght here
þe fayre flowre þat here gan falle
ffrom herowdys kynge he gan vs kalle
he taught vs hom tyll oure halle
A wey by a-nother mere.

2us Rex

¶ I saughe a syght[2]
myn hert is lyght
to wendyn home
god fful of myght
hath us dyght
ffro develys dome.

3us Rex

¶ Oure god I blysse
he sent us i-wys
his Aungel bryght
now we wake[3]
þe wey to[4] take
home full ryght.

Fo. 97v

¶ Symeon Justus

19

I Haue be prest in jherusalem here
And tawth goddys lawe many A ȝere
desyrynge in all my mende
þat þe tyme we[5] neyhand nere
In which goddys son xul Apere

[1] *þe* has been added before *ffadyr* by the reviser.
[2] From line 323 to line 334 three lines are written in one, divided as before.
[3] *be* inserted before *wake* by the Ff. 95–6 hand.
[4] *to* crossed out and *whe* written above the line by the Ff. 95–6 hand.
[5] So in MS. for *were*.

in erthe to take mankende
Or I deyd þ*a*t I myght fynde
my savyo*ur* w*ith* my*n* ey to se
but þ*a*t it is so longe be-hynde
it is grett dysco*m*forte · on to me. 10

¶ Ffor I wax old *and* wante my myght
and be-gynne to fayle my syght
þ*e* more I sorwe þis tyde
save only as I telle ȝow ryght
god of his grace · hath me hyght 15
þ*a*t blysful byrth to byde
Wherfor*e* now her*e* be-syde
to *sancta sanctorum* · wyl I go
to pray god to be my gyde
to comfort me aftyr my wo 20
here Symeon knelyth and *seyth*

¶ A gode god · in trinite
Whow longe xal I abyde the
tyl þ*a*t þ*ou* þ*i* son þ*ou*[1] doth sende
þ*a*t I in erth̄ myght hym se
good lord consydyr to me 25
I drawe fast to An ende
þ*a*t or my strenthis fro me wende
gode lorde send dow[2] þ*i* son
þ*a*t I w*ith* my ful mende
myght wurcheppe hy*m* if I con̄[3]. 30

Fo. 98 ¶ Bothe w*ith* my fete · *and* hondys to
to go to hym *and* handele also
my eyn to se hy*m* in certayn
my tonge for to speke hym to
and all̄ my lemys to werke *and* do 35
In his servyse to be bayn.
Send forth þ*i* son · my lord sovereyn
hastely A-non w*ith*-owte teryenge.
Ffor fro þis world · I wolde be ffayn
It is contrary to my levynge. 40

[1] So in MS. Should one *þ*ⁿ (*þ*ᵘ) be expanded to *þan*?
[2] So in MS. [3] Ang*elus* scribbled here in the margin.

Angelus[1]

¶ Symeon leff þ*i* careful stevene
Ffor þ*i* prayer is herd in hevene
to jher*usal*em ffast now wynne
And þ*er* xalt se ful evene
he þ*at* is goddys son ffor to nemene[2] 45
in þ*e* templ[3] þ*er* þ*ou* dwellyst inne
The dyrknes of orygynal synne
he xal make lyght *and* clarefye
and now þ*e* dede xal be-gynne
whiche hath be spokyn be p*ro*phecye. 50

Symeon

¶ A· I thanke þ*e* lord of grace
þ*at* hath grauntyd me tyme *and* space
to lyve *and* byde thys
and I wyl walk now to þ*e* place
Where I may se þ*i* sonys face 55
which is my joye *and* blys
I was nevyr lyghtere i-wys
to walke · nevyr here be-forn
Ffor a mery tyme now is
Whan god my lord is born. 60

Anna p*ro*phetessa

¶ Al heyl symeon what tydyng*ys* w*ith* ȝow
Why make ȝe al þis myrth now
telle me whedyr ȝe fare.

Symeon

Anne p*ro*phetes · *and* ȝe wyst whov
So xulde ȝe · I make A-vow 65
and all maner men þ*at* Are
Ffor goddys son as I declare
Is born to bye mankende
Fo. 98v Oure savyour is come to sesyn oure care
þ*er*fore haue I grett merth to wende. 70

¶ And þ*at* is þ*e* cawse I hast me
On to þ*e* temple hym to se
and[4] þ*er*fore lett me not good frende.

[1] Some words in another ink are written against *Angelus*, and have been clipped with the margin: *Cum do* . . .
[2] So in MS. [3] So in MS. [4] A miswritten *and* deleted.

Anna

Now blyssyd be god in trinyte
syn þ*at* tyme is come to be 75
and w*ith* ȝow wyl I wende
To se my savyo*ur* ende
and wurcheppe hym Also
w*ith* alł my wyłł *and* my ful mende
As I am bound · now wyl I do. 80
*Et tunc ibu*nt *Ambo ad templu*m. et *prophetissa.*

Symeon

¶ In þ*e* temple of god · who vndyrstod
þis day · xal be offeryd w*ith* mylde mood
which þ*at* is kynge of Alle.
þ*at* xal be skorgyd *and* shedde his blood
And Aftyr dyen on þ*e* rood 85
W*ith*-owtyn cawse to calle
Ffor whos passyon þ*er* xal be-ffalle
Swych a sorwe bothe sharpe *and* smerte
þ*at* as a swerd perce it xalle
ȝevene thorwe his moderys herte. 90

Anna p*rop*[hetissa

¶ ȝa þ*at* xal be as I wel fende
Ffor redempc*io*n of Alł mankende
þ*at* blysse ffor to restor*e*
Whiche hath be lost fro oute of mende
As be our*e* fadyr of our*e* owyn kende 95
Adam *and* Eue be-ffore.

Maria

¶ Joseph my husbond · w*ith*-owtyn mys
ȝe wote þ*at* ffourty days nere is
Sythe my sonys byrth fful ryght
Fo. 99 Wherfor*e* we must to þ*e* temple i-wys 100
þ*er* fore to offre our*e* sone of blys
up to his fadyr i*n* hyght
And I in goddys syght
puryfyed ffor to be
in clene sowle w*ith* al my myght 105
in p*re*sence of þ*e* trinyte[2]

[1] *fende* has been altered to *fynde* by another hand.
[2] Lines 103–6 are written in two long lines divided by black strokes without rubrication.

Joseph

¶ To be purefyed · haue ȝe no nede
ne þi son · to be offeryd · so god me spede
ffor fyrst þou art ful clene
Vndefowlyd in thought *and* dede
and a-nothyr þi son with-owtyn drede
is god *and* man to mene
Wherefore it nedyd not to bene
but to kepe þe lawe · on moyses wyse
Whereffore we xal take us be-twene *Et ibunt ad templum.*
Dowys *and* turtelys ffor sacrefyce.

Symeon

¶ All heyl my kyndely comfortour

Anna prophetissa

All heyl mankyndys creatoure

Symeon

All heyl þou god of myght

Anna prophetissa

All heyl mankyndys savyour

Symeon

All heyl bothe kynge *and* emperour

Anna prophetissa

All heyl As it is ryght

Symeon

All heyl · Also mary bryght

Anna prophetissa

All heyl · Salver of Seknes

Symeon

All heyl lanterne of lyght

Anne prophetissa

All heyl þou modyr of mekenes.

Maria

¶ Symeon · I vndyrstand *and* se
þat bothyn of my sone *and* me
ȝe haue knowynge clere
and also in ȝour compane
my sone desyryth for to be
and þer-ffore haue hym here.

Symeon

Fo. 99v Welcome prynce with-owte pere *et Accipiet Jhesum.*
Welcome goddys owyn sone
Welcome my lord so dere
Welcome with me to wone.

Suscepimus deus m*isericord*iam tuam.

¶ Lord god in mageste
we haue receyvyd þis day of þe
in myddys of þi temple here
thy grett mercy As we may se 140
therfore þi name of grett degre
be wurchepyd in alł manere
Over alł þis werde bothe fer *and* nere
ȝevyn on to þe vnterest ende
Ffor now is man out of daungere 145
and rest *and* pes to Alł man-kende.

Nunc dimittis seruu*m* tuu*m* do*m*ine ,' etcet*era*. *The psalme songyn every vers* and þer *qwyl Symeon pleyth with* þe *child* and *qwhan* þe *psalme is endyd · he seyth*

¶ Now lete me dye lorde *and* hens pace
Ffor I þi servaunt in þis place
haue sen my savyo*ur* dere
Whiche þ*ou* hast ordeyned be-forn þe face 150
Of Al mankynde þis tyme of grace
opynly to Appere
þi lyth is shynand clere
to Alł mankyndys savac*i*on
mary take ȝ*our* childe now here 155
and kepe wel · þis man is savac*i*on.

Anna pr*o*phe*tissa*

Fo. 100 ¶ Ne I rowth nere to dye Also[1]
Ffor more than ffowre skore ȝere *and* to
þis tyme hath bede to se
And sythe þ*at* it is come þer to 160
What goddys wyl is w*ith* me to do
ryght evyn[2] so mot it be.

Joseph

Take here these candelys thre
Mary · Symeon · And Anne
And I xal take þe fowrte to me 165
to offre oure child up thanne.

[1] The words *to Alł mankyndys* are written as beginning of the next line and crossed through.

[2] First written *ȝevyn*, the initial *ȝ* being crossed through.

Maria

¶ Hyest Ffadyr god of powere
ȝ*our* owyn dere son I offre ȝow here
as I to ȝ*our* lawe Am swor̃ñ
Receyve þ*i* childe in glad manere
Ffor he is þe fyrst þis childe so dere
þat of his modyr is born
but þow I offre hy*m* ȝow be-forn
good lord ȝit ȝyf me hym A-ȝen
Ffor my comforte were fully lorn
If we xuld longe A-sondyr ben.

Mari leyth þe childe on þe Autere

Joseph

¶ Sere prest of þe temple now
haue he[1] ffyff pens · vnto ȝow
oure childe Aȝen to take
It is þe lawe As ȝe woot how.

Capellan*us*

Joseph · ȝe An do ryght A-now
As for ȝ*our* childys sake
But othere offerynge ȝett must ȝe make
and þerfore take ȝ*our* sone mary
In meche joye ȝe may A-wake
Whylys he is in ȝ*our* company.

Maria

Fo. 100v ¶ Ther to I am ful glad *and* fayn
Ffor to receyve my childe Agayn
ellys were I to blame.
And Aftere-warde · ffor to be bayn̄
to offre to god in ful certayn̄
As in my sonys name
W*ith* ffowlys bothe wylde *and* tame
Ffor in goddys servyse[2] I xal nevyr irke.

Joseph

Lo mary · haue here tho same
to do þ*i* dewtys · of holy kyrke.

And þer *mary offeryth ffowlys on to* þe *Autere* and *seyth*

Maria

Aff-myghty-fful fadyr · mercyful kynge
Receyvyth now þis lytyl offerynge

[1] So in MS. for *here*.
[2] The word *servyse* omitted and written above the line.

Ffor it is þe fyrst in degre
þat ȝour lytyl childe so ȝynge 200
presentyth to-day be my shewyng
to ȝour hyȝ mageste
Of his sympyl poverte
be his devocion *and* my good wylle
Vpon ȝour Awtere receyve of me 205
ȝour sonys offrynge As it is skylle. 1468.[1]

Remainder of 100v—2⅝ inches—left blank.

Fo. 101 *Tunc respiciens senescallus vadyt ad herodem dicens*

¶ Senescallus

LOrd I haue walkyd be dale *and* hylle
And wayted as it is ȝour wyll.
The kyngys iii. stelyn awey full styll
thorwe bedleem londe
They wyl nevyr so moty the 5
Com̃ in þe lond of Galyle
Ffor to se ȝour fay[2] cete
ne dedys of ȝour honde.

herodes Rex[3]

¶ I ryde on my rowel ryche in my regne 20[4]
Rybbys fful reed with rape xal I rende 10
popetys *and* paphawkys I xal puttyn in peyne
with my spere prevyn̄ pychyn *and* to pende
The gomys with gold crownys ne gete nevyr ageyn̄[5]
to seke þo sottys sondys xal I sende
Do howlott howtyn hoberd *and* heyn̄ 15
Whan here barnys blede vndyr credyl bende
Sharply I xal hem shende
The knaue childeryn þat be
in all israel countre
thei xul haue blody ble 20
ffor on I calde vnkende.

¶ It is tolde in grw
his name xulde be jhesu

[1] This date appears to be in the same hand as the text, and it is looped in red as are the names of the speakers.

[2] So in MS. for *fayr*. [3] Red name loop omitted.

[4] A *1* has been erased here before the *2*. Cf. *Introduction, Numbering of Plays*.

[5] *gomys* has been altered to *gowys* (or *gollys*), *ne* has been crossed through, and *þei* inserted after *gete* in this line in darker ink in another hand.

i-fownde
to haue hy*m* ȝe gon
hewe þe flesch w*ith* þe bon
and gyf hy*m* wownde*
Now kene knyght*ys* kythe[1] ȝour*e* craftys[2]
and kyllyth knaue chylderyn *and* castyth hem in clay
Shewyth on ȝ*our* shulder*ys* scheldys *and* schaftys
Shapyht amonge schel chownys[3] ȧshyrlyng shray[4]
doth rowncys[5] rennyn w*ith* rakynge[6] raftys
tyl rybbys be to-rent w*ith* a reed ray
lete no barne[7] beleve on bete baftys
tyl a beggere blede be bestys baye
Fo. 101v Mahound þ*at* best may
I warne ȝow my knyght*ys*
A barn is born I plyghtys
Wolde clymbyn[8] kynge *and* knyt*ys*[9]
and lett my lordly lay.

¶ knyghtys wyse
chosyn[10] ful chyse
A-ryse a-ryse
and take ȝour*e* tolle
And every page
of ii · ȝer*e* Age
or evyr ȝe swage
sleyth ilke a fool[11].

¶ On of hem alle
was born in stalle
ffolys hym calle

* From line 22 to line 27 three lines are written in one; from line 37 to line 72 two lines are written in one.

[1] Altered by the later hand of Ff. 95–6 to *scharpe*.
[2] Altered by the later hand to *knyvys*.
[3] Altered by the later hand to *schel chowthys*.
[4] *shar* first written and crossed through by scribe.
[5] Crossed out and *your speris* written above by the later hand.
[6] Crossed out and *longe* (?) written above by the later hand.
[7] Crossed out and *chyld* written above by the later hand.
[8] *clyuer* . . . first written and crossed through in red ink.
[9] *ky* before *knytys* crossed through by scribe.
[10] Some miswritten letters (*cof*?) before *chosyn* crossed through in red ink.
[11] Or *foal*.

kynge in crowne[1]
With byttyr galle
he xall down falle
my myght in halle 55
xal nevyr go down.

i^us miles

¶ I xall sle scharlys
And qwenys *with* therlys
here knaue gerlys
I xal steke 60
Fforth wyl I spede
to don hem blede
thow gerlys grede
we xul be wreke.

ij^us miles

¶ Ffor swerdys sharpe 65
as An harpe
quenys xul karpe
and of sorwe synge
barnys ȝonge
they xul be stunge 70
thurwe levyr *and* lunge
we xal hem stynge.

Angelus

¶ A-wake joseph *and* take þ*i* wyff
thy chylde also ryd be-lyff
ffor kynge herowde *with* sharpe knyff 75
his knygh*tys* he doth sende
The fadyr of hevy*n* hath to þe sent
In to Egypte þ*at* þ*ou* be bent
Ffor cruel knygh*tys* þ*i* childe haue me*n*t
with swerde to sle *and* shende. 80

Joseph

¶ Awake good wyff out of ȝ*our* slepe
and of ȝ*our* childe takyght good kepe
Whyl I ȝ*our* clothis ley on hepe
and trus he*m* on þ*e* asse
kynge herowde þ*e* chylde wyl scloo 85
þ*er*fo*re* to Egypte muste we goo

[1] *Ang*elus scribbled here in the margin in a later hand.

An Aungel of god seyd me soo
and þ*er*-fo*re* lete us passe.

Fo. 102 *Tunc ibunt milites ad pueros occidendos* et *dicat prima femina*

i^a femina

¶ Longe lullynge haue I lorn
Alas qwhy was my baron born̄ 90
W*ith* swappynge swerde now is he shorn
þ*e* heed ryght fro þ*e* nekke
Shanke *and* shulderyn is al to torn
Sorwyn I se be-hyndyn *and* be-forn
both mydnyth mydday *and* at morn 95
of my lyff̄ I ne recke.

ij^a femina

¶ Serteynly I sey þ*e* same
gon is all̄ my good game
my lytyll̄ childe lyth all̄ lame
þ*at* lullyd on my pappys 100
my Ffourty wekys gronynge
hath sent me sefne ȝer*e* sorwynge
mykyl is my mornynge
and ryght hard arne my*n* happys.

i^us miles

¶ Lorde in trone 105[1]
makyght no mone
qwenys gyn grone
in werdl aboute[2]
upon my spe*re*
a gerle I be*re* 110
I dar*e* well̄ swe*re*
lett moder*ys* howte.

ij^us miles

¶ Lord we han spad
as ȝe bad
barnis ben blad 115
and lyne in dych̄
Fflesch̄ *and* veyn
han tholyd peyn

[1] From line 105 to line 128 two lines are written in one.

[2] The words *go hom wardys* are written here in the margin in the later hand.

and ȝe xul reyne
euer more rych̄. 120

herodes rex

¶ ȝe xul haue stedys
to ȝour medys
londys *and* ledys
ffryth *and* ffe
wele haue ȝe wrought 125
my ffo is sought
to deth is he brought
now come up to me.

¶ In sete now am I sett as kynge of myghtys most
Alł̄ þis werd ffor þer loue to me xul þei lowt 130
both̄ of hevyn *and* of erth̄ *and* of helle[1] cost
Ffor dygne of my dygnyte þei haue of me dowt
Fo. 102v þer is no lord lyke on lyve to me wurth̄ a toost
nother kyng nor kayser in alł̄ þis worlde abought
If any brybour[2] do bragge or blowe a-ȝens my bost 135
I xal rappe þo rebawdys and rake þem on rought
With my bryght bronde
þer xal be neythey[3] kayser nere knyge
But þat I xal hem down dynge
lesse þan he at my byddynge 140
be buxum to myn honde.

¶ Now my jentyl̄ *and* curteys knyghtys herke to me þis stownde
Good tyme sone me thynkygh̄ at dyner þat we were
Smertly þerfore sett a tabyl̄ a-non here fful sownde
Couerid with a coryous cloth̄ *and* with rych̄ wurthy fare 145
Servyse ffor þe lovelyest lorde þat levynge is on grownde
Beste metys *and* wurthyest wynes loke þat ȝe non spare
þow þat a lytyl pynt xulde coste a Mł̄. pownde
brynge alweye of þe beste for coste take ȝe no care
Anon þat it be done. 150

Senescallus

My lorde þe tabyl is redy dyght
here is watyr now wasch̄ forthryght

[1] *hevyn* first written and crossed through.
[2] *or bragges* first written after *brybour* and crossed through.
[3] So in MS.

now blowe up mynstrall *with* all *ȝour* myght
þe servyse comyth in sone.

herodes rex

¶ Now am I sett at mete
and wurthely servyd at my degre
Com forth knyght*ys* sytt down *and* ete
and be as mery as ȝe kan be.

i^us miles

Lord at ȝowr*e* byddynge we take our*e* sete
with herty wyl obey we the
þer is no lorde of myght so grett
thorwe all þis werde in no cou*n*tr*e*
Fo. 103 In Wurchepp to a-byde.

Herodes

(J quire) I was nevyr meryer her*e* be-forn
Sythe *þat* I was fyrst born
than I am now ryght in þis morn
in joy I gynne to glyde.

Mors

¶ Ow I herde a page make preysyng of p*ri*de
all prynces he passyth he wenyth of powste
he wenyth to be *þe* wurthyest of all þis werde wyde
kynge ovyr All kyng*ys* *þat* page wenyth to be
He sent into bedlem to seke on every syde
Cryst for to qwelle yf þei myght hym se
but of his wykkyd wyl lurdeyn ȝitt he lyede
goddys sone doth lyve *þer* is no lorde but he
Ou*er* all lordys he is kynge
I am deth goddys masanger*e*
All myghty god hath sent me her*e*
ȝon lordeyn to Sle w*ith*-owty*n* dwer*e*
ffor his wykkyd werkynge.

¶ I am sent fro god deth is my name
All thynge *þat* is on grownd I welde at my wylle
both man *and* beste *and* byrdys wylde *and* tame
Whan *þat* I come them to · w*ith* deth I do them kylle
Erbe gres *and* tres stronge · take hem all in same
ȝa *þe* grete myghty Okys · w*ith* my dent I spylle
what man *þat* I wrastele w*ith* he xal ryght sone haue schame
I ȝeve hym such a trepett · *he* xal evyr mor*e* ly stylle

* Ffor deth kan no sporte
wher I smyte þ*er* is no grace 190
Ffor after my strook man hath no space
to make amendys ffor his trespace
but god hym graunt comforte

Fo. 103v ¶ Ow se how prowdely ȝon kaytyff sytt at mete
of deth hath he no dowte he wenyth to leve evyr-more 195
to hym wyl I go *and* ȝeve hy*m* such An hete
þ*at* all þ*e* lechis of þ*e* londe his lyf xul nevyr restore
A-ȝens my dredful dentys it vaylyth nevyr to plete
or I hym part fro I xal hym make ful pore
Alł þ*e* blood of his body I xal hym owt swete 200
Ffor now I go to sle hy*m* w*ith* strokys sad *and* sore
Bothe hym *and* his knyght*ys* alł ¶ þis tyde[1]
I xal hem make to me but thrałł
w*ith* my spere sle hem I xałł. 205
and so cast down his p*r*ide
herodes Rex

¶ Now kende knyght*ys* be mery *and* glad
W*ith* alł good diligens shewe now su*m* myrth
Ffor be gracyous mahound more myrth nev*er* I had
ne nevyr more joye was inne from tyme of my byrth 210
Ffor now my fo is ded *and* prendyd as a padde
aboue me is no kynge · on grownd nere on gerth[2]
merthis þ*er*fore make ȝe *and* be ryght no thynge sadde
spare nother mete nor drynke *and* spare for no dyrthe
of wyne nor of brede 215
Ffor now am I a kynge alone
So wurthy as I · may þ*er* be none
þ*er*fore knyght*ys* be mery echone
ffor now my ffo is dede.
i^{us} miles

¶ Whan þ*e* boys sprawlyd at my sperys hende 220
by sathanas oure syre it was a goodly syght
A good game it was þ*at* boy for to shende
þ*at* wolde abene oure kynge *and* put ȝow fro*m* ȝo*ur* ryght

* Paragraph mark erased here.

[1] Halliwell prints this line before the line *Bothe hym etc.* which is its right place in the stanza.

[2] The *e* written by the scribe over some other letter *r* or ?*y*.

ij^{us} miles

Fo. 104 Now trewly my lorde þe kynge we had ben vn-hende[1]
and nevyr non of us Able for to be a knyght 225
If þat Any of us to hem had ben a frende
and asavyd Any lyff a-ȝen þi mekyl myght
Ffrom deth hem to flytt.

herodes Rex

Amonges all þat grett rowthte
he is ded I haue no dowte 230
þerfore menstrell rownd a-bowte
blowe up a mery fytt.

Hic dum buccinant mors interficiat herodem et duos milites subito et diabolus recipiat eos.

Diabolus

¶ All oure all oure þis catel is myn
I xall hem brynge on to my celle
I xal hem teche pleys fyn 235
and showe such myrthe as is in helle
It were more bettyr Amonges swyn
þat evyr more stynkyn þer be to dwelle
ffor in oure logge is so gret peyn
þat non erthely tonge can telle 240
with ȝow I go my way
I xal ȝow bere forth with me
and shewe ȝow sportys of oure gle
of oure myrthis now xal ȝe se
and evyr synge welawey. 245

Mors

¶ Off kynge herowde all men beware
þat hath rejoycyd in pompe *and* pryde
Ffor all his boste of blysse ful bare
he lyth now ded here on his syde
Ffor whan I come I can not spare 250
Fro me no whyht may hym hyde
now is he ded *and* cast in care
In helle pytt evyr to A-byde
his lordchep is al lorn

[1] Altered by the later hand to *vnkende.*

Fo. 104v Now is he as por*e* as I 255[1]
wormys mete is his body
his sowle in helle ful peynfully
of develis is al to-torn̄.

¶ Att̄ men dwellyng upon þe grownde
Be-war*e* of me be my*n* cou*n*cel 260
Ffor feynt felachep in me is fownde
I kan no curtesy as I ȝow tel
Ffor be a man nevyr so sownde
of helth̄ in herte nevyr so wel
I come sodeynly w*ith*-in a stownde 265
me w*ith*-stande may no castel
my jurnay wyl I spede.
of my comyng no man is war*e*
Ffor when men make most mery far*e*
þan sodeynly I cast he*m* in care 270
and sle þe*m* evy*n* i*n* dede.

¶ Thow I be nakyd *and* por*e* of array
and wurmys knawe me al a-bowte
ȝit loke ȝe drede me nyth *and* day
Ffor whan deth comyth ȝe stande i*n* dowte 275
Evyn lyke to me as I ȝow say
shult̄ att̄ ȝe be her*e* in þis rowte
Whan I ȝow chalange at my day
I xal ȝow make ryght lowe to lowth
and nakyd for to be 280
Among*es* wormys as I ȝow telle
Vndyr þ*e* erth̄ xul ȝe dwelle
and thei xul Etyn both̄ flesch̄ *and* felle
As þei haue don me.

Remainder of Fo. 104v—2¼ inches—and Ff. 105–105v left blank.

[1] From line 255 to line 258 two lines are written in one, divided by black strokes without rubrication.

Modo de doctoribus disputantib*us* cu*m* jh*e*su in templo.[1]

Fo. 106

¶ P*rim*u*s* doctor

¶ Scripture sacre e*sse* dinoscim*ur* doctos
We to · bere þe belle of aꝗ manere clergyse.

ij^us doctor

Velud rosa omniu*m* flor*um* flos
lyke on to us was nevyr clerke so wyse.

i^us doctor

Loke what scyens ȝe kan devyse
of redynge wrytynge *and* trewe ortografye
Amonge*s* aꝗ [2] clerkys we bere þe prysse
of gramer cadens and of p*ro*sodye.

ij^us doctor

¶ No clerke Abyl to bere oure book
of versyfyeng nor of other scyens
of Swete musyke who so wyꝗ look
seke no ferther but to oure presens
Of dyaletyk we haue þe hyȝ excellence
of sophestrye · logyk *and* phylosophye
Ageyn oure argemente [3] is no recystence
In metaphesyk [4] ne astronomye.

i^us doctor

¶ Of calculac*i*on and negremau*n*cye
Also of Augrym̃ *and* of asmatryk
O [5] lynyac*i*on þ*a*t longyth to jematrye
of dyetis *and* domys þ*a*t longyth to phesyk
In aꝗ þis scyens is non us lyke
In caton̄ gryscysme nor doctrynal
and ffor endytynge w*ith* retoryke [6]
þe hyest degre is oure be call [7].

[1] This title is not written in larger form in the MS.

[2] The *ꝗ* is written over some other letter, and some letter (? i) is written small above the *a*.

[3] *argeru* first written and crossed through.

[4] An erasure in the word indicates some difficulty with 'trewe ortografye'.

[5] So in MS. for *Of*.

[6] *retr* first written; some letter written above between *t* and *r* and then the whole crossed through.

[7] *be call* crossed out and *over all* written above by another hand, possibly the Ff. 95–6 scribe.

ij^{us} doctor

¶ In grett canon̄ *and* in Cevyle lawe
Also in scyens of polycye
Fo. 106v Is non to us wurthe An hawe
of all̄ cu*n*nynge we bere *þe* maystrye
Therfor*e* in þis temple we sytt on hye
and of most wurchep kepe *þe* souereynte
þer is on erthe no man so wurthye
þe hyȝ stat to holdyn as we tweyn be.

Jh*es*us

¶ Om*n*is sciencia a d*omi*no deo est
Al wytt *and* wysdam of god it is lent[1]
Of all̄ ȝ*our* lernynge w*ith*-inne ȝ*our* brest
thank hyghly *þat* lord *þat* hath ȝow sent
thorwe bost *and* pryde ȝ*our* soulys may be shent
Of wytt *and* wysdome ȝe haue not so mech̄
but god may make at hese entente
of all̄ ȝ*our* connynge many man ȝow lech̄.

i^{us} doctor

¶ Goo hom lytyl babe *and* sytt on *þi* moder*ys* lappe
and put a mokador[2] a-forn *þi* brest
and pr*ay* *þi* modyr to fede *þe* w*ith* *þe* pappe
of *þe* for to lerne we desyr*e* not to lest[3].

ij^{us} doctor

Go to *þi* dyner for *þat* be-hovyth the best
whan *þou* art a-threste *þan* take *þe* A sowke
Aftyr go to cradyl *þer*-in to take *þi* rest
Ffor *þat* canst *þou* do bettyr *þan* for to loke on book.

Jh*es*us

¶ Stondynge *þat* ȝe be so wytty *and* wyse
Can ȝe owth tellyn how *þis* werde was wrought
how longe xal it laste can ȝe devyse
w*ith* all̄ *þe* cu*n*nyge *þat* ȝe han sought.

i^{us} doctor

Nay all̄ erthely clerkys *þat* telle can nought
it passyth our*e* wytt *þat* for to co*n*tryve

[1] *sent* first written and crossed through.
[2] The *o* in *or* is written over some other letter.
[3] *lyest* first written and crossed through.

Fo. 107 It is not possyble A-bought to be brought
þe worldys endyng no man kan dyscryve [1].

Jhesus

¶ How it was wrought *and* how longe it xal endure
þat I telle [2] be good delyberacion
not only þer-of but of every creature
How it is wrought I knowe þe plasmacion.

ij^us doctor

Of þi wurdys I haue skorne *and* deryson [3]
how schulde a chylde þat nevyr lettyr dude [4] lere
Com to þe wytt of so hyȝ cognysion̄
Of þo grete wurkys þat so wundyrfull [5] were.

Jhesus

¶ All thynge is brought to informacion
be thre personys · oo · god in trynite
and on of þo thre hath take in-carnacion
bothe flesch *and* blood of a mayd ffre
And be þat myght of þo personys thre
hevyn *and* erth *and* all thynge is wrought
and as it plesyth þat hyȝ mageste
all thynge xal leste *and* lenger nowght.

i^us doctor

¶ I grawnt weyl all thynge þat god dyde make
and with-owtyn hym no thynge may be
But o thynge þou seydyst *and* þat I for-sake
þat oo god alone was personys thre
Ryght on-possyble þat is to me
that on is thre I kan not thynke
If þou canst preve it a-non lett se
Ffor in oure hertys it may nevyr synke.

Jhesus

Fo. 107v ¶ In þe sunne · consydyr ȝe thyngys thre
The splendure þe hete *and* þe lyght
as þo thre partys but oo sunne be
Ryght so thre personys be oo god of myght.

[1] The *s* in the word has a double stroke and might be read as *ss*.

[2] The word *can* has been written above the word *telle* by another hand. The tailed *n* (ŋ) of this *can* is found again in a later correction on Fo. 135. It is not found in Ff. 95, 96, or 112.

[3] So in MS. for soñ (= sion).

[4] *dude* has been changed to *dyde* in darker ink and another hand.

[5] The *u* of *full* written by the scribe over the beginning of an *o*.

ij^{us} doctor

In very feyth þis reson is ryght 85
but ȝitt fayr babe oo thynge we pr*a*y ȝow
what do all þo thre p*er*sonys hyght
Vs to enforme · ȝe sey to me now.

Jh*e*sus

¶ The fyrst is calde þe fadyr of myght
þe secunde þe sone of wysdam *and* wytt 90
þe holy gost þe iij^{de}. of grace he is hyght
and in oo substau*n*s all these iij. be knyt.

i^{us} doctor

A-nother questyon[1] I Aske ȝow ȝitt
ȝe seyd on of þese iij. toke flesch *and* blood
and sche a clene mayde I kan not be-leue it 95
clene mayde *and* modyr nevyr ȝit in oo p*er*sone stood.

Jh*e*sus

¶ Lyke as be su*n*ne doth p*er*ysch[2] þe glas
þe glas not hurte of his natur*e*
ryght so þe godhed entryd has
þe virgynes wombe *and* sche mayd pur*e* 100
That maydonys childe xal do grett cur*e*
co*n*victe þe devyl in þe opyn felde
and with his bolde[3] berst fecch hom his creature
mankende to saue his brest xal be þe shelde.

ij^{us} doctor

¶ This childys doctryne doth passe our*e* wytt 105
Su*m* Au*n*gel of hevyn I trowe þ*a*t he be
Fo. 108 But blyssyd babe of oo dowte ȝitt
We pray ȝow enforme us for charyte
Which toke flesch of þe p*er*sonys thre
Ageyn þe fende to holde such batayle. 110

Jh*e*sus

The secunde p*er*sone for sothe is he
xal fray þe fende w*ith*-owte fayle.

i^{us} doctor

¶ Why rather he than Any of þ*a*t other[4]
The fyrst or þe thyrde why come they nowth.

[1] *ȝitt* first written after *questyon* and crossed through.

[2] *perysch* crossed through and *pers* written above in another hand not certainly identifiable with that of Ff. 95, 96.

[3] *bolde* similarly replaced by *bluddy*.

[4] *other* first written and corrected by scribe.

Jhesus

this is þe cawse why sertys *and* non other 115
Ageyn þe secunde þe trespas was wrought
Whan þe serpent adam to synne browth
He temptyd hym nowght be þe faderys myght
Of þe gostys goodnes spak he ryght nowght
but in connynge he temptyd hym ryght. 120

¶ Myght is þe faderys owyn propyrte
to þe gost apperyd is goodnes
in none of these tweyn temptyd he
mankende to synne whan he dede dresse
To þe sone connynge doth longe expres 125
ther with þe serpent dyd Adam A-say
Ete of þis Appyl he seyd no lesse
and þou xalt haue connynge as god verray.

¶ Þus þe secunde person Attrybute
Was only towchyd be temptacion 130
Wherfore hym self wyl hold þe sewte
And kepe his propyrte fro maculacion.

ij^us doctor

Fo. 108v This is An hevynly declaracion
oure naturall wytt it doth excede
so ȝonge a childe of such informacion 135
in al þis werld neuyr er non ȝede.

i^us doctor

¶ We be not worthy to kepe þis sete
Whyll þat oure mayster is in presens
þe maystry of us þis childe doth gete
We must hym wurchep with hyȝ reverens 140
Come forth swete babe of grett excellens
þe whysest clerke þat evyr ȝett was born[1]
to ȝow we ȝeve þe hyȝ resydens
Vs more to teche as ȝe haue done be-forn.

hic adducunt ihesum inter ipsos et in scanno altiori ipsum sedere faciunt ipsis in inferioribus scannis sedentibus et ait ij^us doctor

2us doctor

¶ So ȝonge A chylde suche clergye to reche 145
and so sadly to sey it we woundyr sore

[1] *bar* miswritten before *born* and crossed through.

Who was ȝowr*e* mayster who dede ȝow teche
of what man had ȝe þis wurthy lor*e*.

Jh*es*us

My wytt *and* my lernynge is no ȝonge stor*e*
or þis worde was wrought aꝗ þinge dede I knowe 150
Ffyrst or ȝe wor*e* borne ȝer*ys* many scor*e*
thorwe þ*e* myght of my fadyr my wytt in me dede flowe[1].

i^{us} doctor

¶ Or þ*at* we weryn born nay þ*at* may nat be
þ*e* ȝongest of us tweyn is iij. scor*e* ȝer*e* of Age
and þi-selfe art but a chylde al men may wel se 155
late camst out of cradyl as it semyth be þ*i* vesage.

Jh*es*us

I am of dobyl byrtꝗ *and* of dobyl lenage
Ffyrst be my fadyr I am w*ith*-out gynnynge
Fo. 109 And lyke as he is hendeles in his hyȝ stage
So xal I also neuyr mor haue endynge. 160

¶ Ffor be my ffadyr kynge celestyaꝗ
W*ith*-out begynnyng I am endles
but be my modyr þ*at* is carnaꝗ
I am but xij ȝer*e* of age þ*at* is exp*res*
My body of ȝoughȝ[2] doth shewe wyttnes 165
Which of my modyr her*e* I dude take
but my*n* hyȝ godhede þis is no lesse
aꝗ thynge in þis world for sothe dude I make.

ij^{us} doctor

¶ Be ȝ*our* ffadyr þ*at* endles is
Who is ȝ*our* modyr telle us we p*ray*. 170

Jh*es*us

Be my fadyr þ*e* hyȝ kynge of blys
A modyrles chylde I am veray.

i^{us} doctor

Who was ȝ*our* ffadyr to us than say
be ȝ*our* modyr A woman þ*at* was.

Jh*es*us

I am fadyrles as for þ*at* may 175
of fleschly luste she dude nevyr trespas

[1] The words *dede flowe* are written above the line by the scribe, some miswritten words (*dede flawe ?*) being crossed through.

[2] The *h* appears to have been written over some other letter.

ij^us doctor

¶ Telle us I p*ray* ȝow what is ȝ*our* name
what hyght ȝoure modyr telle us Also.

Jh*es*u

Jh*es*u of Nazareth I am þe same
born of a clene mayd p*ro*phetys seyd so 180
Ysaye seyd þus · Ecce virgo
A mayd xal conceyve in clennes a chylde
ȝitt ageyn nature *and* alkende loo
ffrom alł wem of synne pure *and* vndefylde.

Fo. 109v ¶ Mary þe chylde of Joachym And anne 185
ys þ*at* clene mayd *and* here childe am I
þe frute of here wombe xal saue euery ma*n*ne[1]
Ffrom þe grett dowte of þe ffyndys tormentry.

i^us doctor

Alł þe clerkys of þis worlde trewly
can not brynge this to declarac*i*on 190
lesse þan þei haue of god Almyghty
Su*m* influens of informac*i*on

ij^us doctor

¶ No jentyl jh*es*u we ȝow pray
Whyl þ*at* we stodye a whyle to dwelle
In cas mo dowtys þ*at* we fynde may 195
þe trewtł of hem ȝe may us telle.

Jh*es*u

Goo take ȝ*our* stodye *and* avyse ȝow wełł
And alł ȝ*our* leysere I xal a-byde
If Any dowtys to me ȝe mełł
þe trewtł þerof I xałł vn-hyde. 200

Maria

¶ Alas Alas my*n* hert is wo
My blyssyd babe a-wey is went
I wott nevyr whedyr þ*at* he is go
Alas for sorwe my*n* hert is rent
Jentyl husbond haue[2] hym sent 205
Out on herrande to Any place

[1] MS. mane· (with circumflex and dot over *a*). The *a* and the *n* are run together and the *e* appears an after thought.

[2] *yow* is written above the line after *haue*, probably by the later hand of Ff. 95, 96.

but yf ȝe knowe were her[1] ys bent
my*n* hert for woo A-sondyr wyl race.

Joseph

¶ On my massage I hym not sent
forsothe good[2] wyff in no degre 210
how longe is it þ*a*t he hens went
What tyme dude ȝe ȝ*our* childe last se.

Maria

Fo. 110 Trewly gode spowse not þese days thre
.þ*e*rfor*e* my*n* herte is cast in car*e*
hym for to seke wher so he be 215
in hast good husbonde lete us forth far*e*.

Joseph

¶ Than to hjerusalem[3] lete us streyte wende
Ffor kynred gladly to-gedyr wole gon̄
I hope he þer[4] w*it*h su*m* good ffrende
þ*e*r he hath cosynys rygh̄t many on̄. 220

Maria

I am aferde þ*a*t he hath fon̄
Ffor his grett wytt*ys* *and* werkys good
lyke hym of wytt ffor-soth is non̄
Euery childe w*it*h hy*m* is wroth̄ *and* wood.

¶ Also my babe my blys my blood 225
Whedyr art þ*ou* þus gon fro me.
my sowle my swetyng my frute my*n* ffood
Send me ssu*m* wurd where þ*a*t þ*ou* be
Telle me good ser*ys* for charyte
Jh*e*su my childe þ*a*t babe of blysse 230
Amonge þis companye dude ȝe hy*m* se
Ffor godys hyȝ[5] loue telle where he is.

i^us doctor

¶ Of oo qwestyon̄ I am be-thought
Alh̄ of ȝ*our* modyr þ*a*t blyssyd may
In what governauns is she brought 235
How is sche rewlyd be nyght *and* day.

[1] So in MS. for *he*. [2] *god* first written and crossed through.
[3] The *j* written over an *e*.
[4] *is* is written above the line before *þer* in another ink.
[5] This word is crossed out in different ink.

Jhesu

An old man joseph · as I ȝow say
here weddyd be meracle on to his Wyff
Here for to fede *and* kepe Alway
and bothyn in clennesse be maydonys Olyff. 240

ij^us doctor

Fo. 110v [marked 110] ¶ What nede was it here to be wedde
On to A man of so grett Age
lesse þan þei myght bothe ago to bedde
and kept þe lawe of maryage.

Jhesus

To blynde þe devyl of his knowlache 245
and my byrth from hym to hyde
þat holy wedlok was grett stopage
þe devyl in dowte to do A-byde.

¶ Also whan sche xulde to egypte gon[1]
and fle from herowde for dowte of me 250
be-cawse she xulde nat go Alon
Joseph was ordeyned here make to be
my ffadyr of his hyȝ mageste
here for to comforte in þe way
these be þe cawsys as ȝe may se 255
why joseph weddyd þat holy may.

Maria

¶ A dere childe dere chylde why hast þou þus done
Ffor þe we haue had grett sorwe *and* care
thy ffadyr *and* I thre days haue gone
Wyde þe to seke of blysse[2] ful bare. 260

Jhesus

Why haue ȝe sought me with evy[3] fare
Wete ȝe not wele I muste been[4]

[1] The words *Jhesus adhuc* first written in the margin against this line with red loop as the name of speaker, and the line separating one speech from another drawn above this stanza; words and line cancelled in red ink.

[2] Crossed out and *comfort* roughly written above by a (? that of Ff. 95, 96) later hand.

[3] Altered to *hery* by the later hand of Ff. 95, 96.

[4] Or *ben*. Two *es* seem to be run together. A final *e* is marked with a dot for deletion.

A-monge hem þat is my fader*ys* ware
his gostly catel for to ovyrsen̄.

Maria

¶ ȝ*our* ffader*ys* wyl must nedys be wrought 265
It is most wurthy þat it so be
ȝitt on ȝ*our* modyr haue ȝe su*m* thought
And be nevyr mor*e* so longe fro me
Fo. 111 As to my thynkynge these days thre
þat ȝe Absente haue ben A-way 270
be mor*e* lengere in þ*er* degre
þan Att̄ þe space of xij ȝer*e* day.

Jh*es*us

¶ Now ffor to plese my modyr mylde
I xal ȝow folwe w*ith* obedyence
I am ȝ*our* sone *and* subjecte childe 275
and Owe to do ȝow hyȝ reverence
home w*ith* ȝow I wyl go hens
Of ȝow clerkys my leve I take
Euery childe xulde w*ith* good dyligens
his modyr to plese his owyn wyl forsake [1]. 280

i^us doctor

¶ O blyssyd jh*es*u w*ith* ȝow we wende
of ȝow to haue more informac*i*on
Fful blyssyd is ȝ*our* modyr hende
of whom ȝe toke ȝ*our* incarnac*i*on
we p*ray* ȝow jh*es*u of consolac*i*on 285
At our*e* most nede of ȝow to haue
att̄ þ*at* hath herd þis consu*m*mac*i*on
of þis pagent ȝ*our* grace þem saue.

Amen.

[Remainder of Fo. 111—2⅝ inches—and Fo. 111v—except for title of next play added by reviser and a scribbled name, He. Kinge the yownger,—left blank.]

[1] *to* added above the line between *wyl* and *forsake* in darker ink.

Fo. 111v

Hic Incipyt Joh*ann*es Baptysta.

Fo. 112[1]

Joh*ann*es[2]

Ecce vox clamantes in deserto
I am þ*e* voys of wyldirnese
þ*at* her spekyth *and* prechych yow to
loke ȝe for-sake aſt wrecchidnesse
ffor sake aſt synne þ*at* werkyth woo
And turne to[3] vertu and holynese
Beth clene of levyng in yo*ur* sowle also
Thaꝰ xaſt ȝe be savyd from peynfulnese
Of fyer*e* brynnyng in heſt
If þ*at* ȝe for-sak synne
hevyꝰ blysse xaſt ȝe wyne
Drede ȝe not þ*e* devyll*ys* gynne
w*ith* Angeſt*ys* xaſt yow dweſt.

Penitenci*am* nu*nc* agite
Appropinquabit regnu*m* celor*um*
ffor yo*ur* trespas penaunce do ȝe
and ȝe xaſt wyꝰ hevyꝰ dei deor*um*
In hevyn blyse ye xaſt wyn to be
Among þ*e* blyssyd *com*pany om*n*ium sup*er*nor*um*
þ*er* as is aſt m*er*th joye *and* glee
Inter agmina angelor*um*
In blyse to a-byde
Baptyme I cownceſt yow[4] for to take
And do penaunce for yo*ur* synnys sake
and for yo*ur* offens ame*n*dys ȝe make
yo*ur* synnys for to hyde.

I gyff baptymꝰ in wat*er* puer*e*
þ*at* is callyd flomjordoꝰ
My baptymꝰ is but sygnyfur*e*

[1] This folio is an interpolated leaf of different paper, the wire lines of which correspond with those of Ff. 95 and 96. The handwriting is that found on Ff. 95 and 96. There is no rubrication.

[2] This name stands at the top of the page.

[3] *to* omitted and written over the line.

[4] *take* first written before *for* and crossed through.

Of h*is* baptym̃ þ*at* h*is* lyke hath no*n* 30
Fo. 112v He is a lord̃ of gret valo*ur*
I am not worthy to[1] on̄-bokyll̃ h*is* schon̄
ffor he xall̃ baptyze as[2] seyth scrypto*ur*
þ*at* comyth of hem all̃ eu*er*y-chone
In þ*e* holy goost 35
he may dampne *and* he may save
all̃ goodnese of hem we haue
þ*er* may no[3] ma*n* his werk*ys* deprave
ffor he is lord̃ of myght*ys* most.

hic accedit Jhesus ad Johannem quem intuens Johannes dicat digito demonstrans Jhesum

Ecce Agnus dei qui tollit p*eccat*a mu*n*di 40
Be-holde þ*e* lombe of god̃ is this
þ*at* comyth now here be-forne
þ*e* wich xall̃ wasch̃ þ*e* worldys mys
and saue all̃ þ*at* that was for-lorne
This same lombe for soth it is
þ*at* of a mayd̃ full̃ clene was borne 45
Shamfull̃ deth̃ þ*is* lambe[4] i-wys
xall̃ suffer for us *and* be all̃ to-torne
And rent on̄ a roode
he xall̃ suffer for ma*n*nys sake
lytyll̃ rest *and*[5] moch gret sorow *and*[5] wrake 50
hys bake xall̃ be bowndyn to a stake
And̃ betyn̄ owt all̃ his bloode.

Jhesus

Fo. 113[6] ¶ Joh*an* Baptyste my*n* owyn good ffrende
þ*at* ffeythffully doth̃ prech̃ my wylle
I the thanke w*ith* all̃ my mende 55
Ffor þ*at* good servyse þ*ou* dost me tylle
thy desyre is synne to shende
All̃ synful lyff[7] þ*ou* woldyst spylle.
thyn entente hath a good hende

[1] *v* first written before *on̄* and crossed through.
[2] *scrypture* first written before *seyth* and crossed through.
[3] *no* omitted and written above the line.
[4] The *b* is written over some other letter.
[5] The words from *and* to *and* are written over an erasure.
[6] The original hand of the MS. begins again here. [7] MS. synfullyll̃.

þe lawe of god þou dost fful-fylle
þis tyde
Baptym̃ to take I come to the
and conferme þat sacrement þat newe xal be
In flomjordon þou baptyze me
In water þat is wyde.

Joh*ann*es

¶ My lorde god þis be-hovyth me nought
w*ith* my*n* hondys to baptyze the
I xulde rather of the haue sought
holy baptym þan þou of me.

Jh*es*us

Suffyr now Joh*an* my wyl were wrought
alł ryght-ffułłnes[1] þus ffulfyłł we
me to baptyze take þou no dowth
þe vertu of mekenes here tawth xal be
Euery man to lere
And take ensawmple here by me
How mekely þat I come to þe
Baptym confermyd now xal be
me to baptyze take þou no dwere[2].

Johannes

¶ Alł men may take exaunple[3] lo
of lowly mekenes evyn ryght here
be oure lorde god þat comyth me to
hese pore servaunt *and* his su . . tere[4]
Fo. 113v Euery man[5] lere to worke ryght so
Bothe Kynge *and* Caysere *and* gret Empere[6]
be meke *and* lowe þe pore man to
And put out pryde in alł manere
God doth here þe same
To þi byddynge my lord so dere
I me obey w*ith* gladsu*m* chere
And baptyze the w*ith* watyr clere
Eu*er* halwyd be þi name.

[1] *ffl* first written instead of *fful* and crossed through.
[2] *dwere* crossed through and *fere* written after it by the later hand.
[3] So in MS. [4] Some letter, possibly a *w*, has been erased between *u* and *t*.
[5] *man* omitted and written above the line.
[6] The contraction mark might be expanded Emper*er*—the same being used for (r) *e* and *er*—but the rhyme demands *Empere*.

Spiritus sanctus hic descendat super ipsum et deus ¶ pater celestis dicet in celo.

¶ This is my welbelovyd chylde
Ou*er* whom my spryte doth ou*er* sprede
Clene *and* pur*e* And vndefylyd
of body of sowle ffor thought for[1] dede 95
That he is buxhum meke *and* mylde
I am wel plesyd w*ith*-owtyn drede
Wysly to wysse ȝow ffrom weys wylde
to lysten his lor*e* alł men I rede
And ȝour*e* erys to herke 100
Take good hede what he doth p*re*che
and ffolwyth þ*e* lawys þ*at* he doth teche
Ffor he xal be ȝ*our* Alther-is leche
to saue ȝow from deuelys derke.

Johannes bap[tista

¶ Her*e* I se w*ith* opyn syght 105
The sone of God þat þ*ou* erte[2]
the hooly goost ou*er* the doth lyght
þ*i* fader*ys* voys I her*e* fful smerte
The childe of god as I þ*e* plyght
Fo. 114 þ*at* þ*ou* be whilys I am qwerte 110
I xalł wyttnes to every whyght
and teche it trewly w*ith* alł my*n* hert
To sese it wer*e* grett synne
Ffor goddys sone I wurchypp the
Ffrom hevyn þ*in* hyȝ mageste 115
thu comyst hedyr ffrom dygnite
mannys sowle to wynne.

Jh*esus*

¶ Joh*an* Baptyste þ*ou* be wyttnes
the trewth loke þ*at* þ*ou* nat hyde
Ffor now I passe forth in to wyldernes 120
the holy gost xal be my gyde.

hic ihesus transit in desertum dicens et cetera.

In whylsum place of desertnes

[1] *of* first written and crossed through ; and *for* written above the line.
[2] *arte* first written and corrected by scribe to *erte.*

xl^ti days a terme[1] ful wyde
and ffourty nyght*ys* both more *and* lesse
W*ith*-owtyn bodyly ffode þ*er* to a-byde 125
Ffor man þus do I swynke
In to deserte I[2] passe my way
ffor mannys sake as I ȝow say
xl^ti nyght*ys and* xl^ti day
I xal nowther ete nor drynke. 130

Joh*an* baptyst

¶ In place where I passe · wyttnes I bere
the trewth xal I telle where-so-evyr I go
þ*at* cryst þe sone of god is be-come oure fere
clad in oure clothynge to sofer for us wo
I baptyzid w*ith* myn owyn handys cryst jh*es*u ryght here 135
and now he is to wyldyrnes penawns þ*er* to do
In-formyng so all us þ*at* lord þ*at* hath no pere
to do for oure trespace penawnce here also
Fo. 114v Of penawnce do I p*re*che
In wyttnes ryght be this 140
þ*at* what man for his mys
doth penawns here i-wys
his sowle he doth wel leche.

¶ All men on ground þ*at* be ȝitt on lyue
Ffor ȝ*our* grett offens loke ȝe be repentaunt 145
Of all ȝ*our* venym synne I rede þ*at* ȝe ȝow shryve
Ffor god is ful redy mercy for to graunt
Be contryte[3] for ȝ*our* trespas *and* penauns do be-lyve
Reconsyle ȝ*our*-self *and* be to god plesaunt
W*ith* contrysc*i*on schryffte *and* penau*ns* þe devyl may ȝe dryve 150
Ffor fro ȝ*our* Ffelachep he xal not be erraunt
ȝow for to meve
To penauns[4] *and* synne forsake
Shryfte of mowth loke þ*at* ȝe make

[1] The MS. has been rubbed here, and this word has been written in darker ink over the original word, which may, however, have been the same. The rubbing and subsequent touching up extends down seven lines, but in no other case is the original letter or word unclear.

[2] Inked over by the later hand.

[3] *conty* first written for *contry*[*te*] and crossed through.

[4] *per* first written for *pen*[*auns*] and crossed through.

And þan þe fende[1] in helle so blake 155
he xal ȝow nevyr more greve.

¶ A tre þat is bareyn *and* wyl bere no frute
þe ownere wyl hewe it dowñ *and* cast it on þe fyre
Ryght so it be man þat folwyth þe fowle sute
of þe devyl of helle *and* werkyth his desyre 160
God wyl be vengyd on man þat is both̄ dum *and* mute
þat wyl nevyr be shrevyn but evyr more doth delyre
Clothe the in clennes with vertu be indute
And god with his grace he wyl þe sone inspyre
to Amendynge of þi mys 165
Schryfte of mowthe may best þe saue
Penauns for synne what man wyl haue
Fo. 115 whan þat his body is leyd in grave
His sowle xal[2] go to blys.

¶ Corne þat is good man kepe it ful clene 170
Chaff þat is sympyl is sett wul nere at nought
So good men of levynge to god chosyn bene
Whan synful men be lyke chaff *and* to helle xul be brought
Good penauns ȝow to preche ful hertyly do I mene
Shrylft *and* satysfaccion evyr more to haue in thought 175
What man in good penauns *and* schryfte of mowth̄ be sene
of god he is welbelovyd þat alł þis worlde hath wrought
and alł þinge of nowth dede make
Now haue I tawght ȝow good penauns
god graunt ȝow grace at his plesauns 180
to haue of synne delyverauns
Ffor now my leve I take.

Sathañ

Fo. 116 ¶ Now belyard *and* belzabub ȝe derwurthy devel of helle **23**
And wysest of councel amonges alł þe rowte
herke now what I sey a tale I xalł ȝow telle
þat trobelyth sore my stomak þer of I haue grett dowte.

[1] *felle* first written and corrected to *fende*.
[2] *xal* and *go*, and *good* in the line below, are obscured by a stain on the MS.

Belyall

Syr sathanas oure souereyn syre with þe wol we dwelle 5
All redy at þi byddynge to þe do we lowte
If þou haue Any nede of oure wyse counselle
telle us now þi qwestyon All out *and* oute
sey al þi dowte be-den[e.

Belsabub

ȝa sere telle us þi dowte by *and* by 10
and we xul telle þe so sekyrly
þat þou xalt knowe verryly
What þi dowte doth mene.

Sathan

¶ The dowte þat I haue it is of cryst i-wys
born he was in bedleem as it is seyd 15
And many a man wenyth þat goddys sone he is
born of a woman *and* she a clene mayd
And all þat evyr he prechyth it is of hevyn blys
he wyl lese oure lawe I am ryght sore afrayd
Ffayn wold I knowe who were ffadyr his 20
Ffor of þis grett dowte I am sore dysmayd
in dede
If þat he be goddys childe
and born of a mayd mylde
than be we rygh[1] sore begylde 25
and short xal ben oure[2] spede.

¶ Therfore serys sum what þat ȝe shewe
In þis grett dowth what is best to do
If he be goddys sone he wyl brede a shrewe
and werke us mech wrake both wrech *and* woo 30
Sorwe *and* care he wyl sone strewe
Fo. 116v All oure gode days þan xulde sone be goo
and all oure lore *and* all oure lawe he wyl down hewe
and þan be we all lorn if þat it be soo
he wyll don us all tene 35
he wyll be lorde ouer hevyn *and* helle
and ffeche awey all oure catelle
þerfore shewe now sum good counselle
What comfort may best bene.

[1] So in MS. [2] Some word before *oure* crossed through.

Belyaſſ

¶ The best wytt þat I kan say 40
hym to tempte forsoth it is
with sotyl whylys if þ*at* þ*ou* may
A-say to make hym to don A-mys
If þ*at* he synne þis is no nay
he may nat be kynge of blys 45
hym to tempte go walke þ*i* way
Ffor best cou*n*seſſ I trowe be this
Go forth now *and* assay.

Belsabub

The best wytt I hold it be
hym to tempte in synnys thre 50
the whiche mankende is frelte
doth ſtalle sonest Alway.

Sathan

¶ So Afftyr ȝ*our* wytt now wyſſ I werke
I wyſſ no lenger*e* her*e* a-byde
be he nevyr so wyse a clerke 55
I xal apposyn hym w*ith*-inne A tyde.

Belsabub

now louely lucyfer in helle so derke
Kynge *and* lorde of synne *and* pryde
w*ith* su*m* myst his wyttys to merke
He send þ*e* grace to be þ*i* gyde 60
and evyr more be þ*i* spede.

Belyaſſ

Fo. 117 Aſſ þ*e* deuelys þ*at* ben in helle
shul p*ray* to Mahound as I þ*e* telle
þ*at* þ*ou* mayst spede þis jurney weſſ
and comforte the in þis dede [1]. 65

Jhesus

¶ xl^{ti} days *and* xl^{ti} nyght
now haue I fastyd for ma*n*nys sake
A mor*e* grett hungyr had neuyr no wyght
than I myself be-gynne to take
Ffor hungyr in peyn stronge am I pyght 70
and bred haue I non my*n* hungyr for to slake
A lytel of a loof relese my*n* hungyr myght
but mursele haue I non my comforte for to make

[1] *caas* first written and crossed through.

This suffyr I man for the
Ffor þ*i* glotenye *and* metys wrong 75
I suffyr for þ*e* þis hungyr stronge
I am afferde it wyl be longe
Or þ*ou* do þus for me.

Sathan̄

¶ The sone of god if þ*at* þ*ou* be
be þ*e* grett myght of þ*i* godhede 80
turne these flynt*ys* Anon lett se
ffrom Arde stonys to tendyr brede
mor*e* bettyr it is as I telle the
wysely to werke aftyr my reed
and shewe þ*i* myght of grett majeste 85
than thorwe grett hun̄gyr ffor to be dede
these stonys now bred þ*ou* make
goddys sone if þ*at* þ*ou* be
make these stonys bred lett se
þan mayste þ*ou* ete ryght good plente 90
thy*n* hungyr for to slake.

Jh*esus*

Fo. 117v ¶ Nott only be bred ma*n*nys lyff ȝitt stood
but in þ*e* wurde of god as I þ*e* say
to mannys sowle is neuyr mete so good
As is þ*e* wurd of god þ*at* þ*rechid* is Alway 95
bred materyal doth̄ norch̄ blood
but to ma*n*nys sowle þis is no nay
nevyr mor*e* may be a betyr food
þan þ*e* wurd of god þ*at* lestyth ay
* to her*e* goddys wurde þ*erfore* man loue 100
Thi body doth loue materal brede
W*ith*-oute þ*e* wurde of god þ*i* soule is but dede
to loue p*rechynge* þ*erfore* I rede
If þ*ou* wylt duellyn in blysse a-bove.

Sathan̄

¶ Ffor no grett hungyr þ*at* I kan se 105
In glotony þ*ou* wylt not synne
now to þ*e* temple com forth̄ w*ith* me
and þ*er* xal I shewe þ*e* a praty gynne
Vp to þis py*n*nacle now go we

* Paragraph mark erased here.

I xal þe sett on þe hyȝest pynne 110
ther I preue what þat þou be
Or þat we tweyn part a-twynne
I xal knowe what myght þou haue.

hic ascendit deus pinnaculum templi dum diabolus dicit quod sequitur.

Whan þou art sett upon þe pynnacle
þou xalt þer pleyn a qweynt steracle 115
Or ellys shewe a grett meracle
thy-sself[1] ffrom hurte þou saue.

hic satanas ponit ihesum super pinnaculum dicens

¶ Now If þou be goddys ssone of myght
Fo. 118 Ryght down to þe erth̄ anon þou ffalle
And saue þi-sylf in every plyght 120
Ffrom harm *and* hurte *and* Scappys alle
Ffor it is wretyn[2] with aungelys bryght
þat ben in hevyn þi faderys halle
the to kepe both̄ day *and* nyght
xul be ful redy as þi thralle[3] 125
hurt þat þou non haue
þat þou stomele not a-geyn þe stoñ
and hurt þi fote as þou dost gon
Aungell̄ be redy all̄ everychon
in weys þe to saue. 130

Jhesus

¶ It is wretyn in holy book
þi lorde god þou xalt not tempte
all̄ thynge must obeye to goddys look
out of his myght is[4] non exempt
Out of þi cursydnes *and* cruel crook 135
by godys grace man xal be redempt
whan þou to helle þi brennynge brook
to endles peyne xal evyr be dempt

[1] A dot by the side of the first *s* may be intended to indicate deletion, but cf *ssone* beneath.

[2] The final *n* is in darker ink and may have been added later, but there is the usual space between it and the next word.

[3] MS. *tharalle*, with deleting dot beneath first *a*.

[4] Some word (? *in*) first written.

ther in Alwey to a-byde
Thi lorde god þ*ou* tempt no mor*e* 140
It is nott syttenge to þ*i* lor*e*
I bydde þ*e* sese a-non þ*er*for*e*
And tempte god in no tyde.

Sathan̄

¶ Ow i*n* gloteny nor i*n* veynglory it doth̄ ryght nott a-vayl
Cryst for to tempt it p*r*ofyteth me ryght nought 145
I must now be-gynne to haue a newe travayl[1]
In covetyse to tempt hy*m* it comyth now i*n* my thought
Fo. 118v Ffor If I went þus A-way *and* shrynkyd as a snayle
lorn wer*e* þ*e* labor*e* all̄ þ*at* I haue wrought
þ*er*for*e* in covetyse our*e* syr*e* I xal a-sayle 150
And assay into þ*at* synne yf he may be brought
anon forth̄ ryght
Syr ȝitt onys I pray to the
to þis hyȝ hyl com forth̄ w*ith* me
I xal þ*e* shewe many a cete 155
And many a wurthy syght.

*Tunc-ih*esus *transit cu*m *diabolo su*p*er mont*em et *diabolus dicit*

¶ In to þ*e* northe loke fforth̄ evyn pleyn̄
the towr*e* of babolony þ*er* mayst þ*ou* se
the cete of Jerusalem stondyth þ*er* ageyn
and evyn ffast þ*er* by stondyth Galyle 160
Nazareth naverne *and* þe kyngdom of Spayn̄
zabulon *and* neptalym þ*at* is a rych̄ cou*n*tre
bothe zebee *and* salmana þ*ou* mayst se serteyn
Itayl *and* Archage þ*at* wurthy remys be
bothe januense *and* jurye 165
Rome doth stonde be-for*e* þ*e* ryght
the temple of Salamon as sylver bryght
And her*e* mayst þ*ou* se opynly w*ith* syght
both̄ Ffraunce *and* normandye.

¶ Turne þ*e* now on þis syde *and* se her*e* lumbardye 170
of spycery þ*er* growyth many An C. balys
Archas *and* Aragon̄ *and* grett Almonye

[1] *trav* first written for *trav*[*ayl*] and crossed through.

Parys *and* portyngale *and* þe town of Galys
Pownteys *and* poperynge *and* also pycardye
Erlonde scottlonde *and* þe londe of walys * 175
Fo. 119 Grete pylis *and* castellys þou mayst se *with* eye
ȝa *and* all þe wyd werde *with*-oute mo talys
All þis longygh to me
If þou wylt knele down to þe grownde
and wurchepp me now in þis stownde[1] 180
all þis world[2] þat is so rownd
I xal it gyve to the.

Jhesus

¶ Go Abak þou fowle sathanas
in holy scrypture wretyn it is
thi lorde god to wurchipp in every plas 185
As for his thrall *and* þou servaunt his.

Sathan

Out out harrow Alas Alas
I woundyr sore what is he this
I can not brynge hym to no[3] trespas
nere be no synne to don a-mys 190
he byddyth me gon a-bakke
What þat he is I kan not se
Whethyr god or man what þat he be
I kan not telle in no degre
Ffor sorwe I lete a crakke. 195

hic venient Angeli cantantes et *ministrantes ei* · *Gloria tibi domine* · *dicens*[4]

Jhesus

¶ Now All mankende exaumple[5] take
by these grete werkys þat þou dost se
how þat þe devyll of helle so blake
in synne was besy to tempte me
Ffor all hise maystryes þat he dyd make 200

* Cf. similar lists in the *Castle of Perseverance* and in the Croxton *Play of the Sacrament*.

[1] The final *e* is blotted, possibly meant to be obliterated.

[2] The *l* is corrected from another letter; possibly the scribe first intended *word*.

[3] Some word written before *no* and crossed through.

[4] *dicens* is enclosed in a red loop.

[5] Possibly *exaunple* corrected to *exaumple*.

he is ou*er*com̄ *and* now doth ffle
aſt þis I suffyr ffor ma*n*nys sake
to teche þe how þ*ou* xalt rewle the
Fo. 119v Whan þe devylle dothe the Assayle
loke þ*ou* concente nevyr to synne
For no sleytys ne for no gynne
and þan þe victory xalt þ*ou* wynne
þe devyl xal lesyn aſt his travayl.

¶ To Suffyr temptac*i*on it is grett peyn̄
If þ*ou* w*i*t*h*-stonde it þ*ou* wynnyst grett mede
Of god þe mor*e* grace þ*ou* hast serteyn̄
If þ*ou* w*i*t*h*-sett þe devyl in his dede
thow þ*a*t þe fende tempt þe Ageyn
of his power take þ*ou* no drede
Ffor god hath the ȝovyn both̄ myght *and* mayn
hym for to w*i*t*h*-sytt evyr at nede
þ*ou* hast mor*e* myght than he
Whan þe devyl doth tempte the ,' thoo
Shewe þ*i* myght aȝens þ*i* ffoo
whan þ*i* sowle partyth the froo
In blysse þan xal it be.

Amen

[Remainder of Fo. 119v—3⅝ inches—and Ff. 120, 120v blank (except for scribblings).]

Fo. 121 Hic de muliere in adulterio dep*r*ehensa. ¶ Nolo mortem pe*cc*atoris [1].

¶ Jhesus

MAn for þ*i* synne take repentau*n*ce
If þ*ou* amende þ*a*t is amys
Than hevyn xal be þ*i*n herytau*n*ce
Thow þ*ou* haue don Aȝens god grevauns
ȝett mercy to haske loke þ*ou* be bolde
his mercy doth passe in trewe balauns
Aſt cruel jugement be many folde.

24

[1] This is written in larger but not liturgical script.

A note written very small in the hand of the scribe stands above it: *gyn at nolo morte*[m.

¶ thow *þat* *ȝour* synnys be nevyr so grett
Ffor hem be sad *and* aske mercy
sone of my ffadyr grace ȝe may gett
with þe leste teer wepynge owte of *ȝour* ey
My ffadyr me sent the man to bye
Alł *þi* Raunsom my-sylfe must pay
Ffor loue of *þe* my-sylfe wyl dye
Iff *þou* aske mercy I sey nevyr nay.

¶ In to *þe* erth̄ ffrom hevyn A-bove
þi sorwe to sese *and* joye to restor*e*
man I cam down alł ffor *þi* loue
Loue me ageyn I aske no mor*e*
þow *þou* mys-happe *and* synne ful sor*e*
ȝit turne Aȝen *and* mercy craue
it is *þi* fawte *and* *þou* be lor*e*
haske *þou* mercy *and* *þou* xalt haue.

¶ Vppon *þi* neybor*e* be not vengabyl
Ageyn *þe* lawe if he offende
lyke as he is *þou* art vnstabyl
thyn owyn frelte evyr *þou* attende
'o. 121v (marked 120) Eu*er* mor*e* *þi* neybor*e* helpe to Amende
evyn as *þou* woldyst he xulde *þe*
Ageyn hym wrath̄ if *þou* accende
the same in happ wyłł falle on the.

¶ Eche man to othyr be mercyable
And mercy he xal haue at nede
What man of mercy is not tretable
Whan he Askyth̄ mercy he xal not spede
Mercy to graunt I com in dede
Who so Aske mercy he xal haue grace
Lett no man dowte for his mysdede
But evyr Aske mercy whyl he hath space.

Scriba

¶ Alas Alas our*e* Lawe is lorn
A fals Ypocryte jh*e*su be name
þat of a sheppherdis dowtyr was born
Wyl breke our*e* lawe *and* make it lame
he wyl us werke ryght mekyl shame

HOMERTON COLLEGE LIBRARY

his fals purpos if he up-holde
alł oure lawys he doth defame
þat Stynkynge beggere is woundyr bolde.

Phariseus

¶ Sere scrybe in feyth *þat* ypocryte
wyl turne þis londe al to his lore
*þ*erfore I coun*c*eł · hym to indyte
And chastyse hy*m* ryght wel *þ*erfore.

Scriba

On hym be-leve many A score
In his p*r*echynge he is so gay
Ech man hy*m* ffolwygh ev*er* mor*e* *and* mor*e*
A3ens *þat* he seyth no man seyth nay.

Phariseus

Fo. 122 (K quire) ¶ A Ffals qwarel if we cowde feyne
þat ypocrite to putty*n* in blame
Alł his p*r*echynge xulde sone disteyne
and than his wurchepp xuld turne to shame
w*ith* su*m* falshede to spyllyn his name
lett vs assay his lor*e* to spylle
þe pepyl w*ith* hym yff we cowde grame
than xulde we sone haue al our*e* wyłł.

Açcusator

¶ Herke ser*e* pharysew *and* ser*e* scrybe
A ryght good sporte I kan 3ow telle
I vndyr-take *þat* ryght a good brybe
We alł xul haue to kepe coun*c*eł
A fayr*e* 3onge qwene her*e*-by doth dwelle
both ffresch *and* gay upon to loke
And a talł man w*ith* her*e* doth melle
the wey in to hyr*e* chawmer*e* ryght evyn he toke.

¶ lett us thre now go streyte thedyr
the wey fful evyn I xalł 3ow lede
and we xul take them both to-gedyr
Whyłł *þat* þei do *þat* synful dede.

Scriba

Art *þou* sekyr *þat* we xal spede
Shalł we hym fynde whan we cu*m* there.

Accusator

be my trowth I haue no drede
þe hare fro *þe* fforme we xal a-rere.

Phariseus

¶ We xal haue game *and* þis be trewe
lete us thre werke by on Assent
We wyl here brynge evy*n* be-forn jhesu
and of here lyff þe truth present
Fo. 122v How in advowtrye hyre lyff is lent 85
Than hym be-forn whan she is browth
We xul hym Aske þe trew jugement
What lawfull deth to here is wrouth.

¶ Of grace *and* mercy hevyr he doth preche
And þat no man xulde be vengeable 90
Ageyn þe woman if he sey wrech
Than of his prechynge he is vnstable[1]
And if we fynde hym varyable
of his prechynge þat he hath tawth
than haue we cawse bothe juste *and* Able 95
Ffor a fals man þat he be cawth.

Scriba

¶ Now be[2] grete god ȝe sey fful well
If we hym fyndyn in varyaunce
We haue good reson as ȝe do tell
Hym for to brynge to foule myschauns 100
If he holde stylle his dalyauns
and preche of mercy hi*e* for to saue
than haue we mater of gret substauns
hym for to kylle *and* putt in graue.

¶ Grett reson why I xal ȝow telle 105
Ffor moyses doth bydde in oure lawe
that Euery Advowterere we xuld qwelle
and ȝitt with stonys thei xulde be slawe
Ageyn moyses[3] if þat he drawe
þat Synful woman with grace to helpe 110
he xal nevyr skape out of oure Awe
but he xal dye lyke a dogge whelpe.

[1] *unstabyl* first written and corrected to *vnstable*.
[2] The words *Now be* and *we* in the line below are obscured by a blot.
[3] The words *Ageyn moyses* are written in a slightly larger form.

Accusator

Fo. 123 ¶ ȝe tary ovyr longe ser*ys* I sey ȝow
they wyl sone parte as þ*at* I gesse
þ*er*fore if ȝe wyl haue ȝo*ur* pray now 115
lete us go take them i*n* here whantownnesse.

Phariseus

Goo þ*ou* be-forn þ*e* wey to dresse
We xal þ*e* ffolwe w*ith*-in short whyle
Iff þ*at* we may þ*at* quene dystresse
I hope we xal jh*es*u be-gyle. 120

Scriba

¶ Breke up þ*e* dor*e* *and* go we inne
Sett to þ*e* shuldyr w*ith* all þ*i* myght
We xal hem take evyn in here synne
her*e* owyn trespas shal þem indite.

hic juuenis quidam extra currit indeploydo calligis n*on* *ligat*is et *braccas in manu tenens* et *dicit accusator*

Accusator

Stow þ*at* harlot sum erthely wyght 125
that in advowtrye her*e* is ffownde.

Juuenis

ȝiff Any man stow me þis nyth
I xal hym ȝeve a dedly wownde.

¶ I[1] Any man my wey doth stoppe
or we departe ded xal I be 130
I xal þis daggar*e* putt in his croppe
I xal hym kylle or he xal me.

Phariseus

Grett goddys curse mut go w*ith* the
W*ith* suche a shrewe wyll I not melle.

Juuenes[2]

That same blyssynge I ȝyff ȝow thre 135
And qwhetħ ȝow alle to þ*e* devyl of helle.

* In feyth I was so sor*e* affrayd
Fo. 123v Of ȝone thre shrewys þ*e* sothe to say
my breche be nott ȝett well up-teyd
I had such hast to renne A-way 140

[1] So in MS. for *If*. [2] So in MS.
* No paragraph sign here in MS.

Thei xal nevyr cacche me in such affray
I am full glad þ*at* I am gon
Adewe Adewe a xx^ti^ devyl way
and goddys curse haue ȝe every-chon.

Scriba

¶ Come forth þ*ou* stotte com forth þ*ou* scowte[1] 145
com forth þ*ou* bysmar*e* *and* brothel bolde
com forth þ*ou* hor*e* *and* stynkynge bych clowte
how longe hast þ*ou* such harlotry holde.

Phariseus

Com forth þ*ou* quene com forth þ*ou* scolde
com forth þ*ou* sloveyn com forth þ*ou* slutte 150
we xal the tecche w*ith* carys colde
A lytyl bettyr to kepe þ*i* kutte.

Mulyer

¶ A mercy mercy ser*ys* I ȝow pray
Ffor goddys loue haue mercy on me
of my mys-levynge me not be-wray 155
haue mercy on me for charyte.

Accusator

Aske us no mercy it xal not be
We xul so ordeyn ffor þ*i* lott
þ*at* þ*ou* xalt dye for þ*in* Advowtrye
þ*er*fore com forth þ*ou* stynkynge stott. 160

Mulier

¶ Ser*ys* my wurchepp if ȝe wyl saue
And helpe I haue non opyn shame
bothe gold *and* sylvyr ȝe xul haue
So þ*at* in clennes ȝe kepe my name.

Scriba

Fo. 124 (L quire) Mede ffor to take we were to blame 165
to save suche stottys it xal not be
We xal brynge the to suche a game
þ*at* all advowterer*ys* xul lern be the.

Mulier

¶ Stondynge ȝe wyl not graunt me grace
but for my synne þ*at* I xal dye 170
I pray ȝow kylle me here in þis place
and lete not þ*e* pepyl up-on me crye
If I be sclaundryd opynly

[1] MS. *scowtte*, with deleting dot beneath the first *t*.

to all my frendys it xul be shame
I pray ȝow kylle me p*re*vyly 175
lete not þe pepyl know my defame.

Phariseus

¶ Ffy on þe scowte þe devyl þe qwelle
Ageyn þe lawe xul we þe kyll
Ffyrst xal hange þe · þe devyl of helle
or we such folyes xulde ffulfyll 180
thow it lyke þe nevyr so ill
be-fforn þe pr*o*phete þ*ou* xalt haue lawe
lyke as moyse doth charge ut[1] tyll
W*ith* grett stonys þ*ou* xalt be slawe.

Accusator

¶ Com forth a-pase þ*ou* stynkynge scowte 185
be-fore þe pr*o*phete þ*ou* were þis day
or I xal ȝeue þe such a clowte
þ*at* þ*ou* xalt fall down evyn in þe way.

Scriba

now be grett god *and* I þe pay
Such a buffett I xal þe take 190
þ*at* all þe teth[2] I dare wel say
w*ith*-inne þ*in* heed ffor who xul shake.

Phariseus

Fo. 124v ¶ Herke sere pr*o*phete we all ȝow pray
to gyff trewe dom *and* just sentence
Vpon þis woman which þis same day 195
In synfull advowtery hath don offense.

hic ihesus dum isti accusant mulierem continue debet digito suo scribere in terra.

Accusator

Se we haue brought here to ȝ*our* presens
be-cawse ȝe ben a wyse pr*o*phete
þ*at* ȝe xal telle be consyens
What deth to hyre ȝe thynke most mete. 200

Scriba

¶ In moyses lawe ryght þus we fynde
þ*at* such fals louers xul be slayn
streyte to a stake we xul hem bynde
and w*ith* grett stonys brest out þ*er* brayn

[1] So in MS. for *us*.
[2] This word first miswritten in some way and crossed through.

Of ȝ*our* concyens telle us þ*e* playñ 205
w*ith* þis woman what xal be wrought
Shall we lete her*e* go qwyte agayñ
or to hir*e* deth xal she be brought.

Jhesus nichil respondit sed semper scrybyt in terra.

¶ Now holy p*ro*phete be mercyable Mulier[1]
vpon me wrecch take no vengeaunce 210
Ffor my synnys Abhomynable
In hert I haue grett repentaunce
I am wel wurthy to haue myschau*n*ce
Both bodyly deth *and* werdly shame
but gracyous p*ro*phete of socurraunce 215
þis tyme pray ȝow for goddys name.

Fo. 125 Phariseus

¶ Ageyn þ*e* lawe þ*ou* dedyst offens
þ*er*fore of grace speke[2] þ*ou* no mor*e*
As moyses gevyth in law sentens
þ*ou* xalt be stonyd to deth þ*er*-fore. 220

Accusator

Ha don ser*e* p*ro*phete telle us ȝour*e* lor*e*
xul we þis woman w*ith* stonys kyll
or to hir*e* hous hir*e* home restor*e*
In þis mater tell us ȝ*our* wyll.

Scriba

¶ In a colde stodye me thynkyth ȝe sytt 225
good ser*e* awake telle us ȝ*our* thought
xal she be stonyd telle us ȝ*our* wytt
or in what rewle xal sche[3] be brought.

Jh*es*us

Loke which of ȝow þ*at* nevyr synne wrought
but is of lyff clennere þan she 230
Cast at her*e* stonys *and* spar*e* her*e* nowght
Clene out of synne if þ*at* ȝe be.

hic ihesus iterum se inclinans scribet in terra et *omnes accusatores quasi confusi separatim in tribus locis se disiungent,'*

[1] This word is written in the left-hand margin in another hand, the name of the speaker having been omitted.

[2] *spe* first written, *ke* inserted by the scribe over the line.

[3] Or *scho*—the letter is blotted.

Phariseus

¶ Alas Alas I am ashamyd
I am a-fferde[1] þat I xal deye
ałł my*n* synnys evyn p*ro*pyrly namyd 235
ȝon p*ro*phyte dede wryte be-for my*n* eye
Iff þ*at* my felawys þ*at* dude Aspye
they wyłł telle it bothe ffer *and* wyde
my synfułł levynge if þei out crye
I wot nevyr wher my*n* heed to hyde. 240

Accusator

¶ Alas for sorwe my*n* herte doth blede
Ałł my synnes ȝon man dude wryte
Fo. 125v If þ*at* my felawys to them toke hede
I kan not me ffrom deth Acquyte
I wolde I wor*e* hyd su*m*-wher*e* out of syght 245
þ*at* men xuld me no-wher*e* se ne knowe
Iff I be take I am afflyght
In mekyl shame I xal be throwe.

Scriba

¶ Alas þ*e* tyme þ*at* þis be-tyd
Ryght byttyr care doth me enbrace[2] 250
Ałł my synnys be now vnhyd
ȝon man be-for me hem ałł doth trace
If I wer*e* onys out of þis place
to suffyr deth gret || *and* vengeau*n*s Able[3]
I wyl nevyr come be-for his face 255
þow I xuld dye in a stable.

Mulier

¶ Thow I be wurthy ffor my trespas
to suffyr deth̄ ab-homynable
ȝitt holy p*ro*phete of ȝ*our* hyȝ grace
In ȝ*our* jugement be mercyable 260
I wyl nevyr mor*e* be so vnstable
O holy prophete graunt me mercy
of my*n* synnys vnresonable
W*ith* ałł my*n* hert I am sory.

Jhesus

¶ Wher*e* be þ*i* fomen þ*at* dude þ*e* Accuse 265
Why haue þei left us to Alone.

[1] *fed* first written for *fferde* and crossed through.
[2] So in MS. [3] Line thus divided at *gret* in MS.

Mulier

By-cawse they cowde nat hemself excuse
W*ith* shame they ffled hens Eu*er*y-chone
Fo. 126 But gracyous p*ro*phete lyst to my mone
(I quire) of my sorwe take compassyon̄ 270
now all̄ my*n* enmyes hens be gone
Sey me su*m* wurde of consolac*io*n.

Jh*es*us

¶ Ffor þo synnys þ*at* þ*ou* hast wrought
hath Any man condempnyd the.

Mulier

Nay for-soth̄ þ*at* hath̄ þ*er* nought 275
but in ȝo*ur* grace I putt me.

Jh*es*us

Ffor me þ*ou* xalt nat condempnyd be
go hom A-geyn *and* walk at large
loke þ*at* þ*ou* leve in honeste
and wyl no mor*e* to synne I þ*e* charge. 280

Mulier

¶ I thanke ȝow hyȝly holy p*ro*phete
Of þis grett grace ȝe haue me graunt
all̄ my lewde lyff I xal dou*n* lete
and ffonde to be goddys trewe servaunt.

Jh*es*us[1]

What man of synne be repentaunt 285
of god if he wyl mercy craue
God of mercy is so habundawnt
þ*at* what man haske it he xal it haue.

¶ Whan man is contrite *and* hath wonne grace
God wele not kepe olde wreth̄ in mynde 290
but bettyr loue to hem he has
Very contryte whan he them fynde
Now god þ*at* dyed ffor all̄ mankende
saue all̄ þese pepyl both̄ nyght *and* day
and of our*e* synnys he us vnbynde 295
hyȝe lorde of hevyn þ*at* best may.

Amen̄.

[Ff. 126ᵛ and 127 left blank.]

[1] The word *Jhesus* has been crossed out and the word *Doctor* written below in the same hand as the word *Mulier* earlier in the play.

Fo. 127v hic incipit de suscitac*i*one lazari.

¶ Lazarus

GOd þ*a*t all thynge dede make of nowth
And puttyst eche creatur*e* to his fenaunce
Saue thyn handwerke þ*a*t þ*ou* hast wrought **25**
As þ*ou* art lord of hyȝ substauns
O gracyous god att þ*i* plesau*n*s 5
of my dysese now comforte me
Which þ*ur*owe syknes hath such penawnce
On-ethys ffor heed Ache may I now se.

¶ Systyr Martha *and* Mawdelyn eke
W*i*t*h* hast helpe me in bedde to dresse 10
Ffor trewly I am so woundyrly seke
I may nevyr schape[1] þis grett seknes
My deth is com̃ now I gesse
help in to chawmer*e* þ*a*t I be led
my grett desesse I hope xal lesse 15
If I wer*e* leyd upon a bed.

Martha

¶ Lazarus brother be of good cher
I hope ȝo*ur* syknes ryght wel xal slake
Vpon þis bed rest ȝow rygh[2] her
And a good slep assay to take. 20

Magdalyn̄

Now jentyl brothyr ffor goddys sake
lyfte up ȝowr*e* hert *and* be not feynt
An hevy housholde w*i*t*h* vs ȝe make
If dedly syknes haue ȝow Ateynt.

Lazarus

¶ Ffor sothe der*e* systeryn I may not slepe 25
my seknes so sor*e* doth̄ evyr encrese
of me I pray ȝow take ryght good kepe
tyll þ*a*t my peyne be-gynne relese.

Martha

Fo. 128[3] God graunt grace þ*a*t it may sese
of syknes god make ȝow sownde 30

[1] The *h* is corrected from some other letter. [2] So in MS.
[3] The red loops of the names of the speakers are omitted on Fo. 128.

or ellys oure joy wyth sone dyscres
In so grett peynes if ȝe ly bownde.

Magdalyn

¶ A brothir brothir lyfte up ȝoure herte
ȝ*our* hevy cher doth us grevaunce
If deth from us ȝow xulde de-parte 35
than were we brought in comberaunce
ȝe be oure brothyr syb of Alyaunce
If ȝe wore deed þan had we none
ȝe do us brynge in distemp*er*aunce
Whan ȝe us telle ȝe xal hens gone. 40

1*us* consolator

¶ Dame martha *and* magdalyne
how faryth ȝ*our* brothir*e* lete us hym se.

Martha

he is ryght seke *and* hath grett pyne
I am aferde deed he xal be.

Magdalyn

A man may haue ryght grett pete 45
the fervent hete of hym to fele.

ij*us* consolator

Take ȝe no thought in no degre
I hope þ*at* he xal ffar*e* fful wele.

Martha

¶ he may nat leve his colowr*e* doth chaunge
com to his bed ȝe xal hym se. 50

Magdalyn

Iff he longe leve it wyl be straunge
but as god wole so mut it be
Cher*e* hym gode frendys ffor charyte
Comforte of hym we kan non gete
Fo. 128v Alas Alas what eylight me 55
myne herte for wo is wundyr grete.

3*us* consolator

¶ Al heyl syr lazarus how do ȝe far*e*
how do ȝe ffele ȝow in ȝ*our* herte.

lazarus

I am with syknes all woundyn in car*e*
And loke whan deth me xulde departe. 60

4*us* consolator et nu*n*ci*us*

ȝe xal haue hele *and* leue in qwart
If ȝe wol take to ȝow good chere.

Lazarus

Whan deth[1] on me hath shet his dart
I xal haue hele *and* ly on bere.

1us consolator

¶ Be of good comforte *and* thynke not so
put out of herte þat idyl thought
ȝoure owyn mys-demynge may werke ȝow wo
and cause ȝow sonere to deth be brought.

ij^{us} consolator

With gret syknes þow ȝe be sought
Vpon ȝoure-sylf haue no mystruste
If þat ȝe haue I wundyr ryght nought
Þow ȝe be deed *and* cast in duste.

3us consolator

¶ Many on hath had ryght grett syknesse
And aftyr hath had his hele agayn[2]
and many a man þis is no lesse
With his wantruste hym-sylf hath slayn[3]
ȝe be a man of ryght sad brayn
þow þat ȝour syknes greve ȝow ryght ill
Pluk up ȝour herte with myght *and* mayn
and chere ȝour-sylf with all ȝour wyll.

Lazarus

Fo. 129 ¶ Ageyn my syknes þer is non ese
but jhesu cryst my maystyr dere
If þat he wyst of my dyssese
Ryght sone I trust he wolde ben here.

4us Consolator

I xal go to hym with-outyn dwere
And of ȝour syknes telle hym serteyn
loke þat ȝe be of ryght good chere
Whyll þat I go *and* com ageyn.

Martha

¶ Now jentyl ffrend telle hym ryght thus
he þat he lovyth hath grett syknes
hedyr to come *and* comforte vs
Say þat we prayd hym of his goodnes.

[1] Some correction has been made by the scribe in this word.
[2] *ageyn* first written and corrected to *agayn*.
[3] The *a* here seems also to be written over another letter.

Magdalyn

Recomende us on to his hyȝnes
And telle hym alł oure hertys wo
but he comforte oure hevynes 95
oure werdly[1] joy awey wyl go.

4us consolator et nuncius

¶ The trewth ffor soth alł every dele
as ȝe haue told so xal I say
go to ȝ*our* brothyr *and* cheryse hy*m* wele
Ffor I walke ffortħ streyte in my way. *here goth he his way*[2] 100

Martha

What chere good brothyr telle me I pray
What wele ȝe ete what wele ȝe drynk
loke what is plesynge to ȝ*our* pay
ȝe xal haue what ȝe wole thynke.

Lazarus

¶ My wynde is stoppyd gon is my breth 105
And deth is come to make my*n* ende
to god in hevyn my sowle I qweth
Ffarwełł systeryn ffor hens I wende.

hic lazarus moritur et cetera.

Magdalyn

Fo. 129v Alas ffor wo my*n* here I rende
Myn owyn dere brothyr lyth here now ded 110
now haue we lost a trusty ffrende
þe sybbest blood of oure kynreed.

Martha

¶ Alas alas *and* wele way
now be we tweyn bothe brotherles
Ffor who my hert is colde as clay 115
A hoo xal comforte oure carefulnes
Ther had nevyr woman more doolfulnes
A systyr Magdalyn[3] what is ȝ*our* reed
What whith may helpe oure hevynes
now þ*at* oure brother is gon *and* deed. 120

Magdalyn

¶ Alas dere systyr I cannot telle
þe best comforte þat I can sey[4]

[1] Some letter, possibly *s*, is erased at the end of this word.
[2] Written roughly in the hand of the scribe of Ff. 95, 96.
[3] *Mad* (?) first written for *Mag*[*dalyn*] and crossed through.
[4] *say* first written and corrected to *sey*.

but su*m* man do us sle *and* qwelle
lete us ly down by hy*m* *and* dey
Alas why went he alone awey
If we had deyd w*ith* hy*m* Also
than had our*e* car*e* all turnyd to pley
ther now all joye is turnyd to woo.

i^us^ consolator

¶ Be of good comforte *and* thank god of al
Ffor deth is dew to every man
What tyme þ*at* deth on us xal ffal
non erthely wyght þ*e* our*e* telle can.

Martha

We all xul dye þ*at* is sertan[1]
but ȝit þ*e* blood of kynde nature
Whan deth þ*e* brothyr Awey hath tan
must nedys murne þ*at* sepulture.

ij^us^ consolator

Fo. 130 ¶ Good ffrendys I pray ȝow holde ȝ*our* pes
All ȝ*our* wepynge may not amende itt
of ȝ*our* sorwynge þ*er* fo*re* now ses
And helpe he were buryed in a cley pitt.

Magdalyn

Alas þ*at* wurde my*n* herte doth slytt
þ*at* he must now in cley be graue
I wolde su*m* man my throte wulde kytt
þ*at* I w*ith* hym myght lyne in caue.

iij^us^ consolator

¶ Bothe heed *and* ffoot now he is wounde
in a chete[2] bothe ffayr *and* clene
lete us ber*e* hym streyte to þ*at* grounde
Wher*e* þ*at* ȝe thynke his graue xal bene.

Martha

We be ffull lothe þ*at* pytt to sen
but stondynge it may no bettyr be
The coors take up ȝow thre betwen
w*ith* carefull herte ȝow ffolwe xal we.

*hic portauit corpus ad sepelliendu*m.

[1] The scribe began to write *e* after the *t* (? for *ser(eyn)*).

[2] An initial *s* has been added to this word in a different ink—apparently by a later corrector.

Magdaleyn

¶ Alas comforte I se non othyr
but all of sorwe *and* care *and* woo
We dulfull women must burry oure brothir 155
Alas þat deth me wyl not slo
If I to pitt w*ith* hym myght go
þer-in evyr more w*ith* hym to Abyde
than were my care all went me fro
þer now grett sorwe doth wounde me wyde. 160

i^us consolator

¶ This coors we burry here in þis pytte
all myghty god þe sowle mut haue
and w*ith* þis ston þis graue we shytte
ffro ravenous[1] best*ys* þe body to saue.

Magdalyn

Fo. 130v He is now brought in to his cave[2] 165
my*n* hert ffor wo þis syght doth kylle
Lete us sytt down here by þe grave
or we go hens wepe all oure fylle.

Martha

¶ Vs for to wepe no man may lett
be-forn oure face to se þis syght 170
Alas qwy doth deth us not fett
Vs for to brynge to þis same plyght.

ij^us consolator

Arys for shame ȝe do not ryght
streyth from þis grave ȝe xul go hens
þus for to grugge ageyns godys myght 175
Aȝens hyȝ god ȝe do offens.

Magdalen

¶ Syth I must nedys w*ith* ȝow hens gon
my brother*ys* graue lete me fyrst kys
Alas no whith may helpe my mon
Ffare wel my brothyr fare wel my blys. 180

iij^us consolator

hom to ȝ*our* place we xal ȝow wysse
Ffor goddys loue be of good chere
In dede ȝe do ryght sore amys
so sore to wepe as ȝe do here.

[1] *rauenous* first written, the *u* corrected to *v* by scribe.

[2] This *v* also appears to have been touched up, but there is no trace of an original *u*.

Martha

¶ Lete us go hom than to our*e* place
we pray ȝow alł w*ith* us to Abyde
Vs to comforte w*ith* sum solace
tyl þ*at* our*e* sorwe doth slake *and* sclyde.

1*us* consolator

ȝow for to comforte at every tyde
We xałł dwelle her*e* bothe nyght *and* day
and god þ*at* made þis werd so wyde
be ȝowr*e* comforte þ*at* best may.

*hic iiij*us *consolator* et *nuncius loquitur jhesu dicens*

4*us* consolator

Fo. 131 ¶ Heyl holy p*ro*phete jh*es*u be name
Martha and Mawdelyn þ*o* systeryn too
Recomende hem to ȝ*our* hyȝ fame
And bad me sey to ȝow þus loo
How þ*at* Lazar*e* qwhich þ*at* ȝe lovyd so
w*ith* grett syknes is sor*e* dyssesyd
to hy*m* they prayd ȝow þ*at* ȝe wolde goo
If þ*at* ȝ*our* hyȝnes þ*er*w*ith* wer*e* plesyd.

Jh*es*us

¶ Dedly syknes lazar*e* hath non
but for to shewe goddys grete glorye
Ffor þ*at* syknes is ordeynyd a-lon
þ*e* sone of god to gloryfie.

Nuncius

They be in dowte þ*at* he xal deye
grett syknes hym sor*e* doth holde
Ffor veruent hete his blood doth dr*e*ye
his color*e* chaungyth as they me tolde.

Jh*es*us

¶ Goo hom ageyn And telle hem thus
I xal come to hem whan þ*at* I may.

Nu*n*cius

At ȝ*our* comaundement · O p*ro*phete jh*es*us
I xal hem telle as ȝe do say.

Jh*es*us

Come forth bretheryn walke we our*e* way
In to jurye go we A-non
I cam not ther*e* ful many a day
þ*er*for*e* thedyr now wyl I gon.

Omnes discipuli

¶ The jewys ageyn the were grym *and* grylle
Whan þou were there they wolde þe aslayn
Fo. 131v (marked 130) With stonys they sowte þe ffor to kyll̄
And wylt þou now go thedyr ageyn. 220

Jhesus

xij owrys þe day hath̄ in certeyn
In them to walke both clere *and* bryght
he xal not stomble ageyn hyll̄ nor pleyn
þat goth þe wey whyl it is day lyght.

¶ but if men walke whan it is nyght[1] 225
Sone they offende in þat dyrknes
be-cawse they may haue no cler syght
they hurte there ffete ofte in suche myrkenes
but as ffor this ȝitt nevyr-þe-lesse
the cawse þerfore I thedyr wyl wende 230
is ffor to reyse ffrom bedde expresse
lazare þat slepyth̄ · oure Althere ffrende.

Omnes discipuli

¶ Of his syknes[2] he xal be save
If þat he slepe good sygne it is.

Jhesus

lazare is deed *and* leyd in grave 235
Of his slepynge ȝe deme amys
I was not there ȝe know weyl[3] this
to strengthe ȝoure feyth I am ful glad
þerfore I telle ȝow þe trewthe i-wys
oure ffrende is deed *and* vndyr erth̄ clad. 240

Thomas

¶ Than goo we all̄ ryght evyn streyth thedyr
there as oure ffrende lazare is deed
and lete us deye with hym to-gedyr
þer as he lyth in þe same stede.

Jhesus

Fo. 132 The ffor[4] to deye haue þou no drede 245
the wey streyth thedyr in hast we take

[1] The colour of the ink becomes darker with this line and the writing slightly more angular. This continues more or less to the bottom of Fo. 134.
[2] The *k* is corrected from some other letter.
[3] The *e* omitted and written above the line. [4] So in MS.

be þe grett myght of my*n* godhede
oute of his slepe he xal awake.

Nuncius

¶ Aſt heyl Martha *and* mawdelyn eke
to jh*e*su I haue ȝou*r* massage seyd[1] 250
I tolde hy*m* how þ*a*t ȝ*our* brothyr was seke
and w*ith* grett peyn in his bed leyd
He bad ȝe xulde not be dysmayde
aſt his syknes he xal Askape
he wyſt byn here w*ith*-in a brayde 255
as he me tolde he comyth in rape.

Mawdelyn[2]

¶ That holy p*ro*phete doth come to late
our*e* brothyr is beryed iij days or this
A grett [stone][3] stoppyth þ*e* pyttys gate
there as our*e* brothere beryde is. 260

Nuncius

Is lazare deed now god his sowle blys
ȝit loke ȝe take non hevynes
So longe to wepe ȝe don A-mys
It may not helpe ȝ*our* sorynes.

Martha

¶ Oute of my*n* herte aſt care to lete 265
aſt sorwe *and* wo to caste Away
I xal go forth in þ*e* strete
to mete w*ith* jh*e*su if þ*a*t I may.

ij^us consolator

God be ȝ*our* spede bothe evyr *and* Ay
Ffor w*ith* ȝ*our* sustyr we wyl abyde 270
here to comforte we xal a-say
and aſt here care to caste a-syde.

3us consolator

Fo. 132v ¶ Mary Mawdelyn be of good herte
And wel be-thynke ȝow in ȝ*our* mynde
Eche creature hens must depart 275
þ*er* is no man but hens must wende.

[1] *Sayd* first written and corrected to *seyd*.
[2] The *w* written over a *d*.
[3] This word is written above the line in a different ink, possibly by the scribe of Ff. 95, 96.

Deth to no wyht can be a frende
Alł þinge to erth̄ he wyl dowñ cast
Whan *þat* god wol alł thynge hath ende
lengere than hym lyst no thynge may last. 280

Magdalyñ

¶ I thanke ȝow frendys ffor ȝ*our* good chere
my*n* hed doth ake as it xulde brest
I pray ȝow therfore while[1] ȝe ben here
A lytil whyle *þat* I may rest.

4us co*n*solat*or* nunc[ius

þat lord *þat* made bothe est *and* west 285
graunt ȝow good grace suche rest to take
þat on to hym xulde plese most best
As he þis worlde of nought dyd make.

Martha

¶ A gracyous lord had ȝe ben here
My brother lazare þis tyme had lyved[2] 290
but iiij days gon upon a bere
We dede hym berye whan he was ded
ȝitt now I knowe w*ith*-owtyn drede
What thynge of god *þat* *þou* do craue
þou xalt spede of *þe* hyȝ godhede 295
what so *þou* aske · *þou* xalt it haue.

Jhesus

¶ Thy brothyr lazare Aȝen xal ryse
A levynge man aȝen to be.

Martha

I woot wel *þat* at ·/ *þe* grett last syse
he xal a ryse | *and* also we[3]. 300

Jhesus

Fo. 133 Resurrec*c*ion *þou* mast me se
And hendeles lyff I am also
What man *þat* deyth *and* levyth in me
Ffrom deth to lyve he xal ageyn go.

¶ Eche man in me *þat* feyth̄ful is 305
and ledyth his lyff aftere my lore
of hendeles lyff may he nevyr mys

[1] *wyl* first written and crossed through.
[2] There are traces of correction in this word—possibly *leved* first written.
[3] These line divisions seem intended to mark the extra rhyme.

euere he xal leve *and* deye nevyr more
The body *and* sowle I xal restore
to endeles joye dost *þou* trowe this. 310

Martha

I hope in the || O cryst ful sore[1]
þou art *þe* sone of god in blys.

¶ Thy Ffadyr is god of lyff endeles
þi self is sone of lyff *and* gras
to sese these wordlys wrecchydnes 315
Ffrom hefne to erth *þou* tokè *þe* pas.

Jhesus

Of hevynly myght ryght grett solas
to all þis world me[2] xul sone se
go calle *þi* systyr in to þis plas
byd mary mawdelyn come hedyr to me. 320

Martha

¶ At *þi* byddyng I xal here calle
In hast we were here ȝow be-forn.

Mawdelyn

Alas my mowth is bytter as galle
grett sorwyn my herte on tweyn hath scorn
Now *þat* my brothyr from syth is lorn 325
þer may no myrth my care releve
Alas *þe* tyme *þat* I was born
þe swerde of sorwe my*n* hert doth cleve.

i^us consolator

Fo. 133v ¶ Ffor his dere loue *þat* all hath wrought
Ses su*m* tyme of ȝ*our* wepynge 330
and put all thynge out of thought
in-to þis care *þat* ȝow doth brynge.

2us consolat*or*

ȝe do ȝ*our*-self ryght grett hyndrynge
and short ȝoure lyff or ȝe be ware
Ffor goddys loue ses of ȝ*our* sorwynge 335
and w*ith* good wysdam refreyn ȝ*our* care.

Martha

¶ Sustyr magdalen com out of halle
oure maystyr is com as I ȝow say

[1] The reason for the dividing strokes in this line is not apparent.
[2] So in MS. for *men*.

he sent me hedyr ȝow for to calle
come forth in hast as I ȝow pray. 340

Magdalen

Ha · where hath he ben many a longe day
Alas why cam he no sonere hedyr
In hast I folwe ȝow anon þe way
me thynkyth longe or I come thedyr.

3us consolator

¶ Herke gode ffrendys I ȝow pray 345
Aftyr þis woman in hast we wende
I am a-ferde ryght in good fay
here-self for sorwe þat she wyl shende.

Nuncius

Here brothyr so sore is in hire mende
She may not ete drynke nor slepe 350
streyte to his graue she goth on ende
As a mad woman þer for to wepe.

Magdalen

¶ A Souereyn lord *and* mayster dere
had ȝe with us ben in presens
Fo. 134 Than had my brother on lyue ben here 355
nat ded but qwyk þat now is hens
Ageyn deth is no resystens
Alas myn hert is woundyrly wo
Whan þat I thynke of his Absens
þat ȝe ȝour-self in herte lovyd so. 360

1us consolator

whan we haue mynde of his sore deth
he was to us so gentyl[1] *and* good
þat mend of hym oure hertys sleth
þe losse of hym doth marre oure mood.

ij^us^ [2] consolator

Be bettyr neybore nevyr man stood 365
to euery man he was ryght hende
Vs he dede refresch with drynk *and* ffood
now he is gon | gon is oure frende.

Jhesus

¶ ȝowre grett wepynge doth me constreyne
Ffor my good ffrend to[3] wepe also 370

[1] This initial *g* has been corrected from an original *j* by the scribe.
[2] 3 first written for ii and crossed through.
[3] Some letter (? *w*) before *to* crossed through.

I can not me for wo restreyn
but I must wepe lyke as ȝe do.

hic ihesus fingit se lacrimari. 3us consolat*or*

Be-holde þis pr*o*phete how he doth wepe lo
he louyd lazar*e* ryght woundyrly sor*e*
he wolde not ellys for hy*m* þus wepe so 375
but if þ*at* his loue on hym were þ*e* mor*e*.

Nunci*us*

¶ A straw for þ*i* tale what nedyth hy*m* to wepe
A man born blynde · dede[1] he nat ȝeue syght
Fo. 134v myght he nat thanne his frende on lyve kepe
be the uertu of þ*at* same hyȝ myght. 380

Jh*es*us

Wher*e* is he put telle me anon ryght
brynge me þ*e* weye streyth to his grave.

Martha

lord at ȝo*ur* wylle we xal brynge ȝow tyght
evyn to þ*at* place þ*er* he doth lyne in caue.

Magdalyn̄

Whan þ*at* we had þ*e* massangere sent 385
Or he had fullych̄ half a myle gon̄
deyd[2] my brother *and* up we hy*m* hent
her*e* in þis graue we beryed hy*m* anon̄.

Jh*es*us

þ*e* myght of þ*e* godhed xal glathe[3] ȝow every-chon̄
suche syght xal ȝe se hens or ȝe wende 390
Sett to ȝo*ur* handys take of þ*e* ston̄
A syght lete me haue of lazar*e* my ffrende.

Martha

¶ he stynkygh ryght fowle longe tyme or this
iiij days gon for sothe he was dede
lete hym ly stylle ryght evyn as he is 395
þ*e* stynke of his careyn myght hurte[4] us I drede.

Jh*es*us

As I haue þ*e* tolde syght of þ*e* god-hede
thy-self xuldyst haue · feythful if þ*ou* be

[1] *dede* has been altered to *dyde* in different ink and by a later hand, the *y* being written over the *e*.

[2] *deyd* has been altered to *dyyd* in a different ink and (probably) by a later hand—the *y* written over the *e*.

[3] *glathe* has been altered to *gladd* in a different ink, the *dd* written over the *the*.

[4] *hurf* first written and crossed through.

take of þe ston do Aftyr my rede
þe glorye of þe godhede a-non ȝe xal se. 400

i^us consolator

¶ ȝoure byddynge xal be don a[non][1] ful swyfte
Sett to ȝour handys *and* helpe echon̄
I pray ȝow serys help me to lyfte
I may not reyse it my-self[2] a-lon.

2us consolator

Fo. 135 In feyth it is An hevy ston 405
Ryth sad of weyth *and* hevyof peys.

3us consolator

Thow it were twyes so evy[3] as on[4]
Vndyr vs foure we xal it reyse.

Nuncius

¶ Now is þe ston take ffrom þe caue
here may men se A rewly sygth 410
of þis ded body þat lyth here graue[5]
Wrappyd in a petefful plyght.

Jhesus eleuatis ad celum oculis dicit

Jhesus

I thanke þe fadyr of þin hyȝ myght[6]
þat þou hast herd my prayour þis day
I know ful wel bothe day *and* nyght 415
Euer þou dost graunt þat I do say.

¶ But for þis pepyl þat stondyth about
And be-leue not þe power of þe *and* me
them for to brynge clene out of dowt
this day oure myght thei[7] all xul se. 420

hic Jhesus clamat voce magna dicens

Lazare Lazare my frende so fre
Ffrom þat depe pitt come out a-non̄

[1] MS. *an̄* with *non* written in the left-hand margin.
[2] *seff* first written, imperfectly corrected to *self*.
[3] An initial *h* has been added in different ink by a later corrector.
[4] Some letter has been corrected to *o* in this word and to make it clear an *o* is written above.
[5] The word *in* has been written above the line before *graue* by the later hand.
[6] This line as far as *fadyr* written and crossed through in red ink to make room for the stage direction, then written again beneath.
[7] *thei* has been altered to *they* in a different ink by the later corrector.

be þe grett myght of þe hyȝ mageste
A-lyve þou xalt on erth ageyn gon.

lazarus

¶ At ȝoure comaundement I ryse up ful ryght 425
hevyn helle And erth ȝoure byddyng must obeye
Ffor ȝe be god *and* man *and* lord of most myght
Of lyff *and* of deth ȝe haue both lok *and* keye.

hic resurget lazarus ligatis manibus et pedibus ad modum sepulti[1] *et dicit ihesus*

Jhesus

Fo. 135v Goo forthe bretheryn[2] *and* lazare ȝe vntey
And all his bondys losyth hem asundyr 430
Late hym walke hom *with* ȝow in þe wey
Ageyn god*ys* myght þis meracle is no wundyr.

Petrus

¶ At ȝo*ur* byddynge his bondys we vnbynde
all thynge muste lowte[3] ȝ*our* mageste
be þis grett meracle opynly we fynde 435
Very god *and* man in trewth *þat* ȝe be.

Johannes

þat þ*ou* art very god every man may se
be this meracle so grett *and* so m*er*uayll
all thynge vndyr hevyn must nedys obey þe
whan Aȝens þe powh deth be · he may not p*re*uayll. 440

Omnes consolatores

¶ We All *with* o voys ffor god do þe knowe
and for oure sauyo*ur* we do þe reverens
All oure hool loue now in þe doth growe
O sovereyn lord of most excellens
helpe vs of ȝ*our* grace whan *þat* we go hens 445
Ffor aȝens deth us helpyht not to stryve
but aȝen ȝoure myght is no resistens
oure deth ȝe may A-slake *and* kepe vs[4] stylle on lyve.

Jhesus

¶ Now I haue shewyd in opyn syght
of my godhed þe gret glorye 450

[1] MS. sepulł.
[2] *breryn* first written and crossed through.
[3] The word *vnto* has been written above the line by the later hand between *lowte* and *ȝour*.
[4] *ust* written before *vs* and crossed through.

to-ward my passyon I wyl me dyght
the tyme is nere *þat* I must deye
Ffor alł mankynde his sowle to bye
A crowne of thorn xal p*er*chyn my*n* brayn̄
and on þe mont of caluarye 455
Vpon a cros I xal be slayn.

The Passion Play. I

Fo. 136 ¶ Demon
N quire)[1] ¶[2] I am ȝ*our* lord łucifer *þat* out of helle cam
Prince of þis werd · *and* gret duke of helle **26**
Where*fore* my name is clepyd sere satan
Whech Aperyth among ȝow · A matere to spelle.

¶ I am Norsshere of synne · to þe confusyon of man 5
To bryng hym to my dongeon̄ · *þer* in fyre to dwelle
Ho so evyr serve me so reward hy*m* I kan̄
þat he xal syng wellaway · ev*er* in peynes ffelle.

¶ Lo þus bou*n*tevous A lord þan · now am I
To reward so synners · as my kend is 10
Who so wole folwe my lore · *and* serve me dayly
Of sorwe *and* peyne A-now · he xal nevyr mys.

¶ Ffor I began in hefne synne for to sowe *
Among alł þe Angellys · *þat* weryn þere so bryth
And *þer* fore was I cast out · in to helle ful lowe 15
Not withstandyng I was þe fayrest *and* berere of lyth.

¶ ȝet I drowe in my tayle · of þo Angelys bryth
W*ith* me in to helle · takyth good hed what I say
I lefte but tweyn A-ȝens on · to Abyde þere in lyth
But þe iij^de part come w*ith* me · þis may not be seyd nay. 20

[1] The writing is less regular in this quire and the rubrication of the initial letters of the line more abundant and less careful. On Ff. 136 and 136^v the writing varies noticeably in size from line to line.

[2] The paragraph marking of this Prologue is irregular, as is the scheme of linked quatrains.

* *Wylliam Dere* scribbled in the margin.

¶ Takyth hed to ȝ*our* prince þan̄ · my pepyl euery-chon̄
and seyth what maystryes in hefne · I gan þ*er* do play
To gete A thowsand sowlys in an houre · me thynkyth it but skorn
Syth I wan Adam And Eve · on þ*e* fyrst day.

¶ But now mervelous mendys rennyn in my*n* rememberawns
Of on cryst wiche is clepyd Joseph · *and* maryes sone
Thryes I tempte hym be ryth sotylle instawnce
Aftyr he fast fourty days · ageyns sensual myth or reson
Fo. 136v Ffor of þ*e* stonys to a mad bred · but sone I had conclusyon̄
Þan upon a pynnacle but Angelys were to hy*m* Assystent
His Answerys were mervelous · I knew not his intenc*i*on̄
and at þ*e* last to veyn glory · but nevyr I had my*n* intent.

¶ And now hath he xij dysypulys · to his Attendau*n*s
To eche town̄ *and* cety · he sendyth hem as bedellys
in dyverce place to make ffor hy*m* pu*r*uyau*n*s
The pepyl of hese werkys ful grettly m*er*veyllys
To þe crokyd blynd *and* down̄ · his werkys p*r*ovaylys[1]
Lazar*e* þ*at* four*e* days lay ded · his lyff recuryd
and where I purpose me to tempt · A-non he me Asaylys
Mawdelyn playne remyssyon̄ · Also he hath ensuryd.

¶ Goddys son̄ he p*r*etendyth · *and* to be born of A mayde
and seyth he xal dey · for mannys saluac*i*on
þan xal þe trewth be tryed · *and* no fordere be delayd
Whan þe soule fro þe body · xal make sep*a*rac*i*on
And as for hem þ*at* be vndre · my grett domynac*i*on
He xal fayle of hese intent *and* pu*r*pose[2] Also
Be þis tyxt of holde · remembryd to my*n* intenc*i*on
Quia in inferno nulla est redempc*i*o.

¶ But whan þe tyme xal neyth of his p*er*secuc*i*on
I xal Arere new Engynes · of malyc*i*ous co*n*spiracy
Plente of rep*r*evys · I xal p*r*ovide · to his confusyon̄
þus xal I false þe wordys · þ*at* his pepyl doth testefy
His discipulis xal for-sake hy*m* · *and* here mayst*er* denye
In-novmberabyl xal his wou*n*dys be · of woful grevau*n*s
A tretowr*e* xal cou*n*tyrfe · his deth · to fortyfye
Þe rebukys þ*at* he gyf me · xal t*ur*ne to his displesau*n*s.

[1] So (pvaylys) in MS. [2] MS. pũpose.

Fo. 137 ¶ Some of hese dyscypulys · xal be chef · of þis Ordenawns
Þat xal fortefye þis termn̄ · *þat* in trost is tresoñ
Þus xal I venge · be sotylte al my malycious grevau*n*s
For no thyng may excede · my prudens *and* dyscrec*i*on. 60

¶ Gyff me ȝ*ou*r love · grawnt me my*n* Affec*c*ion
And I wyl vnclose · þ*e* tres*our* of lovys Alyawns
And gyff ȝow ȝour*e* desyrys afft*er*e ȝour*e* intenc*i*on
no poverte xal aproche ȝow · fro plentev*ou*s Abu*n*dau*n*s.

¶ By-holde þe dyvercyte · of my dysgysyd varyau*n*s 65
Eche thyng sett ⁝ of dewe naterall, dysposyc*i*on
and eche parte Acordynge · to his resemblau*n*s
Ffro þe sool of þe ffoot, to þe hyest Asenc*i*on.

¶ Off ffyne cordewañ, A goodly peyr*e* of long pekyd schon
hosyn enclosyd ⁝ of þe most costyous cloth · of Crenseyñ 70
þus a bey to a jentylman · to make comp*ar*ycion
W*ith* two doseyn poyntys of cheverelle ⁝ þe Aglott*ys* of syluer[1] feyñ.

¶ A shert of feyn holond[2] ⁝ but car*e* not for þ*e* payment
A stomachere of clere reynes · þe best may be bowth
Þow poverte be chef ⁝ lete p*r*ide þ*er* be p*r*esent 75
And all þo þ*at* repreff p*r*ide, þ*ou* sette hem at nowth.

¶ Cadace · wolle · or flokkys · where it may be sowth
to stuffe w*ith*-al þ*i* dobbelet, *and* make þe of p*r*oporcyon
two smale legges · And a gret body · þow it ryme nowth
ȝet loke þ*at* þ*ou* desyre · to An þe newe facc*i*on. 80

¶ A gowne of thre ȝerdys loke þ*ou* make comp*ar*isoñ
Vn-to all degrees dayly · þ*at* passe þi*n* astat
A purse w*ith*-outyn mony ⁝ a daggere for devoscyoñ
And þere repref is of synne ⁝ loke þu make debat.

Fo. 137v ¶ W*ith* syde lokkys I schrewe þi*n* here[3] · to þ*i* colere hangyng doñ 85
to herborwe qweke bestys þ*at* tekele men onyth

[1] The initial *s* is written over some other letter (? *þ*).
[2] The word *cloth* first written after *holond* but crossed through.
[3] ou*er* þ*in* eyn *and* þ*in* her*ys* is written by the same hand above the line, as if for an alternate reading for *I schrewe þin here.*

An hey smal bonet · for curyng of þe crowne
And alt beggerys *and* pore pepylt · haue hem on[1] dyspyte
On to þe grete Othys · And lycherye gyf þi delyte
to maynteyn þin astate lete brybory be present 90
And yf þe lawe repreve[2] þe · say þou wylt ffyth
And gadere þe A felachep after þin entent.

¶ Loke þou sett not be precept · nor be comawndement
Both sevyle *and* Canone[3] · sett þou at nowth
Lette no membre of god · but with othys be rent 95
Lo þus þis werd at þis tyme · to myn intent is browth
I Sathan with my felawus · þis werd hath Sowth
And now we han it · at houre plesawns
Ffor synne is not shamfast · but boldnes hath bowth
Þat xal cause hem in helle to han inerytawns. 100

¶ A beggerys dowtere to make gret purvyauns[4]
To cownterfete a jentyl womañ · dysgeysyd as she cañ
And yf mony lakke · þis is þe newe chevesauns
With here prevy plesawns to gett it of sum man
Here colere splayed · *and* furryd with Ermyn calabere or satañ 105
A seyn to selle lechory · to hem þat wyl bey
And þei þat wyl not by it, yet i-now xal þei hañ
And telle hem it is for love · she may it not deney.

¶ I haue browth ȝow newe namys, *and* wyl ȝe se why
Ffor synne is so plesaunt, to ech Mannys intent 110
Fo. 138 ȝe xal kalle pride · oneste · *and* naterall kend lechory
And covetyse wysdam · there tresure is present.

¶ Wreth manhod, *and* envye callyd chastement
Seyse nere sessyoñ · lete perjery be chef
Glotonye · rest · let Abstynawnce · beyn Absent 115
And he þat wole exorte þe to vertu · put hem to repreff.

¶ To rehers al my servauntys · my matere is to breff
But alt þese xal eneryth · þe dyvicion eternal

[1] Or *en*—the letter is blotted.
[2] *ll* written before *repreve* and crossed through. [3] MS. Canoñ.
[4] An *a* (mark of stage direction) crossed through in red ink in the left-hand margin here, and the beginning of the red underlining of stage directions beneath the first word of the line.

Þow cryst by his sotylte · many mat*e*rys meef
In evyr-lastynge peyne · w*ith* me dwellyn þei xal. 120

¶ Remembre our*e* seruaunt*ys* · whoys sowlys ben mortall
Ffor I must remeffe · for mor*e* mat*e*rys to p*r*ovyde
I am w*ith* ȝow at all tymes · whan ȝe to cou*n*cel me call
But for A short · tyme · my-self I devoyde.

Joh*ann*es baptis[ta

¶ I joh*a*n baptyst · to ȝow þus p*r*ophesye
Þat on xal come aftyr me · *and* not tary longe
In many folde mor*e* · strengere þan I
Of whose shon · I am not worthy to lose þe thonge
Where-fore I cou*n*cel þe ȝe reforme all wronge 5
in ȝo*ur* concyens of þe mortall dedys · sevyn
And for to do penawns · loke *þat* ȝe ffonge
Ffor now xal come þe kyngdh*a*m of hevyn.

¶ þe weys of our*e* lord cast ȝow to Aray
And þ*er*-in to walk loke ȝe be Applyande 10
And make his pathys · as ryth as ȝe may
Kepyng ryth forth · *and* be not declinande
Fo. 138v Neyther to fele · on ryth · nor on lefte hande [1]
But in þ*e* myddys · p*ur*pose ȝow · to holde
For *þat* in all wyse is most plesande 15
As ȝe xal her*e* · whan I have tolde.

¶ Of þis wey for to make · moralysacyon [2]
Be þe ryth syde ȝe xal vndyrstonde mercy
And on þe lefte syde · lykkenyd dysp*er*ac*i*on
And þe patthe be-twyn bothyn · *þat* may not wry 20
Schal be hope *and* drede · to walke in p*er*fectly
Declynyng not to fele · for no maner nede
grete cawsys I xal shove ȝow why
þat ȝe xal sowe [3] þe patthe of hope *and* drede.

[1] *syde* first written and crossed through.
[2] Two lines following this are crossed through: *Be þe ryth syde lyknyd disperacion* and *and þe pathe betwyn bothyn.*
[3] Or *sewe*—the vowel is not clear.

¶ On þe mercy of god, to meche ȝe xal not holde 25
as in þis wyse · be-hold[1] · what I mene
Ffor to do synne · be þou no more bolde
In trost þat god wole · mercyful bene
And yf be sensualyte · as it is ofte sene
Synnyst dedly · þou xalt not þerfore dyspeyre 30
but þerfore do penawns *and* confesse þe clene
And of hevyn · þou mayst trost to ben eyre.

¶ Þe pathe þat lyth · to þis blyssyd · enherytawns
Is hope *and* drede · copelyd be conjunccyon̄
Be-twyx þese tweyn̄ · may be no dysseuerawns 35
Ffor hope with-outyn drede · is maner of presumpcion
Fo. 139 And drede · with-owtyn hope · is maner of dysperacion
So these tweyn must be knyt be on Acorde
How ȝe xal aray þe wey · I haue made declararacion[2]
Also þe ryth patthis · Aȝens þe comyng of oure lord. 40

Here xal annas shewyn hym-self in his stage be-seyn after a busshop of þe hoold lawe in a skarlet gowne · and *ouer þat a blew tabbard furryd with whyte* and *a mytere on his hed after þe hoold lawe* · *ij doctorys stondyng by hym in furryd hodys* and *on be-forn hem with his staff of A-stat* and *eche of hem on here hedys a furryd cappe with a gret knop in þe crowne* and *on stondyng be-forn as a sarazyn̄ þe wich xal be his masangere* · *Annas þus seyng*

Annas

¶ As a prelat am I properyd, to provyde pes
And of jewys · jewge · þe lawe to fortefye
I Annas be my powere · xal comawnde dowteles
Þe lawys of moyses · no man xal denye
Hoo excede my comawndement · Anon ȝe certefye
Yf Any eretyk here reyn · to me ȝe compleyn̄
For in me lyth þe powere · all̄ trewthis to trye
And pryncypaly oure lawys · þo must I susteyn̄.

¶ ȝef I may aspey · þe contrary[3] no wheyle xal þei reyn

[1] Or *held*. [2] So in MS.
[3] *contrary* apparently first written, corrected to *contraly*, and again, imperfectly but possibly by the scribe, to *contrary*.

But a-non to me be browth · *and* stonde p*re*sent
Be-fo*re* her*e* jewge · wich xal not feyn
But aft*ere* her*e* trespace · to gef hem jugement
Now serys for A p*ro*se · heryth my*n* intent
There is on jh*es*us of nazareth · þ*at* our*e* lawys doth excede
Yf he p*ro*cede · thus · we xal us all repent
For our*e* lawys · he dystroyt · dayly w*ith* his dede.

Fo. 139v ¶ There fo*re* be ȝ*our* cowncel we must take hede
What is be[1] to p*ro*vyde or do in þis case
Ffor yf we let hym þus go · *and* ferder*e* p*ro*sede
Ageyn sesar*e* *and* our*e* lawe we do trespace.

primus doctor Annas[2]

¶ Sere þis is my*n* Avyse · þ*at* ȝe xal do
Send to cayphas for cowncel knowe his intent
For yf jh*es*u proce[3] *and* þus forth go
Our*e* lawys xal be dystroyd thes se we p*re*sent.

ijus doctor annas[4]

¶ Sere remembre þe gret charge · þ*at* on ȝow is leyd
Þe lawe to ke[5] which may not ffayle[6]
Yf any defawth p*re*vyd of ȝow be seyd
Þe jewys w*ith* trewth · wyl ȝow a-sayl
Tak hed whath cownsayl · may best p*ro*vayl[7]
After rewfyn *and* leyon · I rede[8] þ*at* ȝe sende
They arn temp*er*al jewgys, þ*at* knowyth þe p*ar*ayl
W*ith* ȝoure cosyn cayphas · þis matere to Amende.

Annas

¶ Now surely þis cowncel · revyfe my*n* herte
ȝoure cowncel is best as I can se
Arfexe in hast loke þ*at* þ*ou* styte
And pray cayphas my cosyn come speke w*ith* me.

¶ To rewfyn *and* leon þu go Also
And pray hem þei speke w*ith* me in hast

[1] So in MS. for *best*.
[2] *Annas* written in a different hand, possibly that of Ff. 95, 96.
[3] So in MS. for *procede*.
[4] *annas* written as before.
[5] So in MS. for *kepe*.
[6] The first three letters in this word have been written over others.
[7] So (þvayl) in MS.
[8] With this word the ink becomes darker and the writing is perhaps slightly more angular from here to the bottom of Fo. 141v, where the darker ink ceases.

For A pryncipal matere · *þat* haue to do
Wich must be knowe or þis day be past. 40

Arfexe

¶ My souereyn at *ȝour* intent · I xal gon
In al þe hast *þat* I kan hy
On to Cayphas · rewfyn *and* lyon
And charge ȝoure intent *þat* þei xal ply.

Fo. 140 *here goth þe masangere forth* and *in þe mene tyme cayphas shewyth him-self in his skafhald · Arayd lych to Annas savyng his tabbard xal be red furryd with white ij doctorys with hym arayd with pellys aftyr þe old gyse* and *furryd cappys on here hedys · Cayphas þus seyng·*

Cayphas

¶ As A *pr*imat most preudent · I *pr*esent here sensyble 45
buschopys of *þe* lawe w*ith* al *þe* cyrcu*m*stawns
I Cayphas am jewge · w*ith* powerys possyble
To distroye all erro*ur*is · *þ*at in our*e* lawys make varyawns
All thyng*ys* I convey be reson *and* temp*er*awnce
And all mat*er*is possyble · to me ben palpable 50
Of *þe* lawe of moyses I haue A chef governawns
To seuere[1] ryth *and* wrong in me is t*er*mynable.

¶ but *þer* is on Cryst *þat* our*e* lawys is varyable
he p*er*verte *þe* pepyl w*ith* his p*r*echy*n*g ill
We must seke A mene · on to hym rep*r*evable 55
ffor yf he p*r*ocede our*e* lawys he wyl spyll.

¶ We must take good cowncel in *þ*is case
Of *þe* wysest of *þe* lawe[2] *þ*at kan *þe* trewthe telle
of *þe* jewgys of pharasy · *and* of my cosyn Annas
for yf he p*r*ocede be p*r*ossesse · our*e* lawys he wyl felle. 60

p*rim*us doct[or
Cayfas[3]

¶ My*n* lord plesyt ȝow to pardon me for to say
þe blame in ȝow · is as we fynde[4]
to lete cryst contenue *þ*us day be day
W*ith* his fals wichcraft *þe* pepyl to blynde

[1] *deuere* first written and crossed through.
[2] *ll* written before *lawe* and crossed through.
[3] *Cayfas* written as *Annas* above in the Fo. 95, &c., hand.
[4] *haue fow* first written and crossed through.

Fo. 140v He werkyth fals meraclis Ageyns[1] att kende 65
And makyth oure pepyl to leve hem in
It is ȝ*our* part to take hym *and* do hym bynde
And gyf hy*m* jugement for his gret syn.

ij^us doct*or*
¶ For-sothe sere of trewth this is þe case Cayphas[2]
On to o*ur* lawe ȝe don oppressyon̄ 70
Þ*at* ȝe let cryst from ȝ*ou* pace
and wyl not don on hy*m* correxion̄
Let Annas knowe ȝ*our* intenc*i*on
W*ith* prestys *and* jewg*ys* · of þe lawe
And do cryst fforsake his fals oppynyon 75
Or in to A p*re*son lete he*m* be thrawe.

Cayphas
¶ Wel ser*ys* ȝe sal se w*ith*-inne short whyle
I xal correcte hy*m* · for his trespas
He xal no lenger oure pepyl be-gyle
Out of my*n* dawngere he xal not pas. 80

here comyth þe masangere to cayphas and *in þe mene tyme rewfyn* and *lyon schewyn hem in þe place in ray tabardys furryd* and *ray hodys a-bouth here neckys furryd · þe masangere seyng·*

Masangere
¶ Myn reverent souereyn · *and* it do ȝow plese
Sere annas my lord hath to ȝ*ou* sent
He p*ra*yt ȝ*ou* þ*at* ȝe xal not sese
Tyl þ*at* ȝe ben w*ith* hy*m* p*re*sent.

Cayphas
¶ Sere telle my*n* cosyn I xal not fayl 85
it was my p*ur*pose hy*m* for to se
For serteyn mater*ys* þ*at* wyl[3] p*ro*vayle[4]
Þow he had notwth a sent to me.

Masag*er*
Fo. 141 ¶ I recomende me to ȝ*our* hey degre
on mor*e* massagys I must wende. 90

Cayphas
Ffare wel sere *and* wel ȝe be
gret wel my cosyn *and* my ffrede[5]

[1] *att kende* first written before *Ageyns* and crossed through.
[2] *Cayphas* writtten as *Annas* above in the Fo. 95, &c., hand.
[3] The *l* seems to have been omitted and inserted later.
[4] So in MS. [5] So in MS. for *ffrende*.

here þe masager metyth with þe jewgys sayng Masager

¶ Heyl jewgys of jewry · of reson most prudent
of my massage to *ȝou* · I make relacion
my lord sere Annas hath for *ȝou* sent 95
to se his presens *with*-owth delacion.

Rewfyn

¶ Sere we Are redy At his comawndement
To se sere Annas in his place
it was oure purpose *and* oure intent
to A be *with* hym *with*-inne short space. 100

leyon

¶ We are ful glad his presence to se
sere telle hym · we xal come in hast
no declaracion *þer*-in xal be
but to his presens hye us fast.

masager

¶ I xal telle my lord *ser*is as ȝe say 105
ȝe wyl ful-fylle al his plesawns.

Rewfyn

Sere telle hy*m* · we xal make no delay
but come in hast · at his instawns.

a here þe masangere comyth to Annas þus seyng

masan[ger

¶ My lord *and* it plese *ȝou* to haue intellygens
Ser Cayphas comyth to *ȝou* in hast 110
Rewfyn *and* lyon · wyl se ȝo*ur* presens
And se ȝow here or þis day be past.

Annas

Fo. 141v (marked 140) ¶ Sere I kan þe thank of þ*i* dyligens
Now ageyn my cosyn I wole walk
Serys folwyth me on to his presens 115
Ffor of these mater*ys* we must talk.

a here Annas goth down to mete with cayphas and *in þe mene tyme þus seyng*

Cayphas

¶ Now on to annas let us wende
ech of vs to knowe other*ys* intent
Many mater*ys* I haue in mende
Þe wich to hy*m* I xal present. 120

i^us^ doctor C.

¶ Sere of alł othere thyng remembre þis case
Loke þat ihesus be put to schame.

ij^us^ doctor C.

Whan we come present beforn annas
Whe xal rehers alł his gret blame.

α *here þe buschopys with here clerk*ys and *þe Pharaseus mett* and[1] *þe myd place* and *þe*r *xal be a lytil oratory with stolys* and *cusshonys clenly be-seyn*[2] *lych as it were a cownsel hous* · *Annas* þus *seyng*

Annas

¶ Wel come *ser* Cayphas · *and* ȝe jewgys[3] alłe
now xal ȝe knowe alł myn entent
A wondyr case serys · here is be-falle
On wich we must · gyf jewgement[4]
Lyst þat we aftyre þe case repent
Of on cryst þat goddys sone · som doth hym calle
He shewyth meraclys · *and* sythe present
þat he is prynce of prync*ys* alle.

¶ The pepyl so fast to hym doth falle
Be prevy menys as we a-spye
ȝyf he procede · son sen ȝe xalle
þat oure lawys he wyl dystrye[5].

Fo. 142 ¶ It is oure part þus[6] to deny
What is ȝ*our* cownceł in þis cas.

Cayphas

Be reson þe trewth here may we try
I cannot dem hym with-outh trespace
Be-cause he seyth in every A place
þat he kyng of jewys in every degre
þerfore he is fals knowe wel þe case
Sesar is kyng *and* non but he.

Rewfyn

¶ He is An eretyk[7] *and* a tret*our* bolde
To sesare *and* to oure lawe sertayn

[1] So in MS. for *at*. [2] A mark over *be*, as for be*n*, is probably accidental.
[3] The *g* omitted and written over the line.
[4] *alle* first written after *jewgement* and crossed through.
[5] *þat oure lawys* scribbled beneath in a slightly later hand (?).
[6] *þus* perhaps corrected to *þis*. The MS. is rubbed.
[7] The *t* in this word is blotted and may have been corrected in some way.

Bothe in word *and* in werke *and* ȝe be-holde
He is worthy to dey w*ith* mekyl peyn.

leon̄

¶ þe cawse þ*at* we been her*e* p*re*sent
To fortefye þ*e* lawe *and* trewth to say 150
Jh*es*us ful ner*e* our*e* lawys hath shent
þ*er*fore he is worthy for to day.

i^us^ doctor An[nas

¶ Ser*ys* ȝe þ*at* ben rewelerys of þe lawe
On jh*es*u ȝe must gyf jugement
Let hy*m* fyrst ben hangyn *and* drawe 155
and þanne his body in fyr*e* be brent.

ij^us^ doctor An[nas

¶ Now xal ȝe her*e* þ*e* intent of me
Take jh*es*u þ*at* werke us all̄ gret schame
Put hym to deth · let hym not fle
For þan þ*e* comownys[1] þei wyl ȝow blame. 160

i^us^ doct*or* Cayp[has

Fo. 142v ¶ He werke w*ith* weche[2]-crafte · in eche place
And drawyth þe pepyl to hese intent
Be whar*e* ȝe jewgys let hym not passe
þan be my trewthe[3] ȝe xal repent[4].

ij^us^ doctor Cayphas

¶ Serys takyth hede on to þis case 165
and in ȝo*ur* jewgement be not slawe
þ*er* was nevyr man dyd so gret trespace
As jh*es*u hath don Ageyn our*e* lawe.

Annas

¶ Now bretheryn þan wyl ȝe her*e* my*n* intent
These ix days let us A-byde 170
We may not gyf so hasty jugement
but eche man inqwer*e* on his syde
Send spyes A-bouth þe cou*n*tre wyde
to se *and* recorde *and* testymonye
And þan hese werkys he xal not hyde 175
nor haue no power hem to denye.

Cayphas

This cownceil̄ A-cordyth to my reson.

[1] *Cowm* first written. [2] Some miswritten letter— *?f*—after *weche*.
[3] Or *trowthe*—letter blotted. [4] *repepent* first written.

Annas

And we Alł to þe same.

va-cat. *here Enteryth þe Apostyl petyr* and *johan þe euangelyst with hym.*

petyr seyng

O ȝe pepyl · dyspeyryng be glad
A gret cause ȝe haue · *and* ȝe kan se
þe lord of alł þing · of nowth mad
is comyng ȝ*our* comfort to be
Alł ȝ*our* langorys · salvy*n* xal he
ȝ*our* helthe is mo*re* þan kan wete.[1]

'o. 143[2] ¶ jhesus

¶ Ffrendys be-holde þe tyme of mercy
The whiche is come now w*ith*-owt dowth 180
ma*n*nys sowle in blys now xal edyfy
and þe prynce of þe werd is cast owth.

¶ Go to ȝon castel þat standyth ȝow Ageyn
su*m* of my*n* dyscyplis · go forth ȝe to
þe*re* xul ȝe ffyndyn bestys tweyn 185
An Asse tyed *and* he*re* fole Also
Vn-losne þat Asse · and brynge it to me pleyn
Iff any mas[3] Aske why þat ȝe do so
Sey þat I haue nede to þis best certeyn
and he xal not lett ȝow · ȝ*our* weys for to go 190
þat best brynge ȝe to me.

1[us] Ap*osto*lus

holy p*ro*phete we gon our*e* way
we wyl not ȝour*e* wourd de-lay
Also sone as þat we may
We xal it brynge to the. 195

here þei *ffecch* þe *Asse with* þe *ffole* and *þe burgeys seyth*

Burgensis

¶ Herke ȝe men who ȝaff ȝow leve
Thus þis[4] best ffor to take Away

[1] This passage is cancelled in the MS. by a red stroke through the stage direction, black strokes through the speech, and a red bracket and the word *vacat*, as above, in black.

[2] This folio is interpolated. It is of different paper and the writing (though this is the same as that of the body of the MS.) and rubrication have not the features that distinguish the rest of quire N.

[3] So in MS. [4] *ff* crossed through after *þis*.

but only ffor pore men to releve
this Asse[1] is ordayned as I ȝow say.

Philippus

Good sere take this At no greff[2]
Oure mayster us sent hedyr þis day
he hath grett nede with-owt repreff
þerfore not lett us I þe pray
þis best for to lede.

Burgensis

Fo. 143v Sethyn þat it is so ? þat he hath ȝow sent
werkyth his wyll *and* his intent
take þe beste as ȝe be bent
and evyr wel mote ȝe spede.

jacobus minor

¶ This best is brought ryght now here lo
holy prophete At þin owyn wylle
And with þis cloth Anon Also
þis bestys bak we xal sone hylle.

Philippus

Now mayst þou ryde whedyr þou wylt go
Thyn holy purpos to ffulfylle
thy best fful redy is dyth þe to
bothe meke *and* tame þe best is stylle
And we be redy Also
Iff it be plesynge to þi ssyght
The to helpe A-non forth ryght
Vpon þis best þat þou were dyght
þi jurney ffor to do.

here cryst rydyth out of þe place And he wyl · and *Petyr* and *Johan Abydyn stylle* · And *at þe last whan þei haue don* þer *prechyng þei mete with jhesu,*

Petrus

¶ O ȝe pepyl dyspeyryng[3] · be glad
A grett cawse ȝe haue *and* ȝe kan se
þe lord þat all thynge · of nought mad
is comynge ȝour comforte to be
All ȝour langoris · salvyn xal he

[1] *assa* first written.
[2] *Thus this best to take* first written as the following line and crossed through.
[3] The last *y* in this word is altered from an *e*.

ȝ*our* helthe is mor*e* than ȝe kan wete
Fo. 144[1] He xal cawse þ*e* blynde · þ*at* þei xal se
þ*e* def to here · þ*e* dome · for to speke.

¶ þei þ*at* be crokyd · he xal cause he*m* to goo
In þe wey · þ*at* joh*an* baptyst · of pr*o*phecyed
Sweche A leche · kam ȝow nevyr non too
Wher-fo*re* what he comawndyth · loke ȝe Applyed
þ*at* som of ȝow · be blynd · it[2] may not be[3] denyid[4]
Ffor hy*m* þ*at* is ȝo*ur* make*re* · w*ith* ȝo*ur* gostly ey ȝe xal not knowe
of his comaundement · in ȝow gret necglygens is Aspyed
Where-fo*re* def · fro gostly heryng · clepe ȝow I howe.

¶ And some of ȝow · may not go · ȝe be so crokyd
For of good werkyng · in ȝow · is lytyl habu*n*dawns
Tweyn[5] fete · hen̄ery man xuld haue · *and* it were lokyd
Wyche xuld bere · þe body gostly · most of substawns
Ffyrst is to love god Above all̄ other plesawns
þe secunde · is to love · þ*i* neybore · as þi*n* owyn p*er*sone
and yf þese tweyn · be kepte in p*er*severawn̄s
Into þ*e* celestyal habytac*i*on · ȝe Arn habyl to gone.

¶ Many of ȝow be dome · why · for ȝe wole not redresse
Be Mowthe · ȝ*our* dedys mortal · but þ*er*-in don p*er*dure
Of þe wych̄ · but ȝe haue contrycyon̄ · *and* ȝow confesse
ȝe may not in-heryte hevy*n* · þis I ȝow ensur*e*
And of All̄ þese maladyes · ȝe may haue gostly cure
For þe hevynly leche · is comyng ȝow · for to vicyte
And as for payment · he wole shewe ȝow · no redrur*e*
Ffor w*ith* þ*e* love of ȝowr*e* hertys · he wole be Aqwhyte.

Joh*ann*es apo*stolus*

Fo. 144v Apostolus joh*ann*es

¶ On to my brotherys for-seyd rehersall̄
þat ȝe xuld ȝeve þe more · veray confydens

[1] The less regular writing and less discriminating rubrication of the N quire resumed.

[2] *it* has been written over an erasure in darker ink.

[3] *dey* first written instead of *be* and crossed through and *be* written above the line.

[4] The *id* seems to have been added as a correction.

[5] The contraction mark is a rough blotted stroke.

I come w*ith* hym as testymonyall[1]
Ffor to conferme · *and* fortefye his sentens
þis lord xal come w*ith*-out resystens
On to þe Cety-ward he is now comyng
Where-fore dresse ȝow w*ith* all dew dylygens 260
To honowr*e* hy*m* as ȝ*our* make*re* *and* kyng.

¶ And to fulfylle þe p*ro*phetys p*ro*phese
Vp-on An Asse he wole hedyr ryde
Shewyng ȝow exawmple of humylyte
Devoydyng þ*e* Abhomynable synne of pryde 265
Whech hath ny conqweryd all þe werd wyde
Grettest cause of All ȝ*our* trybulacyon
Vse it ho so wole · for it is þ*e* best gyde
þ*at* ȝe may haue to þ*e* place of dampnacyon.

¶ Now brothyr in god syth we have intellygens 270
þ*at* our*e* lord is ny come to þis cete
To Attend up-on his p*re*cyous p*re*sens
it syttyth to us · as semyth me
Wherfor*e* to mete whit hy*m* now go we
I wold for*e* no thyng we where[2] to late 275
to þ*e* cete-ward fast drawyth he
me semyth he is ny at þ*e* gate[3]

α here spekyth þe iiij ceteseynys · þe fyrst þus seyng[4]

1*us* ciues de jher*usal*em

Fo. 145[5] ¶ Neyborys gret joye in our*e* herte we may make
þ*at* þis hefly kyng · wole vycyte þis cyte.

ij*us* ciues

Yf our*e* eerly kyng · swech a jorne xuld take 280
to don hym hono*ur* *and* worchepe · besy xuld we be.

iij*us* ciues

Meche mor*e* þan to þe hevynly kyng · bownd Ar*e* we
Ffor to do þ*at* xuld be to his p*er*sone reuerens.

[1] The second *l* is written over some blotted letter (? *e*).
[2] The *h* in this word is written over an *e*.
[3] The *g* has been altered from some other letter.
[4] A note in a later hand stands in the left-hand margin at the bottom of this folio : *here entrith þe fyrst prophete.*
[5] Two numbers, 16 and 40, are here crossed out in paler ink.

iiij ci[ues

late vs þan welcome hy*m* w*ith* flowr*ys and* brawnchis of þ*e* tre
Ffor he wole take þ*at* to plesawns · becawse of redolens. 285

α *here þe iiij ceteseynys makyn hem redy for to mete with oure lord goyng barfot* and *barelegged* and *in here shyrt*ys *savyng þei xal haue here gownys cast A-bouth theme* and *qwan þei seen oure lorde þei xal sprede þ*er *clothis be-forn hyme* and *he xal lyth* and *go þ*er *upone* and *þei xal falle downe up-on þ*er *knes alle atonys þe fyrst þ*us *seyng*

i^us^ ciues

¶ Now blyssyd he be þ*at* in our*e* lordys name
to us in Any wyse wole resorte
And we be-leve · veryly · þ*at* þ*ou* dost þ*e* same
For be þ*i* m*er*cy xal sprynng m*an*nys comforte.

α *here cryst passyth forth · þ*er *metyth with hym a serteyn of chylderyn with flowr*ys and *cast be-forn hyme* and *They synggyn Gloria laus* and *be-forn on seyt*[. . .

Thow sone of davyd' þ*ou* be our*e* supporte 290
At our*e* last day whan we xal dye
Where-fo*re* we Alle Atonys to þ*e* exorte
Cryeng mercy mercy mercye.[1]

Jh*e*su[2]

Fo. 145v ¶ Ffrendys be-holde þ*e* tyme of mercy
þ*e* wich is come now · w*ith*-owtyn dowth 295
Mannys sowle in blysse now xal edyfy
And þ*e* prynce of þ*e* werd · is cast owth
As I haue p*re*chyd in placys A-bowth
And shewyd experyence · to man *and* wyf
In to þis werd goddys sone hath sowth 300
Ffor veray loue · ma*n* to revyfe.

¶ The trewthe of trewthis xal now be tryede
and A p*er*fyth of corde be-twyx god *and* man
Wich trewth xal nevyr be dyvide
Confusyon on to þ*e* fynd sathan. 305

i^us^ paup*er* ho*mo*

¶ þ*ou* sone of davyd on vs haue mercye
As we must stedfast be-levyn in þ*e*

[1] A similar note in the same later hand stands at the bottom of Fo. 145 with a reference mark to the close of this speech: *here entreth þe parte of þe ij^de^ prophete.*

[2] The name of the speaker stands as usual at the bottom of the last page, but in this case is written again at the top of Fo. 145v.

þi goodnesse lord lete us be nye
Whech lyth blynd here *and* may not se.

ij^us^ pauper homo

¶ Lord lete þi mercy to us be sewre
and restore to us oure bodyly syth
We know þou may us wel recure
With þe lest poynt · of þi gret myth.

Jhesu

¶ ȝowre be-leve hath mad ȝou for to se
And delyveryd ȝou fro All mortal peyn
blyssyd be All þo þat be-leve on me
And se me not with here bodyly eyn.

α *here Cryst blyssyth here eyn* and *þei may se þe fryst seyng*

Fo. 146

i^us^ pauper homo

* Gromercy[1] lord of þi gret grace
I þat was blynd · now may se.

ij^us^ pauper homo

Here I for-sake al my trespace
And stedfastly wyl be-levyn on þe.

α *here cryst procedyth on fote with his dyscipulys After hym cryst wepyng up-on þe cyte sayng þus* †

Jhesu

¶ O Jherusalem woful is þe ordenawnce
Of þe day of þi gret persecucyon
þou xalt be dystroy[2] with woful grevans
and þi ryalte browth · to trew confusyon
ȝe þat in þe cete · han habytacyon
þei xal course þe tyme þat þei were born
So gret advercyte *and* trybulacion
Xal falle on hem both evyn *and* morwyn.

27

¶ þei þat han most chylderyn sonest xal wayle
and seyn Alas · what may þis meen
Both mete *and* drynk · sodeynly xal fayle
Þe vengeance of god þer xal be seen

* No paragraph sign here in MS.

[1] Somewhat larger capital than usual.

† A large sign in darker coloured ink stands in the left-hand margin against this stage direction and the first lines of the next speech.

[2] So in MS.

þe tyme is comyng · hes woo xal ben
þe day of trobyl *and* gret grevau*n*s 335
Bothe templys *and* towrys they xal down cleen
O cete ffuł woful · is þi*n* Ordenawns.

Petrus

¶ Lord wher*e* wolte þ*ou* kepe þ*i* mau*n*de
I pray þe now lete us haue knowyng
Þ*a*t we may make redy for þe 340
Þ*e* to serve w*i*t*h*-owte · latyng.

Joh*an*nes

Fo. 146 To p*r*ovyde lord · for þi comyng
W*i*t*h* ałł þ*e* obedyens · we kan A-tende
And make redy for þe in Al thyng
In to what place þ*ou* wytl[1] us send. 345

Jhesu

¶ Serys goth to syon *and* ȝe xal mete
A por*e* man in sympyl A-ray
Beryng watyr in þe strete
Telle hy*m* I xal come þ*a*t way
On-to hym mekely · loke þ*a*t ȝe say 350
Þ*a*t hese house I wele come tylle
he wele not onys to ȝow sey nay
But sofre to haue ałł ȝ*our* wylle.

Petrus

¶ At þi wyl lord it xal be don
To seke þ*a*t place we xal us hye. 355

Joh*an*nes

In Ałł þ*e* hast · þ*a*t we may go
þi*n* comawdement[2] nevyr to denye.

α here petyr and *johan gon forth metyng with symon leprows beryng A kan with watyr petyr þus seyng*

Petrus

¶ Good man þe pr*o*phete · our*e* lord jh*e*sus
Þis nyth wyl · rest wyth-in þi*n* halle
On massage to þe he hath sent vs 360
Þ*a*t ffor his sopere ordeyn þ*ou* xalle[3].

[1] So apparently in MS. for *wylt*—Halliwell reads *wyth*; *wyłł* is possible; the letters are not clear.
[2] So in MS. [3] The *e* is written over another letter (? *t*).

Joh*ann*es

ȝa for hym *and* his dyscipulys Alle
Ordeyn þu for his maunde
A paschall lomb what so be-falle
Ffor he wyl kepe his pasch w*ith* the. 365

Symon

Fo. 147 ¶ What wyl my lord · vesyte my plase
Blyssyd be þe tyme of his comyng
I xal ordeyn w*ith*-inne short space
Ffor my good lordys wel comyng
Serys walkyth in at þe begynny*ng* 370
And se what vetaylys · þ*at* I xal take
I am so glad of þis tydyng
I wot nevyr what joye þ*at* I may make.

here þe dyscypulys gon in with Symone to se þe ordenawns and *cryst comyng thedyr-ward þus seyng*

Jh*es*us

¶ Þis path is Calsydon[1] · be goostly ordenawns
Wech xal co*n*uey us wher we xal be 375
I knowe ful redy is þe p*ur*vyaunce
Of my frendys þ*at* lovyn me
Contewnyng[2] in pees · now p*ro*cede we
For mannys love þis wey I take
W*ith* gostly ey I veryly se 380
Þ*at* man ffor man · an hende must make.

here þe dyscipulys come A-geyn to cryst petyr þus seyng

Petrus

¶ All redy lord is our*e* ordenawns
As I hope to ȝow plesyng xal be
Seymon hath don at ȝour*e* instawns
He is ful glad ȝ*our* p*re*sens to se. 385

Joh*ann*es

¶ All thyng we haue lord at our*e* plesyng
Þ*at* longyth to ȝour*e* mawnde w*ith* ful glad cher*e*
Whan he herd telle of ȝ*our* comyng
Gret joye in hy*m* þ*an* dyd Appere.

Fo. 147v *here comyth symon owt of his hous to welcome cryst.*

[1] So in MS. Halliwell prints *cal Sydon.* [2] So in MS.

Symon

¶ Gracyous lord · wel come þu be 390
Reverens be to þe both god *and* man
My poer hous · þat þou wylt se
Weche am þi servaunt as I kan.

Jhesu

¶ There joye of All joyis to þe is sewre
Symon I knowe þi trewe intent 395
þe blysse of hefne · þou xalt recure
Þis rewarde I xal þe grawnt present.

α *here Crist enteryth in-to þe hous with his disciplis* and *ete þe paschal lomb* and *in þe mene tyme þe counsel hous beforn-seyd xal sodeynly onclose schewyng þe buschopys prestys* and *jewgys syttyng in here Astat lyche as it were A convocacyone · Annas seyng þus*

Annas

¶ Be-hold[1] it is nowth al þat we do
In alle houre materys we prophete nowth
Wole[2] ȝe se wech peusawns of pepyl drawyth hym to 400
Ffor þe mervaylys þat he hath wrowth.

¶ Some othyr sotylte · must be sowth[3]
Ffor in no wyse we may not þus hym leve
Than to A schrewde conclusyoñ · we xal be browth
Ffor þe romaynes þan wyl us myscheve. 405

¶ And take oure Astat *and* put us to repreve
and convey all þe pepyl at here owyn request
and þus all þe pepyl in hym xal be-leve
Þerfore I pray ȝow cosyn · say what is þe best.

Cayphas

Fo. 148 Attende now serys to þat I xal seye 410
on-to us all it is most expedyent
þat o man ffor þe pepyl xuld deye
þan all þe pepyl xuld perysch *and* be shent.

¶ Þerfor late us werk wysely þat we us not repent
We must nedys put on hym som fals dede 415
I sey for me I had levyr he were brent
þan he xuld us alle þus ouyr-lede

[1] Larger capital than usual. [2] Or *Wele*.
[3] *wrowth* first written.

þer-fore every man on his party help at þis nede
and cowntyrfete aꝗ þe sotyltes þat ȝe kan
now late se ho kan ȝeve best rede
to ordeyn sum dystrucc*i*on ffor þis man̄. 420

Gamalyel

¶ Late us no lenger make delac*i*on
but do jh*e*su be takyn in hondys fast
And aꝗ her*e* ffolwer*ys* to her*e* confusyon
And in-to a preson do hem be cast 425
Ley on hem yron þ*a*t wol last
Ffor he hath wrouth a-ȝens þe ryth
And sythyn Aftyr we xal iṅ hast
Jewge hym to deth w*ith* gret dyspyth.

Rewfyn̄

¶ Ffor he hath trespacyd A-ȝens our*e* lawe 430
me semyth þis wer*e* best jewgement
W*ith* wyld hors lete hym be drawe
And afftyr in fyr*e* he xal be brent.

leyon̄

¶ Serys o thyng my self herd hym sey
þat he was kyng of jewys Alle 435
þ*a*t is a-now to do hym dey
Ffor treson to sezar we must it calle.

Fo. 148v ¶ He seyd Also to p*er*sonys þ*a*t I know
þat he xuld *and* myth serteyn
þ*e* gret tempyl mythtyly ovyr-throw 440
and þe thrydde day reysynt Ageyn.

¶ Seche materys þe pepyl doth conseyve [1]
To ȝeve credens to his werkys Alle
In hefne he seyth xal be his reyn
Bothe god *and* man he doth hym calle. 445

Rewfyn̄

And Aꝗ þis day we xuld contryve
What shameful deth jh*e*su xuld haue
We may not do hy*m* to meche myscheve
þe worchep of our*e* lawe to save.

leyon

¶ Vp on A jebet lete hym hongyn be 450
þis jugement me semyth it is reson

[1] So in MS. for some word (? *constreyn*) to rhyme with *reyn*.

þat all̃ þe cou*n*tre may hym se
and be war*e* be his gret treson.

Rewfyn̄

¶ ȝet o thyng serys ȝe must A-spye
And make A ryth sotyl ordenawns 455
Be what menys ȝe may come hy*m* bye
For he hath many folwerys at his instawns.

Annas

¶ Serys *þer* of we must have avysement
and ben Acordyd or þan we go
How we xal han hym at our*e* entent 460
Som wey we xal fynd *þer*to.[1]

Mawdelyn

Fo. 149 (O quire)[2] ¶ As a cursyd creatur*e* closyd all̃ in car*e* Maria Magdalen̄
and as a wyckyd wrecche all̃ wrappyd in wo
Of blysse was nevyr no berde so bare
as I my-ſylf *þat* her*e* now go 465
Alas Alas I xal for far*e*
ffor *þo* grete synnys *þat* I haue do
lesse than my lord god sum-del spar*e*
and his grett mercy receyve me to
Mary mavdelyn is my name 470
Now wyl I go to cryst jhesu
ffor he is lord of all̃ vertu
and for su*m* grace I thynke to sew
ffor of my-self I haue grett shame.

¶ A mercy lord *and* salve my synne 475
Maydenys ffloure *þou* wasch̃ me fre
þer was nevyr woman of ma*n*nys kynne
so ful of synne in no cou*n*tre
I haue be ffowlyd be fryth *and* ffenne

[1] At the bottom of the folio below this line stands the stage direction : *α. here judas caryoth comyth in-to þe place,* followed by the word *Jhesus* as name of the next speaker. These are crossed through in black and red ink. Beneath stand three alternative catchwords : *now cownter fedyd*; in the same ink as the text (cf. Fo. 152) crossed through in red ; *myn hert is ryth,* in fainter ink (cf. Fo. 150) crossed through in black ; and *as a cursyd* (cf. Fo. 149), in the same ink as *Mawdelyn,* written as name of speaker under the cancelled *Jhesus.*

[2] Quire O is of different paper from quire N and the following quires P, Q, R ; and the writing and rubrication have not the characteristic features found in quire N and quires P, Q, R.

and sowght synne in many A cete 480
but þ*ou* me borwe lord I xal brenne
w*ith* blake ffendys Ay bowne to be
where fore kynge of grace
W*ith* þis oynement þ*at* is so sote
lete me A-noynte[1] þ*in* holy fote 485
and for my balys þus wyn su*m* bote
and mercy lord ffor my trespace.

Jhesus

¶ Woman ffor þ*i* wepynge wylle
Su*m* socowre god xal þ*e* sende
þ*e* to saue I haue grett skylle 490
ffor sorwefful hert may synne Amende
Att þ*i* pr*ayour* I[2] xal fulfylle
to þ*i* good hert I wul attende
and saue þ*e* fro þi synne so hylle[3]
and fro vij develys I xal[4] þ*e* ffende 495
ffendys fleth 3o*ur* weye
Wyckyd spyritys I 3ow conjowre
Ffleth out of hire bodyly bowre
In my grace she[5] xal evyr fflowre
tyl deth doth here to deye. 500

Maria Magdalene

¶ I thanke þe lorde of þis grett grace
Now þese vij ffendys be fro me fflytt
I xal nevyr fforfett nor do trespace
Fo. 149v in wurd nor dede ne wyl nor wytt[6]
Now I am brought from þe fendys brace 505
In þ*i* grett mercy closyd *and* shytt
I xal nevyr returne to synful trace
þ*at* xulde me dampne to helle pytt
I wurchep the on knes bare
blyssyd be þe tyme þ*at* I hedyr sowth 510

[1] *noyy* first written and crossed through. [2] A miswritten letter crossed through before *I*. [3] *hende* first written and crossed through.
[4] This *x* is written over another letter. [5] An *x* crossed through before *she*.
[6] Spaces are left in the MS. between lines 504 and 505 and between 508 and 509, also between lines 516 and 517 and between 518 and 519, as if the structure of these stanzas was not understood. The tail-verses are not written to the right on fo. 149v, and the last quatrain, ll. 522–5, is more roughly written than the rest.

And þis oynement þat I heydr brought
Ffor now my*n* hert is clensyd from thought
þat Ffyrst was combryd w*ith* care.

Judas

¶ Lord me thynkyth þ*ou* dost ryght ylle
To lete þis oynement so spylle 515
To selle it | yt were more skylle
and bye mete to poer men
The box was worth of good mone
iij C. pens fayr *and* fre
þis myght a bowht mete plente 520
to Ffede oure power ken.

Jhesus

¶ Pore men xul abyde
A-geyn þe woman þ*ou* spekyst wronge
And I passe forth in A tyde
Off mercy is here mornyng songe. 525

here cryst restyth and *etyth A lytyl* and *seyth syttyng to his disciplis* · and *mary mawdelyn*[1]

Fo. 150 ¶ Jhesus

¶ Myn herte is ryght sory And no wondyr is
Too[2] deth I xal go *and* nevyr dyd trespas
But ȝitt most grevyth[3] my*n* hert evyr of this
On of my bretheryn xal werke þis manas
On of ȝow here syttynge my treson xal tras 530
On of ȝow is besy my deth here to dyth
and ȝitt was I nevyr in no synful[4] plas
Where-fore my deth xuld so shamfully be pyght.

Petrus

¶ My dere lord I pray the þe trowth for to telle
Whiche of vs ys he þat treson xal do 535
Whatt traytou*r* is he þat his lord þat wold selle
expresse his name lord þat xal werke þis woo.

Joh*an*nes

If þat þer be on þat wolde selle so
Good mayster telle us now opynly his name

[1] The words *gohth here outh* follow, but have been crossed out in black and red ink.

[2] *Thoo* first written and the *h* marked with deleting dot.

[3] The *g* has been written over some correction.

[4] Some miswritten letters (? *syv*) crossed out before *synful*.

What traytou*r* is hym þat · from þe þat wolde go 540
And wi*th* ffals treson ffullfylle his grett shame.

Andreas

¶ It is right dredfull̄ such̄ tressoñ to thynke
and wel more dredfful to werk þat bad dede
Ffor þat ffals treson to helle he xal synke
In endles peynes grett myscheff to lede. 545

Jacob*us* maior

It is not I lord ffor dowte I haue drede
þis synne to fulfylle cam nevyr *in* my m*e*nde
Iff þat I solde þe · thy blood ffor to blede
In doyng þat treson my sowle xulde I shende.

Matheus

Fo. 150v ¶ Alas my der*e* lord what man is so wood 550
Ffor gold or for sylvyr hym-self so to spylle
he þat þe doth selle ffor gold or for other good
wi*th* his grett Covetyse hym-self he doth kylle.

Bartholome*us*

What man so evyr he be of so wyckyd wylle
der*e* lord among vs · tell vs his name all̄ owt 555
He þat to hym tendyth · þis dede to fulffille
Ffor his grett treson his sowle stondyth *in* dowt.

Philipp*us*

¶ Golde sylver *and* tresoou*r* sone doth̄ passe away
but wi*th*-owtyn ende evyr doth̄ laste þ*i* grace
A lorde who is that · wyll̄ chaffar*e* þe for monay 560
Ffor he þat sellyth his lord to grett is þe trespace.

Jacob*us* mi*n*or

That traytou*r* þat doth̄ þis orryble manace
bothe body *and* sowle I holde he be lorn
Dampnyd to helle pytt fer fro*m* þ*i* face
Amonge All̄ ffowle fyndys to be rent *and* torn. 565

Symoñ

¶ To bad A marchawnt þ*at* traytou*r* he is
and ffor þat monye [1] he may mornyng make
Alas what cawsyth hym to selle þe kyng of blys
Ffor his fals wynnynge þe devyl hym xal take.

Thomas

Ffor his ffals treson þe fendys so blake 570
xal ber*e* his sowle depe dowñ into helle pytt

[1] *may* first written before *monye* and crossed through.

resste xal he no*n* haue · but evyr-mor*e* wake
brennyng in hoot fyr*e* in p*re*son evyr shytt.

Thadeus

¶ I woundyr ryght sor*e* who þ*at* he xuld be
Among*es* vs all bretheryn þ*at* xulde do þis synne 575
Fo. 151 Alas he is lorn þ*er* may no grace be
In depe helle donjeon his sowle he doth py*n*ne.

Jh*es*us

In my dysche he Etyht þis treson xal be-gynne
Wo xal be-tydyn hym for his werke of dred
he may be ryght sory swych ryches to wynne 580
ad[1] whysshe hymself vn-born ffor þ*at* synful ded.

Judas

¶ The trewth wolde I knowe as leff as ȝe
And *þer*fore good ss*er*e þe trewth þ*ou* me telle
whiche of vs All here þ*at* traylto*ur* may be
Am I þ*at* person þ*at* þe now xal selle. 585

Jh*es*us

So seyst þ*i*-selff Take hed att þ*i* spelle
þ*ou* Askyst me now her*e* if þ*ou* xalt do þ*at* treson
Remembyr þ*i*-self A-vyse þe ryght welle
þ*ou* art of grett Age *and* wotysst what is reson.

here judas rysyth prevely and *goth in* þ*e place* and *seyt now counter*[fetyd.

[Remainder of Fo. 151—2¼ inches—and Fo. 151ᵛ left blank.*]

Judas

Fo. 152 ¶ Now cowntyrfetyd I haue A p*re*vy treson 590
(P quire)[2] My Maysterys power for to felle
I judas xal A-say be some[3] encheson
On-to þe jewys hym for to selle
Som mony for hym ȝet wold I telle
Be p*re*vy menys I xal a-say 595
Myn intent I xal fulfylle
No lenger I wole make delay.

[1] So in MS.

* [*John Holand* is scribbled on Fo. 151ᵛ (marked 150), and two and half a third black tapering marks extend from near the top to near the bottom of the page; cf. Fo. 164.]

[2] In this quire the writing and rubrication have the characteristics found in quire N.

[3] A *j* (*i*) first written and crossed through after *some*.

¶ Þe princys of prestys now be present
Vn-to hem now my way I take
I wyl go tellyn hem myn entent 600
I trow ful mery I xal hem make
Mony I wyl non for-sake
And þei profyr to my plesyng
For covetyse I wyl with hem wake
and on-to my maystyr I xal hem bryng. 605

¶ Heyl prynsesse *and* prestys þat ben present
New tydyng*ys* to ȝow I come to telle
ȝyf ȝe wole folwe my*n* intent
my mayster jhesu I wele ȝow selle
Hese intent *and* purpose for to felle 610
For I wole no lenger folwyn his lawe
Late sen what mony þat I xal telle
and late jhesu my maystyr ben hangy*n* *and* drawe.

Gamalye[l

¶ Now welcome judas oure owyn frende
Take hym in serys be þe honde 615
We xal þe both geve *and* lende
And in every qwarel by þe stonde.

Rewfyn

Fo. 152ᵛ ¶ Judas what xal we ffor þi mayster pay*
Þi sylver is redy *and* we A-corde
Þe payment xal haue no delay 620
But be leyde dowñ here At a worde.

Judas

¶ Late þe mony here down be layde
And I xal telle ȝow as I kan
In old termys I haue herd seyde
þat mony makyth schapmañ. 625

Rewfyñ

¶ Here is thretty platys of sylver bryth
Fast knyth with*in*ne þis glove
And we may haue þi mayster þis nyth
Þis xalt þou haue *and* añ oure love.

* The name John Holand is scribbled again in the margin of this page and of Fo. 153ᵛ; on Fo. 155ᵛ Holland Hary; and on Fo. 154ᵛ there are indecipherable scribblings in similar ink and hand.

Judas

¶ ȝe Ar*e* resonable chapmen to bye *and* selle 630
þis bargany w*ith* ȝow now xal I make
Smyth up ȝe xal haue al ȝ*our* wylle
Ffor mony wyl I non for-sake.

leyon

¶ Now þis bargany is mad ful *and* fast
Noyther part may it for-sake 635
But judas þ*ou* must telle us in hast
Be what menys we xal hym take.

Rewfyn

¶ ȝa þ*er* be many þ*at* hym nevyr sowe
Weche we wyl sende to hym in fer*e*
Þ*er*-for be A tokyn we must hy*m* knowe 640
Þ*at* must be p*r*evy be-twyx us her*e*.

leyon

¶ ȝa be war*e* of þ*at* for ony thynge
For o dyscypil is lyche þ*i* mayst*er* in al parayl
Fo. 153 And ȝe go lyche in Aff clothyng
So myth we of our*e* purpose fayl. 645

Judas

¶ As for þ*at* serys haue ȝe no dowth
I xal ordeyn so ȝe xal not mysse
Whan þ*at* ȝe cvm[1] hy*m* Aff A-bowth
Take þ*e* man þ*at* I xal kysse.

¶ I must go to my maystyr A-geyn 650
Dowth not serys þis matere is sur*e* i-now.

Gamalyel

Far*e* wel judas our*e* frend serteyn
þ*i* labo*ur* we xal ryth wel A-low.

Judas

¶ Now wyl I sotely go seke my mayst*er* Ageyn
and make good face as I nowth knew 655
I haue hym solde to wo *and* peyn
I trowe ful sor*e* he xal it rew.

a here judas goth in sotylly wher-as he cam fro.

Annas

¶ Lo serys a part we haue of our*e* entent
For to take jh*e*su now we must p*r*ovyde

[1] The *v* seems to be written over some other letter (? *u*).

A sotyl meny to be p*re*sent 660
Þ*a*t dar*e* fyth *and* wele A-byde.

Gamalye[l

¶ Ordeyn eche man on his party
Cressetys lanternys *and* torchys lyth[1]
And þis nyth to be þ*er* redy
W*ith* exys gleyvis[2] *and* swerdys bryth. 665

Cayphas

¶ No lenger þan make we teryeng
But eche man to his place hy*m* dyth
And ordeyn p*re*uely for þis thyng
Þ*a*t it be don þis same nyth.

Fo. 153v ℭ *Here The Buschopys partyn in þe pLace And eche of*[3] *hem takyn here leve be contenawns resortyng eche man to his place with here meny to make redy to take cryst* and *þan xal þe place* þer *cryst is in xal sodeynly vn-close rownd Abowtyn shewyng cryst syttyng at þe table* and *his dyscypulys eche in ere degre cryst þus seyng*

Jhesu

¶ Brederyn þis lambe þ*at* was set us beforn̄ 670
Þ*a*t we Alle haue etyn in þis nyth
it was comawndyd be my fadyr to moyses *and* Aaron̄
Whan þei weryn w*ith* þ*e* chylderyn of israel in egythp.

¶ And as we w*ith* swete bredys haue it ete
And Also w*ith* þe byttyr sokelyng 675
And as we take þe hed w*ith* þe fete
So dede þei in aꝰ maner thyng.

¶ And as we stodyn so dede þei stond
and here reynes þei gyrdyn veryly
W*ith* schon on here fete *and* stavys in here hond 680
And as we ete it so dede þei hastyly
þis fygure xal sesse A-nothyr xal folwe þ*er*-by
Weche xal be of my body þ*a*t am ȝ*our* hed
weche xal be shewyd to ȝow be A mystery
Of my fflesch *and* blood in forme of bred. 685

¶ And w*ith* fervent desyr*e* of hertys Affeccion̄
I have enterly desyryd to kepe my mawnde

[1] *bryth* first written. [2] The *le* is written over other letters.
[3] The capital letters in this line are unusually tall, and the initial B of the following speech is elaborated.

A-mong ȝow er þan I suffre my passyoñ
For of þis no more to-gedyr suppe xal we
Fo. 154 And as þe pascHal Lomb etyn Haue we [1] 690
(Q quire) In þe old lawe was vsyd for A sacryfyce
So þe newe lomb þat xal be sacryd be me
Xal be vsyd for A sacryfyce most of price.

here xal jhesus take An oble in his hand lokyng vpward in to hefne to þe fadyr þus seyng

¶ Where fore to þe fadyr of hefne þat art eternall
Thankyng *and* honor I ȝeld on to þe 695
To whom be þe godhed I am eqwall
But be my manhod I am of lesse degre
Wherefore I as man · worchep þe deyte
Thankyng þe fadyr þat þou wylt shew þis mystery
And þus þurwe þi myth fadyr *and* blyssyng of me 700
Of þis þat was bred is mad my body.

here xal he spekyn ageyn to his dyscipulys þus seyng

¶ bretheryn be þe [vertu] of þese wordys þat [re]hercyd be [2]
Þis þat shewyth as bred to ȝour Apparens [3]
Is mad þe very flesche *and* blod of me
To þe weche þei þat wole be savyd must ȝeve credens. 705

¶ And as in þe olde lawe it was comawndyd *and* precepte
To ete þis lomb to þe dystruccyoñ of pharao vn-kende
So to dystroy ȝour gostly [4] enmye · þis xal be kepte
ffor ȝour paschal lombe in-to þe werdys ende.

¶ Ffor þis is þe very lombe with-owte spot of synne 710
Of weche Johan þe baptyst dede prophesy
Whan þis prophesye he dede be-gynne
Seyng · Ecce agnus dey.

¶ *And* how ȝe xal ete þis lombe I xal ȝeve infformacion
In þe same forme as þe eld lawe doth specyfye 715

[1] The capital letters in this top line are again unusually tall.
[2] This line is omitted and written in the margin, where it is partly cut away, and again in large form in red ink at the bottom of the page.
[3] *apperens* first written and corrected to *apparens*.
[4] Perhaps corrected from *goostly*.

As I shewe be gostly interpretacyon
Per-fore to þat I xal sey ȝour wyttys[1] loke ȝe replye.

Fo. 154v[2] ¶ With no byttyr bred þis bred ete xal[3] be
þat is to say with no byttyrnesse of hate *and* envye
But with þe suete bred of loue *and* charyte 720
Weche ffortefyet þe soule gretlye.

¶ And it schuld ben etyn with þe byttyr sokelyng
Þat is to mene ȝyf A man[4] be of synful dyspocycion
Hath led his lyff here with mys-levyng
Perfore in his hert he xal haue byttyr contrycion. 725

¶ Also þe hed with þe feet ete xal ȝe
Be þe hed ȝe xal vndyr-stand my godhed
And be þe feet ȝe xal take myn humanyte
Þese tweyn ȝe xal receyve to-gedyr in dede.

¶ This immaculat lombe þat I xal ȝow ȝeve 730
Is not only þe godhed A-lone
But bothe god *and* man þus must ȝe beleve
Þus þe hed with þe feet ȝe xal receyve ech-on.

¶ Of þis lombe vn-ete · yf owth be levyth i-wys
Yt xuld be cast in þe clere fyre *and* brent 735
Weche is to mene yf þou vndyrstande nowth al þis
Put þi feyth in god *and* þan þou xalt not be shent.

¶ The gyrdyl þat was comawndyd here reynes to sprede
Xal be þe gyrdyl of clennes *and* chastyte
þat is to sayn to be contynent in word thought *and* dede 740
and all leccherous levyng · Cast ȝow for to fle.

¶ And þe schon þat xal be ȝour feet vp-on
Is not ellys but exawnpyl of vertuis levyng
Of ȝour form faderys ȝou be-forn
With þese schon my steppys ȝe xal be sewyng. 745

[1] The *tt*ys of this word blotted—*ȝe* first written after it.

[2] On Ff. 154v and 155 each paragraph begins with a larger and more definitely rubricated capital than has been usual hitherto in the MS. and these larger capitals continue more or less for some pages. They appear sparsely in quires N and P.

[3] *ȝe* first written after *xal*. [4] Some word before *man* crossed through.

Fo. 155 ¶ And þe staf þat in ȝour Handys ȝe xal holde
Is not ellys · but þe exawmplys to other men teche
Hold fast ȝour stauys in ȝour handys *and* beth bolde
To every creature myn precepttys for to preche.

¶ Also ȝe must ete þis paschall lombe hastyly 750
Of weche sentens þis is þe very entent
At every oure *and* tyme ȝe xal be redy
Ffor to fulfylle my cowmawndement.

¶ Ffor þow ȝe leve þis day · ȝe are not sure
Whedyr ȝe xal leve to-morwe or nowth 755
þer-for hastyly every oure do ȝoure besy cure
To kepe my preceptys *and* þan þar ȝe not dowth.

¶ Now haue I lernyd ȝow how ȝe xal ete
ȝour paschal lombe þat is my precyous body
Now I wyl fede ȝow all with Awngellys mete 760
Wherfore to reseyve it · come fforth seryattly [1].

Petrus

¶ Lord ffor to receyve þis gostly sustenawns
In dewe forme it excedyth myn intellygens
Ffor no man of hym-self may have substawns
To receyve it with to meche reverens. 765

¶ Ffor with more delycyous mete lord þou may us not fede
þan with þin owyn precyous body
Wherfore what I haue trespacyd in word thought or dede
With byttyr contrycion · lord I haske þe mercy.

whan oure lord ȝyvyth his body to his dyscypulys he xal sey to eche of hem · except to judas

Fo. 155v ¶ This is my body Fflesch *and* blode 770
þat for þe xal dey up-on þe rode.

α *And whan judas comyth last oure lord xal sey to hym*

¶ Judas art þou Avysyd what þou xalt take.

Judas

Lord þi body I wyl not for-sake.

α *And sythyn oure lord xal sey on-to judas*

[1] *sey* miswritten before this word.

Jhesu

¶ My*n* body to þe I wole not denye
Sythyn þ*ou* wylt p*re*sume þ*er*-upon
Yt xal be þi dampnacyon verylye
I ȝeve þ*e* warnyng now be-forn.

α *And aftyr þat judas hath reseyvyd he xal syt þ*er *he was cryst seyng*

¶ On of ȝow hath be-trayd me
Þ*at* at my borde w*ith* me hath ete
Bettyr it hadde hym for to A be
Bothe vn-born *and* vn-begete.

α *Than eche dyscypyl xal loke on other* and *petyr xal sey*

Petrus

¶ Lord it is not I. and *so alle xul seyn tyl þei comyn at judas weche xal sey*

Judas

Is it owth I lord *þan jhesus xal sey*

Jhesu

Judas þ*ou* seyst þ*at* word̄
Me þ*ou* ast solde þ*at* was þ*i* ffrend
Þ*at* þ*ou* hast be-gonne brenge to An ende.

þan judas xal gon A-geyn to þe jewys · And yf men wolne xal mete with hym and *sey þis spec̄h folwyng | or levynt whether þei wyl · þe devyl þus seyng*

Demon

Fo. 156 (R quire) ¶ A · A · Judas Derlyng myn
Þ*ou* art þe best to me þ*at* evyr was bor*e*
Þ*ou* xalt be crownyd in helle peyn
and þ*er*-of þ*ou* xalt be sekyr for evyr-mor*e*.

¶ Thow hast solde þ*i* maystyr *and* etyn hy*m* also
I wolde þ*ou* kowdyst bryngyn hy*m* to helle every del
But ȝet I fer*e* he xuld do þ*er* su*m* sorwe *and* wo
Þ*at* all̄ helle xal crye out on me þ*at* sel.

¶ Sped up þ*i* mater*e* þ*at* þ*ou* hast be-gonne
I xal to helle for þe to mak redy
Anon þ*ou* xalt come wher þ*ou* xalt wonne
In fyr*e* *and* stynk þ*ou* xalt sytt me by.

Jhesu

¶ Now þe sone of god claryfyed is
and god in hym is claryfyed also 800
I am sory þat judas hath lost his blysse
Weche xal turne hym to sorwe *and* wo.

¶ But now in þe memory of my passyon
To ben partabyl with me in my reyn above
ȝe xal drynk myn blood with gret devocyon̄ 805
Wheche xal be xad ffor mannys love.

¶ Takyth þese chalys of þe newe testament
And kepyth þis evyr in ȝour mende
As oftyn as ȝe do þis with trewe intent
It xal defende ȝow fro þe ffende. 810

α *Than xal þe dysciplys com* and *take þe blod · Jhesus seyng*

Fo. 156v Þis is my blood þat for mannys synne
Outh of myn herte it xal renne.

α *And þe dyscipulys xul sett þem Aȝen þer þei wore* · and *jhesus xal seyn*

¶ Takyth hed now bretheryn what I haue do
With my flesch *and* blood I haue ȝow fed
Ffor mannys love I may do no mo 815
Þan for love of man to be ded.

¶ Werfore petyr *and* ȝe every-chon̄
ȝyf ȝe loue me fede my schep
Þat for fawth of techyng þei go not wrong
But evyr to hem takyth good kep. 820

¶ ȝevyth hem my body as I haue to ȝow
Qweche xal be sacryd be my worde
And evyr I xal þus A-byde with ȝow
In-to þe ende of þe werde.

¶ Ho so etyth my body *and* drynkyth my blood 825
Hol god *and* man he xal me take
It xal hym defende from þe deuyl wood
and at his deth I xal hym nowth for-sake.

¶ *and* ho so not ete my body nor drynke my blood
Lyf in hym is nevyr A dele 830

Kepe wel þis in mende for ȝ*our* good
and every man save hym-self wele.

℞ *here jhesus takyth a basyn with watyr* and *towaly gyrt abowtyn hym* and *fallyth beforn petyr on his o kne.*

Fo. 157 ¶ Jhesus[1]

A-nother exawmpyl I xal ȝow showe
How ȝe xal leve in charyte
Syt here down at[2] wordys fewe 835
and q*w*at I do ȝe sofre me.

here he takyth þe basyn and *þe towaly* and *doth as þe roberych seyth beforn.*

Petrus

¶ Lord what wylt þ*ou* w*ith* me do
Þis servyce of þe I wyl for-sake
To wassche my feet þ*ou* xal not so
I am not worthy it of þe to take. 840

Jhesu

¶ Petyr *and* þ*ou* for-sake my seruyce[3] alle
Þe weche to ȝow þ*at* I xal do
No part w*ith* me haue þ*ou* xal
And nevyr com my blysse on-to.

Petrus

Þ*at* part lord we wyl not for-go 845
We xal A-bey his comawndement
Wasche hed *and* hond we p*ray* þe so
We wyl don After þin entent.

℞ *here jhesus wasshyth his dyscipulys feet by* and *by* and *whypyth hem And kyssyth hem mekely* and *sythym settyth hym down þus seyng*

Jhesu

ffrendys þis wasshyng xal now p*re*vayll
ȝour*e* lord *and* mayster ȝe do me calle 850
And so I am w*ith*-owtyn fayl
ȝet I haue wasschyd ȝow Alle
Fo. 157v A memory of þis haue ȝe xall
Þ*at* eche of ȝow xal do to othyr
W*ith* vmbyl hert submyt egal 855
As eche of ȝow were otherys brother.

[1] The name of the next speaker (possibly *Petrus*) has been erased at the bottom of Fo. 156v and *Jhesus* thus written at the top of the page.

[2] *fe* written before *wordys* and crossed through. [3] MS. *seruyce̅* (? *seruyces*).

¶ No thyng serys so wele plesyth me
Nor no lyf *þat* man may lede
As þei *þat* levyn in charyte
In efne I xal reward her*e* mede 860
Þe day is come I must pr*o*cede
Ffor to fulfylle þe pr*o*phecy
Þis nyth for me ȝe xal han drede
Whan novmbyr of pepyl xal on me cry.

¶ Ffor *þe* pr*o*phetys spoke of me 865
And seydyn of deth *þat* I xuld take
Fro whech deth I wole not fle
But for ma*nn*ys synne A-mendys make.

¶ This nyth fro ȝow be led I xal
and ȝe for fer fro me xal fle 870
Not onys dur speke whan I ȝow call
And some of ȝow for-sake me.

¶ Ffor ȝow xal I dey *and* ryse ageyn
Vn þe thrydde day ȝe xal me se
Be-forn ȝow all walkyng playn 875
In þe lond of Galyle.

Petrus

¶ Lord I wyl þe nevyr for-sake
Nor for no p*er*ellys fro þe fle
Fo. 158 I wyl rather my deth take
Þan onys lord for-sake þe. 880

Jh*es*u

¶ Petyr yn[1] ferther*e* þan þu doyst[2] knowe
As for *þat* pr*o*mese loke *þou* not make
For or þe cok hath twyes crowe
Thryes *þou* xal me for-sake.

¶ But all my frendys *þat* Arn me der*e* 885
Late us go þe tyme drawyth ny
We may no lenger*e* A-bydyn her*e*
Ffor I must walke to betany.

[1] Halliwell reads *þu*, the forms of *yn* and *þu* being indistinguishable in this MS.

[2] The *o* is written over another letter.

¶ þe tyme is come þe day drawyth nere
On-to my deth I must in hast
Now petyr make hall þi felawys chere
My flesch for fere is qwakyng[1] fast.

℟ *here jhesus goth to betany ward* and *his dyscipulys folwyng with sad contenawns jhesus seyng*

¶ Now my dere frendys *and* bretheryn echon
Remembyr þe wordys þat I xal sey **28**
þe tyme is come þat I must gon
For to fulfylle þe prophesey.

þat is seyd of me þat I xal dey
þe fendys power fro ȝow to flem
Weche deth I wole not deney
Mannys sowle my spovse[2] for to redem.

¶ þe oyle of mercy is grawntyd playn
Be þis jorne þat I xal take
Be my fadyr I am sent sertayn
Be-twyx god *and* man An ende to make.

Fo. 158v ¶ Man for my brother may I not for-sake
Nor shewe hym vn-kendenesse be no wey
In peynys for hym my body schal schake
And for love of man man xal dey.

℟ *here jhesus* and *his discipulys go toward þe mount of olyvet* and *whan he comyth a lytyl þer be-syde in a place lych to A park he byddyt his dyscipulys A-byde hym* þer and *seyth to petyr or he goth*

¶ Petyr with þi felawys · here xalt þou A-byde
and weche tyl I come A-geyn
I must make my prayere here ȝou be-syde
My flesch qwakyth sore for fere *and* peyn.

Petrus

Lord þi request doth me constreyn
In þis place I xal Abyde stylle
Not remeve tyl þat þou comyst Ageyn
In confermyng lord of þi wylle.

℟ *here jhesu goth to olyvet* and *settyth hym down one his knes* and *prayth to his fadyr þus seyng*

[1] The *q* is roughly written, as if over some other letter.
[2] The *v* is written over an original *u*.

Jhesu

¶ O ffadyr fadyr for my sake
Þis gret passyon̄ þou take fro me
Wech arn ordeyned þat I xal take
ȝyf mannys sowle savyd may be 920
And ȝyf it be-hove fadyr for me
To save mannys sowle þat xuld spylle
I am redy in eche degre
Þe vyl of þe for to fulfylle.

α here jhesus goth to his dyscipulis and *fyndyth hem sclepyng jhesus þus seyng to petyr*

Fo. 159 ¶ Petyr petyr þou slepyst fast 925
A-wake þi felawys *and* sclepe no more
Of my deth ȝe Are not Agast
ȝe take ȝour rest *and* I peyn sore.

α here cryst goth Ageyn þe second tyme to olyvet and *seyth knelyng*

¶ Ffadyr in hevyn I be-seche þe
Remeve my [1] peynes be þi gret grace 930
and lete me fro þis deth fle
As I dede nevyr no trespace
The Watyr *and* blood owth of my face
Dystyllyth for peynes þat I xal take
My flesche qwakyth in ferful case 935
As þow þe joyntys A-sondre xuld schake.

α here jhesus goth A-ȝen to his discipulis and *fyndyth hem Asclepe Jhesus þus seyng latyng hem lyne*

¶ Ffadyr þe thrydde tyme I come A-geyn
Ffulleche myn erdon for to spede
Delyuere me fadyr fro þis peyn
Weche is reducyd with ful gret dred 940
On to þi sone fadyr take hede
Þou wotyst I dede nevyr dede but good
It is not for me þis peyn I lede
But for man I swete bothe watyr *and* blode.

α here An Aungel descendyth to jhesus and *bryngyth to hym A chalys with An host* þer *in.*

[1] *þi* first written before *my*.

Angelus

Fo. 159v ¶ Heyl bothe god *and* man in dede 945
The ffadyr hath sent þe þis present
He bad *þat* þ*ou* xuldyst not drede
But fulfylle his intent
As þe parlement of hefne hath ment
Þ*at* mannys sowle xal now redemyd be 950
Ffrom hefne to herd lord þ*ou* wor*e* sent
Þ*at* dede Appendyth on-to þe.

¶ Þis chalys ys þ*i* blood þis bred is þ*i* body
Ffor ma*n*nys synne evyr offeryd xal be
To þe fadyr of heffne þ*at* is al-mythty 955
Þi dyscip*u*lis *and* aſt presthood xal offer*e* for*e* the.

a here þe Aungel Ascendyth A-ȝen sodeynly.

Jhesu

Ffadyr þ*i* wyl ffulfyllyd xal be
It is nowth to say A-ȝens þe case
I xal fulfylle þe p*ro*phesye
and sofr*e* deth ffor mannys trespace. 960

a here goth cryst A-geyn to his dyscipulys and *fyndyth hem sclepyng stylle.*

¶ A-wake petyr þi rest is ful long
of sclep þu wylt make no delay
Judas is redy w*ith* pepyl strong
and doth his part me to be-tray
Ryse up serys I ȝ*ou* pray 965
On-close ȝ*our* eyne for my sake
We xal walke in to þe way
and sen hem com þ*at* xul me take.

Fo. 160 ¶ Petyr whan þ*ou* seyst I Am for-sake
Amonge my*n* frendys *and* stond Alone 970
Aſt þe cher þat þ*ou* kanst make
Geve to þ*i* bretheryn every-chone.

a here jhesus with his dyscipulis goth in-to þe place and *þer xal come in A x personys weyl be-seen in white Arneys* and *bregander*ys and *some dysgysed in odyr garment*ys *with swerdys gleyvys* and *other* *straunge wepone*[1] *as cressettys with feyr* and *lanternys* and *torchis lyth* and *judas formest of Al conveyng hem to jhesu be contenawns*

Jhesus þus s[eyng

¶ Serys[2] in ȝour way ȝe haue gret hast
To seke hym þat wyl not fle
Of ȝow I am ryth nowth A-gast 975
Telle me serys whom seke ȝe.

leyon̄

¶ Whom we seke here I telle þe now
A tretour is worthy to suffer deth
We knowe he is here A-mong ȝow
His name is jhesus of nazareth. 980

Jhesu

¶ Serys I am here þat wyl not fle
Do to me all þat ȝe kan
For sothe I telle ȝow I am he
Jhesus of nazareth þat same man.

here alle þe jewys falle sodeynly to þe Erde whan þei here cryst speke and *qwan byddyth hem rysyn þei rysyn Aȝen cryst þus seyng*

¶ A-ryse serys whom seke ȝe fast haue ȝe gon̄ 985
Is howth ȝour comyng hedyr for me
I stond be-forn ȝow here echon̄
Þat ȝe may me bothe knowe *and* se.

Rufyne

Fo. 160v ¶ Jhesus of Nazareth[3] we seke[4]
And we myth hym here A-spye. 990

Jhesu

I told ȝow now with wordys meke
Be-forn ȝou Aħ þat it was I.

Judas

¶ Welcome jhesu my mayster dere
I haue þe sowth in many A place
I am ful glad I fynd þe here 995
For I wyst nevyr wher[5] þou wace.

here judas kyssyth jhesus and *A-noon Alle þe jewys come A-bowth hym* and *ley handys on hym* and *pullyn hym as þei were wode* and *makyn on hym A gret cry All At-onys* and *Aftyr þis Petyr seyth*

[1] MS. wepon̄, not the usual n̄ (= *ne*).
[2] *What* written at the beginning of the line before *Serys* and crossed through.
[3] A more elaborate capital than usual. [4] *and we* first written after *seke*.
[5] *h* omitted and written above the line—cf. *werfore*, Ff. 156v and 160v.

Petrus

¶ I drawe my swerd now þis sel
Xal I smyte mayster fayn wolde I wete[1].

α and *forth-with he smytyth of Malcheus here* and *he cryeth help myn here myn here* · and *cryst blyssyth it* and *tys hol.*

Jhesus

Put þi swerd in þe shede fayr *and* wel
Ffor he þat smyth with swerd · with swerd xal be smete. 1000

¶ A judas þis treson cowntyrfetyd hast þou
And þat þou xalt ful sore repent
Þou haddyst be bettyr A ben vn-born now
Þi body *and* sowle þou hast shent.

Gamalyel

¶ Lo jhesus þou mayst not þe cace refuse 1005
Bothe treson *and* eresye in þe is fownde
Stody now fast on þin excuse
Whylys þat þou gost in cordys bownde
Þou kallyst þe kyng of þis werd rownde
Now lete me se þi gret power 1010
And saue þi-self here hool *and* sownde
And brynge þe out of þis dawngere.

leyon

Fo. 161 Bryng forth þis tretoure spare hym nowth
On-to Cayphas þi jewge we xal þe ledde
In many A place we haue þe sowth 1015
And to þi werkys take good hede.

Rufyne

¶ Come on jhesus *and* folwe me
I am ful glad þat I þe haue
Þou xalt ben hangyn up-on A tre
A[2] melyon of gold xal þe not save. 1020

leyon

¶ Lete me leyn hand on hym in heye
On to his deth I xal hym bryng
Shewe forth þi wyche-crafte *and* nygramansye
What helpyth þe now Al þi fals werkyng.

[1] *Petyr put þi s* written as the beginning of the next line and crossed through.
[2] *m* written at the beginning of the line before *A* and crossed through.

Jhesu

¶ Ffrendys take hede ȝe don vn-ryth 1025
So vn-kendely *with* cordys to bynd me here
And þus to falle on me be nyth
As thow I were A thevys fere
Many tyme be-forn ȝow I dede A-pere
With-inne þe[1] temple sen me ȝe have 1030
Þe lawys of god to teche *and* lere
To hem þat wele here sowlys sawe[2].

¶ Why dede ȝe not me dysprave
And herd me preche bothe lowd *and* lowe
But now as woodmen ȝe gynne to Rave 1035
And do thyng þat ȝe notwth knove[3].

Gamaly[el

¶ Serys I charge ȝow not o word more þis nyth
But on-to Cayphas in hast loke ȝe hym lede
Fo. 161v [marked 160] Have hym forth *With*[4] gret dyspyte
And to his wordys take ȝe non hede. 1040

a here þe jewys lede cryst outh of þe place with gret cry and *noyse some drawyng cryst forward* and *some bakwarde* and *so ledyng forth with here weponys A-lofte* and *lytys brennyng* and *in þe mene tyme maryę magdalene xal rennyn to oure lady* and *telle here of oure lordys takyng þus seyng*

Maria Magdelene

¶ O in-maculate modyr of all women most meke
O devowtest in holy medytacion evyr A-bydyng
Þe cawse lady þat I to ȝour person seke
Is to wetyn yf ȝe heryn ony tydyng.

¶ Of ȝour swete sone *and* my reverent lord' jhesu 1045
Þat was ȝour dayly solas ȝour gostly consolacyon.

Marya

I wold ȝe xuld telle me mawdelyn *and* ȝe knew
Ffor to here of hym it is All myn Affeccyon.

Maria Magd[alen

¶ I wold fayn telle lady *and* I myth for wepyng
For sothe lady to þe jewys he is solde 1050

[1] *þis* first written, corrected to *þe*. [2] So in MS.
[3] So in MS. [4] W unusually tall.

W*ith* cordys þei haue hym bownde *and* haue hy*m* in kepyng
Þei hy*m* bety spetously[1] *and* haue hy*m* fast in holde[1].
Maria uirgo
A · A · A · how my*n* hert is colde
A hert hard as ston how mayst þ*ou* lest
Whan þese sorweful tydyngys Ar*e* þ*e* told 1055
So wold to god hert þ*at* þ*ou* mytyst brest.

¶ A jhesu · jhesu · jhesu · jhesu ·
Why xuld ȝe sofer*e* þis trybulacyon *and* Advercyte
How may thei fynd in her*e* hertys · ȝow to pursewe
Þ*at* nevyr trespacyd in no maner degre 1060
For nevyr thyng but þ*at* was good thowth ȝe
Fo. 162 Where-fore þan xuld ȝe sofer þis gret peyn̄
I suppoce veryly it is for þe tresspace of me
And I wyst þ*at* my*n* hert xuld cleve on tweyn̄.

¶ Ffor þese langowrys may I susteyn̄[2] 1065
Þe swerd of sorwe hath so thyrlyd my meende
Alas what may I do · Alas what may I seyn̄
Þese prongys myn herte A-sondẏr þei do rende.

¶ O ffadyr of hẽfne wher ben Al þi be-hestys
Þ*at* þ*ou* pr*o*mysyst me whan A modyr þ*ou* me made 1070
Þ*i* blyssyd sone I bare be-twyx tweyn bestys
And now þ*e* bryth colo*ur* of his face doth fade.

¶ A good fadyr why woldyst þ*at* þin owyn der*e* sone xal sofre Al þis
And dede he nevyr Aȝens þ*i* pr*e*cept b*ut*[3] evyr was obedyent
And to every creatur*e* most petyful most jentyl *and* beny*ng* i-wys 1075
And now for all̃ þese kendnessys is now most shameful schent.

¶ Why wolt þ*ou* gracyous fadyr þ*at* it xal be so
May man not ellys be savyd be non other kende
ȝet lord fadyr þan · þ*at* xal comforte my*n*[4] wo 1079
Whan man is savyd be my chylde *and* browth to A good ende.

[1] The *sl* is written over some correction, as is also the *d* in *holde*.
[2] So in MS., *not* omitted (?). [3] MS b^t^.
[4] Omitted and written over the line.

¶ Now dere sone syn þou hast evyr be so ful of mercy
Þat wylt not spare þi-self · for þe love þou hast to man
On All man-kend now haue þou pety
And Also thynk on þi modyr þat hevy woman.

[Remainder of Fo. 162—2¼ inches—and Fo. 162ᵛ left blank.]

Fo. 163 i doctor

[1] O Thou Altitude · of Al gostly Ryches
O þou incomperhensibele · of grete excyllence
O þou luminarye · of pure lyghtnes
Shete out þi bemys · on-tyl þis Audyens.

2 doctor

O Ffily Altissimi · clepyd by eternalyte
Hele þis congregacion · with þe salve of þi passyon̄
And we prey þe · spiritus paraclyte
With þe ffyre of þi love · to slake All detraccion.

i doctor

To þe pepyl not lernyd · I stonde as A techer
Of þis processyon · to ȝeve informacion
And to them þat be lernyd · As A gostly precher
that in my rehersayl · they may haue delectacion.

2

Welcome of þe apostelys · þe gloryous qwere
Ffyrst Petyr ȝour prynce · And eke ȝour presydent
And Andrewe ȝour half brother · to-gedyr in ffere
that Ffyrst ffolwyd Cryst · be on Assent.

1

O ȝe tweyn luminaryes · Jamys · *and* Jhon
Contynualy brennyng · As bryght As þe sonnbem
With þe chene of charyte · bothe knyt in on
And offeryd of ȝour modyr · to cryst in jherusalem.

2

Fo. 163ᵛ Welcome Phelypp · þat conuertyd samaryan
And conuertyd þe tresorere · of þe qwene cavdas
With Jamys · þe lesser · that Apud jherosolyman
Was mad fyrst patryarke · by þe ordenauns of Cephas [2].

[1] No number is attached to this section, which stands by itself in the MS. between blank folios. It is rubricated on a different plan. Each speech has an initial red capital, and the lines connecting the rhyme words are red instead of black as elsewhere in the MS. except four lines on Fo. 214ᵛ. No paragraph marks are used.

[2] † *Petyr* written over *Cephas*.

i^us 1

Heyl mathew the Apostel · *and* Also ev*a*ngelyst
that was clepyd to þ*e* fflok · of gostly conu*e*rsac*i*on
Ffrom thyrknes of concyens · þ*a*t ȝe were in ffest
w*i*th Bertylmew þ*a*t ffled · Ałł carnałł temptac*i*on.

2us 1

Heyl Symeon zelotes · þus be ȝo*ur* name
And Judas þ*a*t bothe · wel lovyd our*e* lord
Ther*e* ffor*e* ȝe haue bothe joye *and* game
Wher nevyr is sstryff · but good A-corde.

i^us

Heyl poul grett docto*ur* · of þ*e* ffeyth
And vessel chosyn be trewe Elecc*i*on
heyl Thomas · of whom þ*e* gospel seyth
in crystys wounde · was ȝo*ur* refecc*i*on.

2us

Heyl Joh*a*n baptyst · most sovereyn creatur*e*
that evyr was born · be naturałł conseyvyng
And hyest of pr*o*phetys · As wytnessyth scryptur*e*
heyl joys [2] þ*a*t in desert · was Ałł-wey cryeng.

1us 3

[Ff. 164 and 164ᵛ originally left blank. Fo. 164 is the first of quire S, and is discoloured, having apparently been an outside leaf at some time. It has on it a tapering stain which corresponds with similar stains lying in an opposite direction on Fo. 151ᵛ, the last leaf of quire O. On Fo. 164 stands the second signature of R. Hegge. At the top of the page in book-hand are the words :

In n*o*mj*n*e Dei · Amen.

In the middle of the page in cursive hand :

ego R. H. Dunelmensis
possideo
οὐ κτησις ἀλλὰ χρησις.*]

[1] These contractions are added in paler ink. [2] So in MS. for *voys*.

[3] This stands at the bottom of the folio—the last of quire R—and suggests that the Doctors' proclamation is incomplete. It is, however, a more roughly made figure than the others.

* This *τ* is inserted in different ink.

The Passion Play. II **29**[1]

Fo. 165 *What tyme þat processyon is enteryd in to þe place* and *þe herowdys takyn his schaffalde* · and *pylat* and *annas* and *cayphas here schaffaldys Also þan come þer An exposytour in doctorys wede þus seyng*

Contemplacio

¶ Sofreynes *and* frendys · ȝe mut alle be gret w*ith* gode
Grace love *and* charyte · evyr be ȝ*ou* A-mong
þe maydenys sone p*re*serve ȝ*ou* · þ*at* for ma*n* deyd on rode
he þ*at* is o god · in p*er*sonys thre · defende ȝ*ou* fro ȝ*our* fon.

¶ Be þe leue *and* soferauns of aH-mythty god
We intendyn[2] to p*ro*cede þe mater*e* þ*at* we lefte þe last ȝer*e*
Wher*e* fore we be-seche ȝow þ*at* ȝ*our* wyllys be good
to kepe þe passyon in ȝ*our* mende þ*at* xal be shewyd her*e*.

¶ The last[3] ȝer*e* we shewyd her*e* how our*e* lord for love of ma*n*
Cam to þ*e* cety of jherusalem · mekely his deth to take
And how he made his mawnde · his body ȝevyng þan̄
to his Apostelys evyr w*ith* us · to A-bydyn for ma*n*nys sake.

¶ In þ*at* mawnde he was betrayd · of judas þ*at* hy*m* solde
to þe jewys for xxx^ti^ platys · to delyvyr hy*m* þ*at* nyth
W*ith* swerdys *and* gleyvys · toke[4] jhesu · they come w*ith* þe tret*our* bolde
And toke hy*m* amonge*s* his Apostelys · About myd nyth.

¶ Now wold we p*ro*cede[5] how he was browth þan
be-forn Annas *and* cayphas · *and* syth̄ beforn pylate
And so forth in his passyon · how mekely he toke it for man
besekyng ȝ*ou* · for mede of ȝ*our* soulys · to take good hede þ*er* Atte.

*α here þe herowndys xal shewe hy*m*self* and *speke.*

herowdys

Fo. 165v

¶ herodes Rex[6]

¶ Now sees of ȝ*our* talkyng · And gevyth lordly Audyence
Not o word I charge ȝ*ou* þ*at* ben her*e* p*re*sent

[1] This is the only number written at the head of the page and not in the margin.
[2] The *d* is written over an original *t*.
[3] *lass* miswritten before *last* and crossed out.
[4] So in MS. for *to*.
[5] Some word has been erased before *how*.
[6] Name in larger form.

noon so hardy to p*re*sume · in my hey p*re*sence
to on-lose hese lyppys Ageyn̄ myn intent
I am herowde of jewys · kyng most reverent
þe lawys of Mahownde · my power*e* xal fortefye
reverens to þ*a*t lord of grace · moost excyllent
Ffor be his power*e* all̄ þinge doth multyplye.

¶ ȝef ony crystyn be so hardy · his feyth to denye
Or onys to erre Ageyns · his lawe
On gebettys w*ith* cheynes · I xal hangyn hym heye
and w*ith* wylde hors þ*o* traytorys · xal I drawe
To kylle A thowsand crystyn · I gyf not An hawe
to se hem hangyn or brent · to me is very plesauns
to dryvyn hem in to doongenys · dragonys to knawe
And to rend her*e* flesche *and* bonys · on-to her*e* sustenau*n*s.

¶ Joh*a*n þe baptyst crystenyd cryst · *and* so he dede many on̄
þer-fo*r*e my-self · dede hy*m* bryng o dawe
It is I · þ*a*t dede hym kylle · I telle ȝ*ou* every-chon̄
Ffor And he had go forth · he xuld A dystroyd o*ur* lawe
Wher*e* as crystyn Apperyth · to me is gret grevau*n*s
It peynyth my*n* hert of tho tretowrys to her*e*
Ffor þe lawys[1] of mahownde I have in governawns
þe which I wele kepe · þ*a*t lord hath no pere
Fo. 166 Ffor he is god most prudent
Now I charge ȝ*ou* my lordys þ*a*t ben her*e*
Yf Any crysty*n* doggys · her*e*[2] doth Aper*e*
Bryng þo tretor*ys* · to my hey power*e*
And þei xal haue sone jewgement.

i^us miles

¶ My sovereyn lord heyest of excillens
In ȝ*ou* all̄ jewgement · is termynabyle
all̄ crystyn dogg*ys* · þ*a*t do not her*e* dyligens
ȝe put hem to peynes þ*a*t ben inportable.

ij^us miles

No þi*n*g in ȝ*ou* may be mor*e* comendable
as to dysstroye þ*o* traytor*ys* þ*a*t erre
ageyn our*e* lawys · þ*a*t ben most p*ro*fytable
be rythwysnesse þ*a*t lawe ȝe must p*ro*ferr*e*.

[1] *aw* is written over some other letters.
[2] The *r* is written over some other letter and the word is not clear.

Rex Herow[. . . .

¶ Now be gloryous mahownd · my sovereyn savyo*ur*
these p*ro*messys I make · as I am trewe knyth
þoo þ*at* excede his lawys be ony erro*ur* 40
to þe most xamefullest deth I xal hem dyth
But o thyng is sor*e* in my gret delyte
þer*e* is on jh*es*us of nazareth · as me*n* me tellyth
of þ*at* man I desyr*e* to han A sythte
ffor w*ith* many gret wondrys · our*e* lawe he fellyth. 45

¶ The son of god hy*m*-self · he callyth
And kyng of jewys he seyth is he
Fo. 166v And many woundrys of hym be-fallyth
My hert desyryth hym for to se
Ser*ys* yf þ*at* he come in þis cowntr*e* 50
W*ith* our*e* jurresdycc*i*on loke ȝe A-spye
And a-non þ*at* he be brouth on-to me
And þe trewth myself þan xal trye.

i^us^ miles

¶ To-morwe my jorne · I xal begynne
to seke jh*es*us w*ith* my dew dilygens 55
ȝyf he come ȝo*ur* provynce w*ith*-inne
he xal not a-skape ȝo*ur* hey p*res*ens.

ij^us^ miles

¶ Myn sovereyn þis my councel þ*at* ȝe xal take
A man þ*at* is bothe wyse *and* stronge
thurwe all galyle A serge to make 60
yf jh*es*u be enteryd ȝo*ur* pepyl A-mong
Correcte hese dedys þ*at* be do wronge
Ffor his body is vndyr ȝo*ur* bayle
as men talkyn hem among
þat he was born in galyle. 65

Rex

¶ Thanne of þese materys serys take hede
Ffor A whyle I wele me rest
Appetyde requyryth me so in dede
And ffesyk tellyth me it is þe best.

a here xal A massanger com in-to þe place rennyng and *criyng Tydyngys tydyngys* · and *so rownd Abowth þe place · jhesus of nazareth is take · Jhesus of nazareth is take* · and *forth-with heylyng þe prynces þus seyng*

Massanger

Fo. 167 ¶ All heyle my lordys · princys of prestys
Sere cayphas *and* sere Annas lordys of þe lawe
tydyng*ys* I brynge ȝ*ou* · reseyve þe*m* in ȝ*our* brestys
Jhesus of nazareth is take · þ*er*-of ȝe may be fawe.

¶ he xal be browth hedyr to ȝ*ou* A-non
I telle ȝ*ou* trewly w*ith* A gret rowth
whan he was take · I was hem Among
and þ*er* was I ner to kachyd a clowte.

¶ Malcus bar A lanterne · *and* put hym in pres
A-noon he had A towche *and* of went his ere
Jhesus bad his dyscyple put up his swerd *and* ces
And sett malcus ere Ageyn · as hool as it was ere.

¶ So moty the methowut it was A strawnge syth
Whan we cam fyrst to hy*m* he cam vs A-geyn
And haskyd whom we sowth þ*at* tyme of nyth
We seyd jhesus of nazareth · we wolde haue hy*m* fayn.

¶ And he seyd it is I · þ*at* am here in ȝ*our* syth
W*ith* þ*at* word we ovyr-throwyn · bakward every-chon
and some on here bakkys · lyeng up-ryth
but standyng up on fote manly · þ*er* was not on.

¶ Cryst stod on his fete · As meke as A lom
and we loyn stylle lyche ded men · tyl he bad us ryse
Whan we were up · fast handys we leyd hy*m* up-on
but ȝet me thought I was not plesyd w*ith* þe newe gyse.

¶ Ther-fore takyth now ȝ*our* cowncel · *and* A-vyse ȝ*ou* ryth weyl
Fo. 167v And beth ryth ware · þ*at* he make ȝ*ou* not A-mat
Ffor be my thryfte I dare sweryn at þis seyl
ȝe xal fynde hym A strawnge watt.

α *here bryng þei jhesus be-forn Annas* and C. and *on xal seyn þus*

Lo · Lo · Lordys here is þe man
þ*at* ȝe sent us fore.

Annas

þerfore we cone ȝ*ou* thanke than
and reward ȝe xal haue þe more.

¶ Jhesus þou Art welcome hedyr to oure presens
ful oftyn-tymes we han þe besyly do sowth
we payd to þi dyscyple for þe thretty pens
and as an ox or an hors · we trewly þe bowth. 105

¶ þer-fore now art oure · as þou standyst us be-fore
Sey why þou ast trobelyd us · *and* subuertyd oure lawe
þou hast ofte concludyd us · *and* so þou hast do more
Where-fore it were ful nedful to bryng þe A dawe.

Cayphas

¶ What arn þi dysciplys · þat folwyn þe A-boute 110
and what is þi dottryne[1] þat þou dost preche
telle me now some-whath *and* bryng us out of doute
þat we may to othere men þi prechyng forth teche. .

jhes[us][2]

¶ Al tymes þat I haue prechyd · opyn it was don̄
in þe synagog or in þe temple · where þat all jewys com̄ 115
Aske hem what I haue seyd · and Also what I haue don̄
þei con telle þe my wordys · Aske hem everychon̄.

i^us judeus

Fo. 168 ¶ What þou fela to whom spekyst þou
xalt þou so speke to A buschop
þou xalt haue on þe cheke · I make A vow 120
and ȝet þer-to A knok.

α here he xal smyte jhesus on þe cheke.

jhesus

¶ yf I haue seyd Amys
þer-of wytnesse þou mayst bere
and yf I haue seyd but weyl in þis
þou dost Amys me to dere. 125

Annas

¶ Serys takyth hed now to þis man̄
þat he dystroye not oure lawe
and brynge ȝe wytnesse · A-ȝens hym þat ȝe can
so þat he may be browt of dawe.

i^us doctor

¶ Sere þis I herd hym · with his owyn mowth seyn 130
brekyth down þis temple · with-out delay
and I xal settynt up Ageyn
as hool as it was be þe thrydde day.

[1] Or *doctryne*; *correcte* on Fo. 166ᵛ is also doubtful.
[2] The last letters of the contraction are blotted.

ij^{us} doctor

¶ ȝa ser and I herd hym seyn Also
þat he was þe sone of god 135
and ȝet many A fole wenyth so
I durst leyn þ*er*-on my*n* hod [1].

iij^{us} doctor

¶ ȝa . ȝa . *and* I herd hy*m* pr*e*che meche þi*ng*
And A-ȝens oure lawe every del
Of wheche it were longe to make rekenyng 140
to tellyn aſt at þis seel.

Cayphas

¶ What seyst now jh*esus* whi Answeryst not
Fo. 168^{v} heryst not what is seyd A-ȝens þe
Spek man spek · spek þ*ou* fop
hast þ*ou* scorn to speke to me 145
Heryst not in how many thyng*ys* þei þ*e* Acuse [2]
α[3] Now I charge þe *and* con*j*ure be þe sonne *and* þe mone
þ*at* þ*ou* telle us *and* þ*ou* be goddys sone.

Jh*esus*

¶ Goddys sone I am I sey not nay to þe
And þ*at* ȝe aſt xal se At domys-day 150
Whan þe sone xal come · in gr*e*t power*e* *and* majeste
And deme þe qweke *and* dede as I þe say.

Cayphas

¶ A · out out Allas what is þis
heryth ȝe not how he blasfemyth god
what nedyth us to haue mor*e* wytness 155
her*e* ȝe han herd Aſt his owyn word
α Thynk ȝe not he is worthy to dey. *et clamab*unt *omnes*
ȝys · ȝys · ȝys · Aſt we seye · he is worthy to dey · ȝa ȝa ȝa.

Annas

α Takyth hym to ȝow · *and* betyth hy*m* som̄ del
ffor hese blasfemyng at þis sel. 160

α *her*e *þei xal bete jh*esus *A-bout þe hed* and *þe body* and *spyttyn in his face* and *pullyn hym down* and *settyn hy*m *on A stol* and *castyn A cloth ouyr his face* · and *þe fyrst xal seyn*

[1] *hed* first written, corrected to *hod*. [2] This line is extra-metrical.

[3] This and the following similar signs affixed to the rhyming couplet on Ff. 168^{v}, 169, 169^{v}, and again on Fo. 171^{v} on Fo. 177 and 178 and 178^{v} and 179 and 183^{v} are in red. With a very few exceptions (Fo. 74), such red signs only occur otherwise in the play of the Assumption, Ff. 215^{v} to 222. Those affixed to stage directions are, as heretofore in the MS., black.

i^us judeus

α A felawys be ware what ȝe do to þis man
ffor he prophecye weyl kan.

ij^us judeus

α Þat xal be A-sayd be þis batte
what þou jhesus ho ȝaff þe þat.

et percuciet super caput[1]

iij^us judeus

'o. 169α Whar whar now wole I
Wetyn how he can prophecy
·ho was þat.

iiij^us judeus

α A *and* now wole I a newe game begynne
þat we mon pley at all þat arn here-inne.

α whele *and* pylle · whele *and* pylle
comyth to halle ho so wylle
ho was þat.

α *here xal þe woman come to jewys* and *seyn*

i^a ancilla

α What serys how take ȝe on with þis man
se ȝe not on of hese dysciplys · how he be-heldyth ȝou þan.

α *here xal þe tother woman seyn to petyr*

ij^a ancille

α A · good man me semyth be þe
þat þou on of his dysciplys xulde be.

Petrus

α A woman I sey nevyr er þis man
syn þat þis werd fyrst be-gan.

et cantabit gallus.

i^a Ancilla

α What þou mayst not sey nay þou art on of hese men
be þi face wel we may þe ken.

Petrus

α Woman þou seyst A-mys of me
I knowe hym not so mote I the.

i^us judeus

α A fela myn wel met
for my cosynys ere pou of smet

[1] This stage direction is not underlined in red.

α Whan we þi mayster in þe ȝerd toke
þan alł þi ffelawys hym for-soke
α *and* now þou mayst not hym for-sake
for þou art of galyle I vndyr-take.

Petrus

¶ Sere I knowe hym not be hym þat made me
and ȝe wole me be-leve ffor An oth
Fo. 169v I take record of alł þis companye
þat I sey to ȝow is soth *et cantabit gallus.*

α *And þan jhesus xal lokyn on petyr* · and *petyr xal wepyn* and *þan he xal gon out* and *seyn*

¶ A weel a-way · weel away · fals hert why whylt þou not brest
Syn þi maystyr so cowardly · þou hast forsake
Alas qwher xal I now on erthe rest
tyl he of his mercy to grace wole me take.

¶ I haue for-sake my mayster *and* my lord jhesu
thre tymes as he tolde me · þat I xulde do þe same
Wherfore I may not haue sorwe A-now
I synful creature · am so mech to blame.

¶ Whan I herd þe cok crowyn he kest on me A loke
As who seyth be-thynke þe · what I seyd be-fore
Alas þe tyme þat I evyr hym for-soke
And so wyl I thynkyn from hens evyr more.

Cayphas **30**

α Massangere Massangere.

Massangere

here lord here.

Cayphas

¶ Massanger to pylat · in hast þou xalt gon
and sey hym we comawnde us in word *and* in dede
and prey hym þat he be at þe mothalle A-noon
ffor we han A[1] gret matere · þat he must nedys spede.

¶ In hast now go þi way
And loke þou tery nowth.

[1] This *A* is in darker ink—perhaps written over another letter later.

Massang*er*

it xal be do lord be þis day
I am as whyt as thought.

α *here pylat syttyth in his skaffald* and þe *massanger knelyth to hym*
Fo. 170 *þus seyng*

¶ Al heyl sere pylat þat semly is to se 215
Prynce of al þis jure · *and* kepere of þe lawe
My lord busshop cayphas comawndyd hy*m* to þe
And prayd the to be At þe mothalle by þe day dawe.

Pylat

¶ Go þ*i* way p*ra*ty masanger · *and* comawnde me Also
I xal be þere in hast *and* so þ*ou* mayst say 220
be þ*e* oure of pr*i*me I xal comy*n* hem to
I tery no longer · no[1] make no delay.

α *here þe massanger comith*[2] *A*ȝen and *bryngith*[2] *An Ansuere þus seyng*

Massanger

¶ Al heyl my*n* lordys · *and* buschoppys · *and* pr*i*ncys of þ*e* lawe
Ser pylat comawndyth hy*m* to ȝ*ou* · *and* bad me to ȝ*ou* say
He wole be at þe mothalle i*n* hast sone After þ*e* day dawe 225
He wold ȝe xuld be þ*er* be pr*i*me · w*ith*-outh lenger de-lay.

Cayphas

α Now weyl mote þ*ou* fare my good page
take þ*ou* þis for þi massage.

α *here enteryth judas on to þe juwys þus seyng*

judas

¶ I judas haue synyd · *and* treson haue don
ffor I haue be-trayd þis rythful blood 230
here is ȝ*our* mony A-ȝen Alł And som̃
Ffor sorwe *and* thowth · I am wax wood.

Annas

¶ What is þ*a*t to us A-vyse þe now
þ*ou* dedyst w*ith* us co*u*nawnt[3] make
þ*ou* soldyst hy*m* us · as hors or kow 235
þ*er*fore þi*n* owyn dedys · þ*ou* must take.

Fo. 170ᵛ *þan judas castyth down þe mony* · and *goth* and *hangyth hym-self.*

[1] *and* first written and crossed through before *no*.
[2] MS. com*t* and brȳg*t*; cf. waxit, Fo. 204ᵛ, seyt (saith), Fo. 151.
[3] MS. c̃onawnt. Halliwell prints *cornawnt*.

Fo. 170v

Cayphas

¶ Now serys þe nyth is passyd þe day is come
it were tyme þis man had his jewgement
And pylat Abydyth in þe mothalle Alone
tyl we xuld þis man present. 240

¶ *And* þer-fore go we now forth with hym in hast.

i^us judeus

it xal be don *and* þat in short spas.

ij^us judeus

ȝa but loke yf he be bownd ryth wel *and* fast.

iij^us judeus

he is saff A-now · go we ryth A good pas.

a here þei ledyn jhesu A-bowt þe place · tyl þei come to þe halle.

Cayphas

¶ Sere pylat takyht hede to þis thyng 245
jhesus we han be-forn þe browth
wheche oure lawe doth down bryng
and mekyl schame he hath us wrowth.

Annas

¶ ffrom þis cetye[1] · in to þe lond of Galyle
he hath browth oure lawys · neyr in-to confusyon 250
with hese craftys wrowth be nygramancye
shewyth to þe pepyl · be fals symulacyon.

i^us doctor

¶ ȝa ȝet ser A-nother · *and* werst of Alle
Aȝens sesare · oure emperour þat is so fre
kyng of jewys · he doth hym calle 255
So oure emperourys power · nowth xulde be.

ij^us doctor

¶ Sere pylat we kan not telle half þe blame
þat jhesus in oure countre · hath wrowth
þerfore we charge þe in þe emperorys name
þat he to þe deth · in hast be browth. 260

Pylat

Fo. 171 ¶ What seyst to these compleyntys jhesu
these pepyl hath þe sore acusyd
be-cause þou bryngyst up lawys newe
þat in oure days were not vsyd.

[1] *ceyt* first written and crossed through,

Jhesus

¶ Of here A-cusyng me rowth nowth 265
so þat þei hurt not here soulys · ne non mo
I haue nowth ȝet founde · þat I haue sowth
ffor my faderys wyl · fforth must I go.

Pylat

¶ Jhesus be þis þan I trowe · þou art A kyng
and þe sone of god · þou art Also 270
lord of erth · *and* of All þing
telle me þe trowth · if it be so.

Jhesus

¶ In hefne is knowyn · my faderys intent
And in þis werlde · I was born
be my fadyr · I was hedyr sent 275
for to seke · þat was for-lorn.

¶ Alle þat me heryn · And in me belevyn
And kepyn here feyth stedfastly
þow þei weryn dede · I xal þem recuryn
and xal þem bryng to blysse · endlesly. 280

Pilate

¶ Lo serys now ȝe An erde þis man · how thynk ȝe
thynke ȝe not All be ȝoure reson
but as he seyth it may wel be
and þat xulde be be þis incheson.

¶ I fynde in hym · non obecyon 285
of error*our* nor treson · ne of no maner gylt
Fo. 171v The lawe wele · in no conclusyon
W*ith*-owte defawth he xuld be spylt.

i^us doctor

¶ Sere pylat þe law restyth in þe
and we knowe veryly his gret trespas 290
to þe emperour · þis mater told xal be
yf þou lete jhesus · þus from þe pas.

Pylat

¶ Serys þan telle me o thyng
What xal be his A-cusyng.

Annas

α Sere we telle þe al to-gedyr 295
Ffor his evyl werkys · we browth hym hedyr
α And yf he had not An evyl doere be
we xuld not Abrowth hym to þe.

Pylat

ɑ Takyth hym þan Aftyr ȝ*our* sawe
and demyth hy*m* Aftyr ȝ*our* lawe. 300

Cayphas

ɑ it is not lefful to vs ȝe[1] seyn
no maner man for to slen.
ɑ þe cawse why · we bryng hy*m* to þe
þ*at* he xuld not our*e* kyng be.
ɑ weyl þ*ou* knowyst kyng we haue non 305
but our*e* emp*erour* Alon.
ɑ Jhesu · þ*ou* Art kyng of jur*e*. Pylat

Jhesus

So þ*ou* seyst now to me.

Pylat

ɑ Tel me þan 310
wher*e* is[2] þ*i* kyngh*a*m.

Jhesus

ɑ my kyngh*a*m is not in þis werld
I telle þ*e* At o word.
ɑ yf my kyngh*a*m her*e* had be
I xuld not A be delyveryd to þ*e*[3]

Pylat

ɑ Ser*ys* A-vyse ȝow as ȝe kan 315
Fo. 172 I can fynde no defawth in þis man.

Annas

¶ Ser*e* her*e* is a gret record · take hed þ*er* to
and knowyng gret myschef in þis man
And not only in o day or to
it is many ȝerys · syn he began 320
We kan telle þe tyme where *and* whan
þ*at* many A thowsand turnyd hath he
As Aꝗ þis pepyꝗ record weyl kan
from hens in-to þe lond of galyle et *clamabunt* ȝa. ȝa. ȝa.

Pilat

¶ Serys of o thyng than · gyf me relacyoꝑ 325
if jhesus were out born in þe lond of galelye
ffor we han no poer ne no jurediccyoꝑ

[1] *we* first written, crossed through and *ȝe* written over the line.
[2] *is* omitted and written over the line.
[3] From line 309 to line 314 two lines are written in one.

of no man of *þat* contre
Ther-fore þe trewth · ȝe telle me
And A-nother wey I xal *pro*vyde 330
if jh*es*us were born · in *þat* cou*n*tre
þe jugement of herowdys he must A-byde.

Cayphas

¶ Ser*e* as I am to þe lawe trewly sworn
to telle þe trewth I have no fer
in galelye · I know *þat* he was born 335
I can telle in what place · *and* where
Aȝens þis no man may Answere
for he was born in bedlem jude
and þis ȝe knowe now Alł *and* haue don her*e*
þat it stant in þe lond of galelye. 340

Pylat

Fo. 172v [marked 170] ¶ Weyl serys syn *þat* I knowe · *þat* it is so
þe trewth of þis I must nedys se
I vndyrstand ryth now · what is to do
þe Jugement of jh*es*u · lyth not to me
herowde is kyng of *þat* cou*n*tre 345
to jewge *þat* regyon in lenth *and* i*n* brede
þe jurysdyccyon of jh*es*u · now han must he
þer fore · jh*es*u in hast · to hy*m* ȝe lede
In hałł þe hast *þat* ȝe may spede
lede hy*m* to *þe* herownde · A-non p*re*sent 350
and sey I comawnde me wi*th* worde *and* dede
And jh*es*u to hy*m* *þat* I haue sent.

i^us doctor

¶ This erand in hast sped xal be
in alł þe hast *þat* we can do
We xal not tary in no degre 355
tyl þe herowdys presens · we com̄ to

α here þei take jhesu and *lede hym* · *in gret hast to þe herowde* · And *þe herowdys scafald xal vn-close shewyng herowdes*[1] *in astat* alł *þe jewys knelying* · *except Annas* and *cayphas þei xal stondy*n *etcetera*

i^us doctor

¶ heyl herowde most excyllent kyng
we Arn comawndyd · to *þin* p*re*sens

[1] The contraction for (d)*es* is here probably a slip for the contraction for (d)*e*.

pylat sendyth þe · be us gretyng
And chargyth · us · be ou*re* Obedyens.

ij^us doctor

þ*at* we xuld do ou*re* dylygens
to bryng jhe*sus* of nazareth on-to þ*e*
And chargyth us to make no resystens
be-cawse he was born in þis cou*n*tre.

Annas

Fo. 173 ¶ we knowe he hath wrowth · gret fole
A-geyns þe lawe · shewyd present
Ther-fo*re* pylat sent hy*m* on to þe
þ*at* þ*ou* xuldyst gyf hy*m* jugement.

Herowde Rex

¶ Now be Mahound my god of grace
of pylat þis is A dede ful kende
I for-gyf hym now his gret trespace
And schal be his frend w*ith*-owtyn ende.

¶ Jhe*sus* to me þ*at* he wole sende
I desyred ful so*re* hy*m* for to se
gret ese in þis pylat xal fynde
and jhe*sus* þ*ou* art welcome to me.

i^us jude*us*

¶ My sovereyn lord þis is þe case
þe gret falsnesse of jhe*su* · is opynly knawe
þer was nevyr man dede so gret trespas
ffor he hath · Al-most · dystroyd ou*re* lawe.

ij^us judeus

¶ ȝa be fals crafte of soserye
wrowth opynly to þe pepyll Alle
and be sotyl poynt*ys* of nygramancye
many thowsandys fro ou*re* lawe be falle.

Cayphas

¶ Most excellent [1] kyng · ȝe must take hede
he wol dystroye all þis cou*n*tre · both elde *and* ȝyng
yf he ten monthis mo*re* procede
be his meraclys *and* fals prechyng
he bryngyth þe pepyl in gret fonnyng
And seyth dayly A-mong hem Alle

[1] *excelyng* first written, and *yng* crossed through.

Fo. 173v That he is lord *and* of þe jewys kyng
and þe sone of god he doth hym calle.

Rex Her*owde*

¶ Serys Alle þese materys · I haue herd sayd
and meche more þan ȝe me telle
Alle to-gedyr þei xal be layde 395
And I wyl take þ*er* on cow*n*celle.

¶ Jhesus þ*ou* Art wel-come to me
I kan pylat gret thank for his sendying
I haue desyryd ful longe þe to se
and of þi meracles · to haue knowyng. 400

¶ It is told me þ*ou* dost many A wondyr thyng
Crokyd to gon *and* blynd men to sen
and þei þ*at* ben dede gevyst hem levyng
And makyst lepers · fayr*e* *and* hool to ben.

¶ These Arn wondyr werkys · wrougth of þe 405
be what wey · I wolde knowe þe trew sentens
Now jhesu I p*ray* the · lete me se
O meracle wrougth in my p*re*sens.

¶ In hast now · do þi dylygens
And p*er*-Aventure · I wyl shew favo*ur* to the 410
for now þ*ou* art in my p*re*sens
thyn lyf *and* deth · her*e* lyth in me.

α *And here jhesus xal not speke no word to þe herowde.*

¶ Jhesus why spekyst not to þi kyng
what is þe cawse þ*ou* stondyst so stylle
þ*ou* knowyst I may deme Alle thyng 415
thyn lyf *and* deth · lyth At my wylle.

Fo. 174 ¶ What spek jhesus · *and* telle me why
þis pepyl do þe so · her*e* Acuse
spar*e* not but telle me now · on hey
how þ*ou* canst þ*i*-self · excuse. 420

Cayphas

¶ loo serys þis is of hym · A false sotylte
he wyl not speke · but whan he lyst
þus he dysceyvyth þ*e* pepyl in eche degre
he is ful fals · ȝe veryly tryst.

Rex her*owde*

What þ*ou* on-hangyd harlot · why wylt þ*ou* not speke 425
hast þ*ou* skorne to speke on to þ*i* kyng
be-cawse þ*ou* dost our*e* lawys breke
I trowe þ*ou* art A-ferd · of our*e* talkyng.

Annas

¶ Nay he is not Aferde · but of A fals wyle
be-cawse we xuld not hym A-cuse 430
if þ*at* he Answerd ȝow on-tylle
he knowyth he can not hym[1]-self excuse.

Rex her*owde*

¶ what spek I say · þ*ou* foulyng · evyl mote þ*ou* far*e*
loke up · þ*e* devyl mote þ*e* cheke
Ser*ys* bete his body w*ith* scorg*ys* bar*e* 435
And A-say to make hy*m* for to speke.

i^us^ jude*us*

¶ it xal be do w*ith*-outyn teryeng
come on þ*ou* tret*our* evyl mot þ*ou* þ*e*
whylt þ*ou* not speke on-to our*e* kyng
A new lesson we xal ler*e* þ*e*. 440

α here þei pulle of jhesus clothis and *betyn hy*m *with whyppys.*

ij^us^ jude*us*

Fo. 174^v^ ¶ Jhesus þ*i* bonys we xal not breke
but we xal make þ*e* to skyppe
þ*ou* hast lost þ*i* tonge · þ*ou* mayst not speke
þ*ou* xalt a-say now of þis whippe.

iij^us^ jude*us*

¶ Serys take þese whyppys in ȝ*our* hande 445
and spar*e* not whyl þei last
and bete þis tretour*e* þ*at* her*e* doth stonde
I trowe þ*at* he wyl speke in hast

α and qwan *þei han betyn hy*m *tyl he is alle blody þan þe herownde seyth*

¶ Sees ser*ys* · I comawnde ȝ*ou* · be name · of þ*e* devyl of helle
Jhesus thynkyst þis · good game 450
þ*ou* Art strong · to suffyr schame
þ*ou* haddyst levyr be betyn lame
þan þ*i* defawtys for to telle.

[1] This *y* is written over an *e*.

¶ But I wyl not þi body · Aſt spyl
nor put it here · in-to more peyn̄ 455
Serys takyth jhesus · At ȝour owyn wyl
and lede hym to pylat hom Ageyn̄
Grete hym weyl *and* telle hym serteyn̄
Aſt my good frenchep xal he haue
I gyf hym powere of jhesus · þus ȝe hym seyn 460
Whether he wole hym dampne or save.

i^us doctor

¶ Sere at ȝour request it xal be do
we xal lede jhesus at ȝour demawde[1]
and delyver hym Pylat · on-to
And telle hym aſt · as ȝe comawnde. 465

Fo. 175 *here enteryth Satan in to þe place in þe most orryble wyse* · and *qwyl* α *þat he pleyth þei xal don on jhesus clothis* · and *ouerest A whyte clothe* and *ledyn hym A-bowth þe place* and *þan to pylat be þe tyme þat hese wyf hath pleyd.*

Sathan

¶ Thus I reyne As A rochand · *with* A rynggyng rowth
As A devyl most dowty · dred is my dynt[2]
Many A thowsand develys · to me do þei lowth
brennyng in flamys · as fyre out of flynt
ho so serve me sathan · to sorwe is he sent 470
with dragonys in dou*n*genys · *and* develys fu[3] derke
in bras *and* in bronston · þe brethellys be brent
þat wone in þis werd · my wyl for to werke.

31

¶ *with* myschef on moolde · here membrys I merke
þat japyn *with* jhesus þat judas solde 475
be he nevyr so crafty · nor conyng clerke
I harry þem to helle · as tretou*r* bolde.

¶ But þ*er* is o thyng · þ*at* grevyth me sore[4]
Of A p*ro*phete · þ*at* jhesu men calle
he peynyth me every day · more *and* more 480
with his holy meraclis *and* werkys Alle.

¶ I had hym onys · in A temptacyon
with glotenye *with* covetyse · *and* veyn glorye

[1] So in MS. [2] *dyth* first written. [3] So in MS.
[4] *myn hert* first written instead of *me sore*.

I ha-sayd hym be Aſt weys *þat* I cownde[1] don
and vttyrly he refusyd hem · *and* gan me defye. 485

¶ *Þat* rebuke *þat* he gaf me · xal not be vn-qwyt
Som̃ what I haue be-gonne · *and* more xal be do
Fo. 175v ffor Aſt his barfot goyng · fro me xal he not skyp
but my derk dongeon I xal bryngyn hy*m* to.

¶ I haue do made redy his cros · *þat* he xal dye up-on̄ 490
And thre nayles to takke hym w*ith* þat he xal not styrte
be he nevyr so holy he xal not fro me gon
but w*ith* a sharpe spere · he xal be smet to *þe* herte.

¶ And sythyn he xal come to helle · be he nevyr so stowte
And ȝet I am Aferd *and* he come · he wole do som wrake 495
*þer*fore I xal go warnyn helle · *þat* þei loke A-bowte
þat þei make redy chenys · to bynd hy*m* w*ith* in lake.

¶ Helle helle · make redy · for here xal come A gest
Hedyr xal come jh*es*us · *þat* is clepyd goddys sone
And he xal ben here · be *þe* oure of none 500
And w*ith* þe here · he xal wone
And han ful shrewyd rest.

α here xal A devyl spekyn in helle demon̄
Out upon þe · we co*n*jure þe
þat nevyr in helle we may hy*m* se
ffor And he onys in helle be 505
he xal oure power brest.

Sathan̄
¶ A · A · than haue I go to ferre
but som̃ wyle help I haue a shrewde torne
My game is wers *þa*n I wend here
I may seyn · my game is lorne. 510

¶ lo A wyle ȝet haue I kast
if I myth jh*es*us lyf s*a*ve
helle gatys xal be sperd fast
And kepe stylle aſt þo I haue.

¶ to pylatys wyff I wele now go 515
Fo. 176 *And* sche is A-slepe A-bed ful fast

[1] So in MS.

And byd here *with*-owtyn wordys mo
to pylat þat sche send in hast.

¶ I xal A-say *and* þis wol be
to bryng pylat in belef 520
with-inne A whyle ȝe xal se
how my craft I wole go pref.

α here xal þe devyl gon to pylatys wyf · þe corteyn drawyn as she lyth in bedde and *he xal no dene make but she xal sone After þat he is come in · makyn a rewly noyse · comyng* and *rennyng of þe schaffald* and *here shert* · and *here kyrtyl in here hand* · and *sche xal come beforn pylat leke A mad woman · seyng þus*

Vx*or* pilaty

¶ Pylat I charge þe · þat þ*ou* take hede
deme not jh*es*u · but be his frende
ȝyf þ*ou* jewge hy*m* to be dede 525
þ*ou* art da*m*pnyd *with*-owtyn ende.

¶ A fend · Aperyd me beforn
As I lay in my bed slepyng fast
Sethyn þ*e* tyme þ*at* I was born
was I nevyr so sore A-gast. 530

¶ As wylde fyr*e* *and* thondyr blast
he cam cryeng on to me
he seyd þei þ*at* bete jh*es*u · or bownd hy*m* fast
with-owtyn ende dampnyd xal be.

¶ þ*er*-fore A wey here-in þ*ou* se 535
and lete jh*es*u from þe cler*e* pace
þe jewys þei wole be-gyle þe
and put on þe Aff þe trespace.

Pylat

Fo. 176v ¶ Gramercy my*n* wyf for evyr ȝe be trewe
ȝ*our* cowncel is good *and* evyr hath be 540
now to ȝ*our* chawmer ȝe do sewe
and aff xal be weyl dame as ȝe xal se.

α here þe jewys bryng jhesus Aȝen to pylat.

i*us* doctor

¶ Sere pylat gode tydandys · þ*ou* here of me
Of herowd þe kyng þ*ou* hast good wyl

And jhesus he sendyth Azen to the
And byddyth þe chese · hym to save or spylle.

ij^us doctor

¶ ȝa *ser* all þe poer lyth now in þe
and þou knowyst oure feyth · he hath ner schent
þou knowyst what myschef · þer of may be
we charge þe · to gyf hym jwgement.

Pylat

¶ Serys trewly ȝe be · to blame
Jhesus þus to bete · dyspoyle or bynde
or put hym to so gret schame
ffor no defawth · in hym I fynde.

¶ Ne herowdys nother · to whom I sent ȝow
defawte in hym cowde fynde ryth non
but sent hym Azen · to me be ȝow
As ȝe knowe wel everychon.

* Ther fore vndyrstande what I xal say
ȝe knowe þe custom is in þis londe
of ȝour pasche day · þat is ner honde
what þeff or tretore · be in bonde
with-out Any price
for worchep of þat day[1] · xal go fre Away
Fo. 177 Now þan me thynkyth · it were ryth
to lete jhesus · now go qwyte
And do to hym no mo dyspyte
serys þis is myn A-vyse
℟[2] I wolde wete what ȝe say.
℟ *here Alle þei xul cryen* || Nay nay nay.

i^us doctor

℟ delyvere us · þe þeff barabas
þat for mansclawth presonde was.

Pylat

℟ What xal I þan with jhesu do
Whethyr xal he A-byde or go.

ij^us doctor

℟ Jhesus xal on þe cros be don
Crucifigatur we crye echon.

* Paragraph mark omitted here.

[1] Some miswritten letters (*ff* ?) crossed out before *xal*.

[2] This and the following similar signs against the couplets are in red ink.

pylat

α Ser*ys* what hath jhesus don A-mys
Crucifigatur · we sey At onys[1]. *populus clamabit.*

Pylat

¶ Serys syn Al gatys · ȝe wolyn so
puttyn jhesu to wo *and* peyn 580
jhesus A wyle w*ith* me xal go
I wole hy*m* examyne betwyx us tweyn.

α *here pylat takyth jhesu* and *ledyth hym in to þe cowncel hous* and *seyth*

¶ Jhesus what seyst now · lete se
This matere now þ*ou* vndyrstonde
In pes þ*ou* myth be for me 585
but for þ*i* pepyl of þ*i* londe.

¶ Busshoppys *and* prestys · of þe lawe
þei love þ*e* not as þ*ou* mayst se
and þ*e* comon[2] pepyl A-ȝens þ*e* drawe
In pes þ*ou* myth A be for me 590
þis I telle þ*e* pleyn.

Fo. 177v ¶[3] What seyst jhesus · whi spekyst not[4] me to
knowyst not I haue power on þe cros þe to do
And also I haue power to lete þe forth go
what kanst þ*ou* here to seyn. 595

Jhesus

¶ On me poer þ*ou* hast ryth non
but þ*at* my fadyr hath grawntyd be-forn
m̄ I cam my faderys wyl to fuł-fylle
þ*at* mankynd xuld not spylle
m̄ he þ*at* hath betrayd me · to þ*e* at þis tyme 600
his trespas is more þ*an* is þine.

i^us doctor

¶ ȝe pryncys *and* mayster*ys* · takyth hed *and* se
how pylat in þis matere is favorabyl
and þus oure lawys dystroyd myth be
And to vs Alle vn-recurabyl. 605

α *here pylat letyth jhesus A-lone* and *goth in to þe jewys* and *seyth*

[1] *Omnes* written in the right-hand margin and crossed through in red ink.
[2] MS. comon̄. [3] This paragraph sign seems inserted by mistake.
[4] *spekyst not* written twice and crossed through in red ink; *to me* written first, corrected, and finally *me to* written above the line.

Pylat

¶ Ser*ys* what wole ȝe now · w*ith* jhesu do
I can fynde in hym but good
it is my cownce[1] ȝe lete hy*m* go
it is rewthe to spylle his blood.

Cayphas

¶ Pylat me thynkyth · þ*ou* dost gret wrong 610
Aȝens[2] our*e* lawe þus to fortefye
and þe pepyl her*e* is so strong
bryngyng þ*e* lawful · testymonye.

Annas

¶ ȝa *and* þ*ou* lete jhesu fro us pace
þis we welyn up-holdyn Alle 615
þ*ou* xalt Answer*e* for his trespas
and tret*our* to þe emp*erour* we xal þ*e* kalle.

Pylat

Fo. 178α Now þan syn ȝe wolne · non other weye
but in Alwyse þ*at* jhesus must deye
α Artyse bryng me watyr I pray þe 620
and what I wole do ȝe xal se *hic vnus afferet aquam.*
As I wasche w*ith* watyr my handys clene
so gyltles of hese deth I mut ben.

i^us doctor

α þe blod of hy*m* mut ben on vs
and on our*e* chyldyr Aftyr vs } *et clamabunt* ȝa · ȝa · ȝa ·

α *þan pylat goth A-ȝen to jhesu* and *bryngith*[3] *hym þus seyng*

Pylat

α lo ser*ys* I bryng hy*m* her*e* to ȝ*our* presens 626
þ*at* ȝe may knowe I fynde in hy*m* non offens.

ij^us doctor

α dylyuer*e* hym · delyuer*e* hym · *and* lete us go
on þ*e* crosse þ*at* he were do.

pilat

α Serys wolde ȝe ȝ*our* kyng · I xulde on þ*e* cros don. 630

iij^us doctor

α Ser*e* we seyn þ*at* we haue no kyng but þ*e* emp*erour* a-lon.

pilat

α Ser*ys* syn Al-gatys it must be so
We must syt *and* o*ur* offyce do

[1] So in MS. [2] The *A* is written over some other letter.
[3] The contraction is not clear. The scribe apparently first wrote *bringys* and then crossed the top of the contraction without making any erasure.

α brynge forth to þe barre þat Arn to be dempt
and þei xal haue here jugement. 635

α *here þei xal brynge barabas · to þe barre* and *jhesu* and *ij þewys*[1] *in here shertys bare leggyd* and *jhesus standyng at þe barre be-twyx them* · and *annas* and *cayphas xal gon in to þe cowncelle hous qwan pylat sytty[th.*

Pylat

α Barabas hold up þ*i* hond
Fo. 178v for here at þ*i* delyvere[2] dost þ*ou* stond and *he halt up his honde*
α Serys qwhat sey ȝe of barabas · thef *and* tret*our* bold
xal he go fre or he xal be kept in holde.

i^us doctor

α Ser*e* for þe solennyte[3] of our*e* pasche day 640
be our*e* lawe he xal go fre A-way.

Pylat

α Barabas þan I dymysse[4] þe
and ȝeve þe lycens to go fre *et curret.*[5]
α dysmas *and* jesmas · ther as ȝe stondys
þe lawe comawndyth ȝ*ou* · to hald up ȝ*our* hondys 645
α Ser*e* what sey ȝe of þese thevys tweyn.

ij^us doctor

Ser*e* þei ben both gylty we seyn.

Pylat

α And what sey ȝe · of jh*e*su of nazareth.

i^us doctor

Ser*e* we sey he xal be put to deth.

Pylat

α And kone ȝe put A-ȝens hy*m* no trespas. 650

ij^us doctor

Ser*e* we wyl All þ*at* he xal be put upon þe crosse.[6]

α *et clamabunt omnes · voce magna dicentes* ȝa · ȝa · ȝa ·

Pylat

α jh*e*su þ*in* owyn pepyl han dysprevyd
Al þ*at* I haue for þe seyd or mevyd.

¶ I charge ȝ*ou* All at þe be-gynny*n*g
as ȝe wole Answer*e* me be-forn 655

[1] So in MS. for *þerys*. Halliwell prints *Jewys*, but the initial letter is clear.
[2] The word *he* (?) has been erased before *dost*. [3] Or possibly *solonnyte*
[4] So in MS. [5] Not underlined in red.
[6] This speech is underlined in red as if it were a stage direction.

þat þer be no man · xal towch *ȝour* kyng
but yf he be knyght · or jentylman born.

¶ fyrst his clothis ȝe xal of don
and makyn hym nakyd · for to be
bynde hym to A pelere · as sore as ȝe mon 660
þan skorge hym with qwyppys · þat al men may se
Fo. 179α Whan he is betyn · crowne hym for ȝour kyng
and þan to þe cros ȝe xal hym bryng
α *And* to þe crosse þou xalt be fest
And on thre naylys þi body xal rest 665
α On xal thorwe þi ryth hand go
Anothyr thorwe þi lyfte hand Also
α þe thred xal be smet thour bothe þi feet
Whech nayl þer-to be mad ful mete
α *And* ȝet þou xalt not hange A-lone 670
but on eyther syde of þe xal be on
α Dysmas now I deme þe
þat on hese ryth hand þou xalt be
α *And* Jesmas on þe left hand hangyd xal ben
on þe mownth[1] of caluerye þat men may sen. 675

α *here pylat xal rysyn* and *gon to his schaffalde*, and *þe busshoppys with hym* and *þe jewys xul crye for joy with a gret voys* · and *Arryn hym* and *pullyn of his clothis* and *byndyn hym to A pelere* and *skorgyn hym on seyng þus*

i^us judeus

Doth gladly oure kyng
for þis is ȝour fyrst begynnyng.

α *and qwan he is skorgyd · þei put upon hym A cloth of sylk* and *settyn hym on a stol* and *puttyn A kroune of þornys on hese hed with forkys* and *þe jewys knelyng to cryst takyng hym A septer* and *skornyng hym · and þan þei xal pullyn of þe purpyl cloth* and *don on A-geyn his owyn clothis* and *leyn þe crosse in hese necke to berynt* and *drawyn hym forth with ropys · And þan xal come to women wepyng* and *with here handys wryngyn seyng þus*

i^us [2]mulier

Fo. 179v ¶ Allas jhesus · Allas jhesus · wo is me **32**
þat þou art þus dyspoylyd Allas

[1] MS. *mowth* (with a stroke over the *th*).

[2] So undoubtedly in MS., though on Fo. 169, p. 277, the distinction in the contraction marks is observed.

and ȝet nevyr defawth · was fownd in the 680
but evyr þou hast be fole of grace.

ij^us mulier

¶ A here is a rewful syth · of jhesu so good
þat he xal þus dye Aȝens þe ryth
A wykkyd men ȝe be more þan wood
to do þat good lord so gret dyspyte. 685

α here jhesus turnyth a-ȝen to þe women with his crosse þus seyng

jhesus

¶ Dowterys of hierusalem · for me wepyth nowth
but for ȝour-self wepyth · *and* for ȝour chyldyr Also
for þe days xal come · þat þei han aftyr sowth
here synne *and* here blyndnesse xal turne hem to wo.

¶ þan xal be sayd · blyssyd be þe wombys þat bareyn be 690
and wo to þe tetys · tho days · þat do ȝevyn sokyng
and to here faderys þei xul seyn · wo to þe tyme þat þou be-gat me
and to here moderys · Allas · wher xal be oure dwellyng.

¶ þan to þe hyllys and mownteynes · they xal crye *and* calle
oppyn *and* hyde us from þe face · of hym syttyng in trone 695
or ellys ovyr-throwyth · *and* on us now come falle
þat we may be hyd · from oure sorweful mone.

α here jhesus turnyth fro þe women and *goth forth* and *þer þei metyn with symonem · in þe place þe jewys seyng to hym*

i^us jude[*us*

¶ Sere to þe A word of good
a man is here þou mayst se
beryth hevy of A Rode 700
where-on he xal hangyd be.

Fo. 180 (T quire) ¶ Therefore we prey all the *þer fore we prey*[1]
þou take þe crosse of þe man
bere it with vs to kalvarye
and ryth gret thank þou xalt han. 705

Symon

¶ Serys I may not in no degre
I haue gret errandys for to do
þerfore I pray ȝow excuse me
and on my herand · lete me go.

[1] Written at the bottom of Fo. 179^v as catchword.

ij^us judeus

¶ what harlot hast þou skorne
to bere þe tre whan we þe preye
þou xalt berynt haddyst þou sworn̄
And yt were ten tyme þe weye.

Symon

¶ serys I prey ȝou dysplese ȝou nowth
I wole help to bere þe[1] tre
in to þe place it xal be browth
where ȝe wole comawnde me.

here symon takyth þe cros of jhesus and *beryth it forth.*

Veronica

¶ A ȝe synful pepyl why fare þus
Ffor swet *and* blood he may not se
Allas holy prophete cryst jhesus

and *sche whypyth his face with here kerchy.*

Careful is myn hert for the.

Jhesus[2]

¶ veronyca þi whipyng doth me ese
my face is clene · þat was blak to se
I xal þem kepe from all̄ mys-ese
þat lokyn on þi kerchy · *and* remembyr me.

Fo. 180v *þan xul þei pulle jhesu out of his clothis* and *leyn them togedyr* and *þer þei xul pullyn hym down* and *leyn hym Along on þe cros* · and *after þat naylyn hym þeron.*

i^us Judeus

¶ Come on now here we xal A-say
Yf þe cros for þe be mete
Cast hym down̄ here in þe devyl way
how long xal he standyn on his fete.

ij^us judeus

¶ pul hym down evyl mote he the
And gyf me his arm in hast
And A-non we xal se
hese good days þei xul be past.

iij^us judeus

¶ Gef hese other Arm to me
A-nother take hed to hese feet

[1] Some word may have been written above *þe* in darker ink and smudged out.
[2] *Jh* written as correction over *ve*.

And A-non we xal se
yf þe borys be for hy*m* meet.
iiij^{us} jude*us*

¶ þis is mete take good hede
pulle out þ*a*t Arm to þ*e* sor*e*.
i^{us} jude*us*

þis is short þe deuyl hy*m* sped
be a large fote *and* more.
ij^{us} jude*us*

¶ Ffest on A Rop · *and* pulle hy*m* long
and I xal drawe þe A-geyn
spar*e* we not þese ropys strong
þow we brest both flesch *and* veyn.
ij^{us} jude*us*

¶ dryve in þ*e* nayl a-non · lete se
And loke *and* þe flesch *and* senues well last.
iiij^{us} jude*us*

þ*a*t I graunt so mote I the
lo þis nayl is dreve ryth wel *and* fast.*
i^{us} jude*us*

Fo. 181 ¶ Ffest A rop þ*a*n to his feet
And drawe him do n long Anow.
ij^{us} jude*us*

her*e* is a nayl for both good *and* greet[1]
I xal dryve[2] it thorwe I make A vow.

α here xule þei leve of and *dawncyn a-bowte þe cros shortly.*
iij^{us} jude*us*

¶ Lo fela here A lythe · takkyd on A tre.
iiij^{us} jude*us*[3]

ȝa *and* I trowe þ*ou* art A worthy kyng.
i^{us} jude*us*

A good ser*e* telle me now what helpyth[4] þ*i* prophecy þ*e*.
ij^{us} jude*us*

ȝa or Any of þ*i* ffals p*r*echyng.
iij^{us} jude*us*

¶ Ser*ys* set up þ*e* cros on þe hende
þ*at* we may loke hy*m* in þe face.

* The first six words of this line are scribbled again in the bottom margin in a slightly later hand.

[1] The *g* written over an *r*. [2] The *r* written over a *y*.

[3] A piece of the original edge has been torn away here.

[4] The *h* (initial) is written over some other letter.

iiij^{us} jude*us*

ȝa *and* we xal knelyn on to our*e* kyng so kend' 760
and preyn hym of his gret grace.

here qwa*n* *þei han set hym up þei xuln gon be-fore hyne*[1] *seyng eche affter other þus*

i^{us} jude*us*

¶ heyl kyng[2] of jewys · yf þ*ou* be.

ij^{us} jude*us*

ȝa · ȝa · sere as þ*ou* hangyst þere flesche *and* bonys.

iij^{us} jude*us*

Com now down of þ*at* tre.

iiij^{us} jude*us*

And we wole worchepe þe all Atonys. 765

here xul poer comonys stand and *loke upon þe jewys iiij or v* and *þe jewys xul come to theme* and *do theme hange þe þevys.*

i^{us} judeus

¶ come on ȝe knavys · *and* set up þise[3] ij crosses ryth
and hange up þese to thevys Anon.

ij^{us} jud[eus

Fo. 181v ȝa *and* in þe worchep of þis worthy knyth
on eche syde of hym xal hangyn on.

here þe sympyl men xul settyn up þese ij crossys and *hangyn up þe thevys be þe Armys* and *þer whylys xal þe jewys cast dyce for his clothis* and *fytyn* and *stryvyn* and *in þe mene tyme xal oure lady come with iij maryes with here* and *sen Johan with heme settyng hem down A-syde A-fore þe cros · oure lady swuonyng*[4] and *mornyng* and *leysere seyng*

Maria

¶ A my good lord my sone so swete 770
what hast þ*ou* don why hangyst now þ*us* here
is þ*er* non other deth to þe now mete
but þe most shamful deth · Among þese thevys fere.

¶ A out on my hert whi brest þ*ou* nowth 774
And þ*ou* art maydyn *and* modyr *and* seyst þus þ*i* childe spylle

[1] So in MS. for *hyme*.

[2] *kyng* is written as correction over some erased word.

[3] The medial vowel is not clear and might be *e*, but is perhaps to be read as an *i* with an unusually marked dot.

[4] There is a blotted stroke between the *w* and the *u*.

how mayst *þou* a-byde þis sorwe *and* þis woful powth
A deth · deth · deth · why wylt *þou* not me kylle.

α here oure lady xal swonge A-ȝen and *ore lord xal seyn þ*us Jhesus

¶ O Ffadyr Al-mythy · makere of man̄
Ffor-gyff þese jewys *þat* don me wo
Ffor-geve hem fadyr for-geve hem þan 780
Ffor thei wete notwh[1] what þei do.

i^us Judeus

¶ ȝa vath̄[2] · vath̄ · now · here is he
þat bad us dystroye oure tempyl on A day
and with-inne days thre
he xulde reysynt A-ȝen in good A-ray. 785

ij^us Judeus

Fo. 182 ¶ Now *and* *þou* kan do swech̄ A dede
help now þi-self yf *þat* *þou* kan
and we xal be-levyn on þe with-outyn drede
and seyn *þou* art a mythty man.

iij^us Judeus

ȝa yf þu be goddys sone · as *þou* dedyst teche 790
ffrom *þe* cros come now down̄
þan of mercy we xal þe beseche
and seyn *þou* art a lord of gret renown.

jestes

¶ Yf *þou* be goddys sone as *þou* dedyst seye
helpe here now both *þe* *and* vs 795
but I fynde it not al in my feye
þat *þou* xuldyst be cryst goddys sone jhesus.

Dysmas

¶ Do wey fool why seyst *þou* so
he is þe sone of god I be-leve it wel
and synne dede he nevyr lo 800
þat he xuld be put þis deth tyl.

¶ but we ful mech̄ wrong han wrowth
he[3] dede nevyr *þing* A-mys
now mercy good lord mercy · *and* for-gete me nowth
whan *þou* comyst to *þi* kyngham *and* to þi blysse. 805

[1] So in MS. [2] *w* written before *vath̄* and crossed out.
[3] Some word before *dede* erased and *he* written beyond the line in the margin.

Jhesus

¶ Amen Amen · þou art ful wyse
þat þou hast Askyd I grawnt þe
þis same day in paradyse
with me þi god þou xalt þer be.

Maria

Fo. 182v [marked 180—some older mark ?100 or 180 smudged out.]

¶ O my sone my sone · my derlyng dere
what haue I defendyd þe
þou hast spoke to alle þo · þat ben here
and not o word þou spekyst to me.

¶ To þe jewys þou art ful kende
þou hast for-gove al here mysdede
and þe thef · þou hast in mende
for onys haskyng mercy · hefne is his mede.

¶ A my sovereyn lord why whylt þou not speke
to me þat am þi modyr · in peyn for þi wrong
A hert hert why whylt þou not breke
þat I were out of þis sorwe so stronge.

Jhesus

¶ A woman woman · be-hold þer þi sone
And þou jon take[1] her for þi modyr
I charge þe to kepe here as besyly as þou kone
þou A clene mayde xal kepe A-nother.

¶ And woman þou knowyst þat my fadyr of hefne me sent
to take þis manhod of þe · Adam ys rawnsom to pay
Ffor þis is þe wyl · *and* my faderys intent
þat I xal þus deye · to delyuere man · fro þe develys pray.

¶ Now syn it is þe wyl of my fadyr · it xuld þus be
Why xuld it dysplese þe modyr · now my deth so sore
And for to suffre Al þis for man · I was born of the
to þe blys þat man had lost · man A-ȝen to restore.

α her oure lady xal ryse and *renne* and *halse þe crosse.*

Maria Magdalena

Fo. 183 ¶ A good lady why do ȝe þus
ȝour dolfol cher · now cheuith[2] us sore
And for þe peyne of my swete lord jhesus
þat he seyth in ȝou it peyneth hym more.

[1] This *a* is corrected from another letter.

[2] MS. *cheut*.

Maria virgo

¶ I pray ȝow Alle lete me ben here
and hang me up here on þis tre
be my frend *and* sone þat me is so dere 840
ffor þer he is þer wold I be.

Johannes

¶ Jentyl lady now leve ȝour mornyng
and go with us now we ȝou pray
And comfort oure lord at hese departyng
ffor he is Al-most redy to go his way. 845

here þei xal take oure lady from þe crosse · and *here xal pylat come down from his shaffald with Cayphas* and *Annas* and *att here mene* and *xul come* and *lokyn on Cryst* and *annas* and *cayphas xul skornfully sey[n*

Cayphas

¶ lo serys lo · be-heldyth *and* se
here hangyth he þat halpe many A man
And now yf he goddys sone be
helpe now hym-self yf þat he kan.

Annas

¶ ȝa *and* yf þou kyng of israel be 850
come down of þe cros Among us alle
And lete þi god now delyuere the
and þan oure kyng we wole þe calle.

here xal pylat Askyn penne and *inke* and *A tabyl xal be take hym wretyn A-fore* · *hic est jhesus nazarenus rex judeorum.*

Fo. 183v and *he xal make hym to wryte* and *þan gon up on A leddere* and *settyn þe tabyl abovyn crystys hed* and *þan cayphas xal makyn hym to redyn* and *seyn*

Cayphas

Sere pylat we merveylyth of þis
þat ȝe wryte hym to be kyng of jewys 855
þerfore we wolde þat ȝe xuld wryte þus
þat he namyd hymself kyng of jewus.

Pylat

þat I haue wretyn · wretyn it is
and so it xal be for me i-wys.

and *so forth att þei xal gon Aȝen to þe skaffalde* and *Jhesus xal cryen*[1]

[1] *Jhesus* is enclosed in a red loop as the name of a speaker, and the red underlining of the stage direction is continued under *xal cryen*.

HOMERTON COLLEGE LIBRARY

¶ heloy · heloy · lamaȝabathany ‖ 860
my fadyr in hevyn on hy
why dost þ*ou* me for-sake[1]
The frelte of my mankende
W*ith* stronge peyn yt gynnyth to peynde
ha der*e* fadyr haue me in mende 865
and lete deth my sorwe slake.

ij^{us} judeus

¶ me thynkyth he this doth calle hely
lete us go ner*e* *and* a-spy
and loke yf he come p*re*uely
from cros[2] hy*m* down to reve. 870

Jh*es*us

So grett a thrust dede nevyr man take
as I haue man now for þ*i* sake
for thrust a-sundyr my lyppys gy*n* crake
for drynes þ*ei* do cleve.

iij^{us} jude*us*

¶ ȝ*our* thrust ser*e* hoberd for to slake 875
eyȝil *and* galle her*e* I þ*e* take
what me thynkyth a mowe ȝe make
is not þis good drynk
to crye for drynke ȝe had gret hast
and now it semyth it is but wast 880
is not þis drynk of good tast
now telle me how ȝe thynk.

4us jude*us*

¶ on lofte ser*e* hoberd now ȝe be sett[3]
we wyl no le*n*ger w*ith* ȝ*ou* lett
w*e* grete ȝ*ou* wel on þ*e* newe gett 885
and make on ȝ*ou* a mowe.

i^{us} judeus

w*e* grete ȝ*ou* wel w*ith* a scorn
and p*ra*y ȝ*ou* bothe evy*n* *and* morn

[1] The ink changes and the writing becomes rather smaller with this line, which is written as one with the preceding line. Note also that the next speaker is 'ii^{us} judeus' where 'i^{us}' might be expected.

[2] This *r* is written over an *o*.

[3] From line 883 to line 890 two lines are written in one.

take good eyd to our*e* corn
and char*e* awey þ*e* crowe. 890

Jh*esus*

Fo. 184[1] ¶ In manus tuas domine
holy fadyr in hefly se
I comende my spyryte to þe
for her*e* now hendyth my fest
I xal go sle þ*e* fende þ*at* freke 895
ffor now my*n* herte be-gynnyth to breke
wurdys mo xal I non speke
Nu*n*c co*n*summatu*m* est.

Maria

¶ Alas Alas I leve to longe
to se my swete sone w*ith* peynes stronge 900
As a theff on cros doth honge
And nevyr ȝet dede he synne
Alas my der*e* chyld to deth is dressyd
now is my car*e* wel mor*e* in-cressyd
A myn herte w*ith* peyn is pressyd 905
Ffor sorwe my*n* hert doth twynne.

Joh*ann*es

¶ A blyssyd mayde chaunge ȝ*our* thought
Ffor þow ȝ*our* sone w*ith* sorwe be sought
ȝitt by his owyn wyl þis werk is wrought
And wylfully his deth to take 910
ȝow to kepe he chargyd me her*e*
I am ȝ*our* servaunt my lady der*e*
wherfor*e* I p*ray* ȝow be of good cher*e*
And merthis þ*at* ȝe make.

Maria

¶ Thow he had nevyr of me be born 915
And I sey his flesch þus al to torn
on bak be-hyndyn on brest be-forn
Rent w*ith* wou*n*dys wyde
Fo. 184v Nedys I must wonyn in woo
to se my ffrende w*ith* many a ffo 920
all to rent from top to too
his flesch w*ith*-owtyn hyde.

[1] This and the following folio (185) are interpolated in this quire, being of different paper. The writing on them is firmer and more regular.

Joh*ann*es

¶ A blyssyd lady as I ȝow telle
had he not deyd we xuld to helle
Among*es* ffendys þ*er* evyr to dwelle 925
In peynes þ*at* ben smert
he sufferyth deth for our*e* trespace
and thorwe his deth we xal haue grace
to dwelle w*ith* hy*m* in hevyn place
þ*er*fore beth mery in hert. 930

Maria

¶ A der*e* ffrende weel woot I this
þ*at* he doth bye us to his blys
but ȝitt of myrth evyr mor I mys
whan I se þis syght.

Joh*ann*es

Now der*e* lady þ*er*fore I ȝow p*ray* 935
Ffro þis dolful dolo*ur* wende we our*e* way
Ffor whan þis syght ȝe se nought may
ȝour*e* care may waxe mor*e* lyght.

Maria

¶ Now sythe I must parte hym fro
ȝit lete me kysse or þ*at* I go 940
his blyssyd ffeyt þ*at* sufferyn wo
naylid on þis tre
So cruelly w*ith* grett dyspyte
þus · shamfully was nevyr man dyghte
þ*er*fore in peyn my*n* hert is pyghte 945
al joye departyth fro me.

hic quasi semi nortua[1] *cadat prona in terra*m et *dic*it *joh*ann*es*

Fo. 185 Joh*ann*es

¶ Now blyssyd mayd com*e* forth w*ith* me
no lengere þis syght þ*at* ȝe se
I xal ȝow gyde in þis countre
where þ*at* it plesyth ȝow best. 950

Maria

Now jentyl Joh*an* my sonys derlyng
to goddys temple þ*ou* me brynge
þat I may prey god w*ith* sor*e* wepynge
And mornynge þ*at* is prest.

[1] So in MS. (seminor tua).

Joh*ann*es

¶ Alł ȝ*our* desyr*e* xal be wrought 955
w*ith* herty wyłł I werke ȝ*our* thought
now blyssyd mayde taryeth nowth
In þ*e* temple þ*at* ȝe ware
Ffor holy prayer*e* may chaunge ȝ*our* mood
and cawse ȝ*our* cher*e* to be mor*e* good 960
whan ȝe se notȝ ȝ*our* childys blood
þ*e* lasse may be ȝ*our* car*e*.

*Tunc transiet maria ad templu*m *cu*m *Io*hanne et *ce*tera.

Maria

¶ Her*e* in þis temple my lyff I lede
And serue my lord god w*ith* hertyly drede
now xal wepynge me fode *and* fede 965
Som̄ comforte tyłł god sende
A my lord god I þ*e* pray
Whan my childe ryseth þ*e* iij^de^ day
Comforte thanne thy*n* hand-may
my car*e* for to Amende. 970

33[1] a*nim*a Chr*isti*

¶ Now alł mankende in herte be glad
w*ith* alł merthis þ*at* may be had
ffor ma*nn*ys sowle þ*at* was be-stad
in þ*e* logge of helle:
Fo. 185v now xal I ryse to lyve agayn 975
from peyn to pleys of p*ar*adyse pleyn
þ*er*fore man in hert be fayn
in m*er*the now xalt þ*ou* dwelle.

¶ I am þe sowle of cryst jh*es*u
þ*e* which is kynge of alł vertu 980
my body is ded þ*e* jewys it slew
þ*at* hangyth ȝitt on þ*e* rode
rent *and* torn̄ all blody red
ffor ma*nn*ys sake my body is deed
ffor ma*nn*ys helpe my body is bred 985
And sowle drynk my bodyes blode.

¶ þow my body be now sclayn

[1] This number is written partly against the preceding speech.

þe thrydde day *þis* is certayn
I xal reyse my body a-gayn
to lyve as I ȝow say
now wole I go streyth to helle 990
and feche from *þe* fendys felle
aſſ my frendys *þat* *þer*-in dwelle
to blysse *þat* lestyth Ay.

* *Nota anima latronis*

The sowle goth to helle gatys and *seyth*

℣ [2] Attollite portas *pr*incipes *vest*ras et eleuam*ini* porte et*er*nales et *in*troib*it* rex glorie.

¶ Ondothe ȝour*e* ȝatys of sorwatorie
On ma*n*nys sowle I haue memorie 995
her*e* comyth now *þe* kynge of glorye
these gat*ys* for to breke
ȝe develys *þat* arn her*e* wi*th*-inne
helle gatys ȝe xal vn-pynne
I xal delyver*e* mannys kynne 1000
ffrom wo I wole he*m* wreke.

Belyaſſ

Alas Alas out *and* harrow
Onto þi byddynge must we bow
þat *þou* art god now do we know
Of *þe* had we grett dowte 1005
Aȝens *þe* may no thynge stonde
Aſſ thynge obeyth to thyn honde
bothe hevyn *and* helle watyr *and* londe
Aſſ thynge must to *þe* lowte.

A*nim*a Chr*isti*

Fo. 186 ¶ Aȝens me it wor*e* but wast 1010
to holdyn or to stondyn fast
helle logge may not last
Aȝens · *þe* kynge of glorye
þi derke dor*e* down I throwe
My fayr ffrendys now wele I knowe 1015

* These words are written in the margin apparently in the hand of the scribe of Ff. 95, 96.

[1] This contraction for *Versus* is otherwise only used to mark the Latin versicles in the Assumption play.

I xal he*m* brynge reknyd be rowe
Out of here p*ur*catorye.

Centurio

¶ In trewth̄ now I knowe w*ith* ful opyn syght
that goddys dere sone is naylid on tre
these wundyrful tokenys Ap*re*vyn ful ryght 1020
q*uo*d vere filius dei erat iste.

Ali*us* miles 2 **34**

The very childe of god I suppose þ*at* he be
and so it semyth wele be his wundyrful werk
þe erth̄ sore qwakyth *and* þ*at* agresyth me
W*ith* myst *and* grett wedyr it is wo*u*ndyr dyrk. 1025

Ali*us* miles 3

¶ Soch̄ merveylis shewe may non erthely man̄
þe eyr is ryght derke þ*at* fyrst was ryght clere
The Erth̄-qwave is grett þe clowdys waxe wha*n̄*
those tokenys p*re*ve hym a lorde w*ith*-out Any[1] pere.

Centurio

his fadyr is pereles kyng of most empere 1030
bothe lorde of þis world *and* kynge of hevyn hyȝe
ȝit out of all̄ sy*n*ne to brynge us owt of daungere
he soferyth his dere sone for us all̄ to dye.

Nichodem*us*

¶ Alas Alas what syght is this 1035
to se þe lorde *and* kynge of blys
þ*at* nevyr synnyd ne dede Amys
þus naylid vpon a rode
Alas ȝewys what haue ȝe wrought
A ȝe wyckyd wytys what was ȝo*ur* thought 1040
Why haue ȝe bobbyd *and* þus betyn owth
all̄ his blyssyd blood.

Fo. 186v Senturyo[2]

¶ A now trewly telle weyl I kan
þ*at* þis was goddys owyn sone
I knowe he is both god *and* man 1045
be þis wark þ*at* here is done.

¶ þ*er* was nevyr man but god · þ*at* cowde make þis werk
þ*at* evyr was of woman born

[1] A large *A* seems to have been written over a small one.

[2] This name stands at the top of the page. No name of next speaker is given at the bottom of Fo. 186.

were he nevyr so gret A clerk
it passeth hem alł þow þei had sworn. 1050

¶ hese lawe was trewe I dare wel saye
þat he tawth us here A-monge
þerfore I rede ȝe turne ȝ*our* faye
and amende *þat* ȝe han do wronge.

Joseph of Ara . . [[1]

¶ O good lord jh*esu* · *þat* deyst now here on rode 1055
haue mercy on me · *and* for-gyf me my mys
I wold þe worchep here w*ith* my good
þat I may come to þi blysse.

¶ to pylat now wole I goon[2]
and aske þe body of my lord jh*esu* 1060
to bery *þat* now wold I soon
in my grave *þat* is so new.

¶ heyl sere pylat *þat* syttyth in sete
heyl justyce of jewys men do þe calle
heyl w*ith* helth · I do þe grete 1065
I p*ray* þe of A bone what so befalle.

¶ to bery jh*esu* is body I wole *þe* p*ray*
þat he were out of mennys syth
for to-morwyn xal be oure holyday
þan wole no man hy*m* bery I *þe* plyth. 1070

Fo. 187 ¶ And yf we lete hym hange *þer* stylle
Some wolde seyn *þer* of A-now
þe pepyl *þer*-of wold seyn ful ylle
þat nother xuld be ȝ*our* worchep nor prow.

Pylat

¶ Sere joseph of baramathie · I graunt þe 1075
W*ith* jhesu is body do þi*n* intent
but fyrst I wole wete *þat* he ded be
as it was his jugement.

¶ Sere knytys I comawnd ȝow *þat* ȝe go
in hast w*ith* josepht of baramathie 1080

[1] Strokes of *m* lost in hole in paper—rest of the word or mark of contraction cut off with margin.

[2] *god* first written and crossed through.

and loke ȝe take good hede þ*er*-to
þ*at* jh*es*u suerly ded be.

¶ se þ*at* þis comawndement ȝe fulfylle
w*ith*-out wordys ony mo
and þan lete joseph do his wylle 1085
what þ*at* he wyl w*ith* jh*es*u do.

α *here come to knyt*ys *be-forn pylat At-onys þus seyng*

i^us^ miles

¶ Ser*e* we xal do our*e* dylygens
w*ith* joseph goyng to Caluerye
be we out of þi p*re*sens
sone þe trewth we xal aspye. 1090

Joseph Ab[. . . .

¶ Gramercy pylat of ȝ*our* jentylnesse
þ*at* ȝe han grawntyd me my lyst
Any thyng in my p*ro*vince
ȝe xal haue at ȝ*our* resquest [1]

Pylat

Fo. 187v ¶ Ser*e* at̄ ȝ*our* lest ȝe xal haue 1095
w*ith* jh*es*u is body do ȝ*our* intent
whethyr ȝe bery hy*m* in pyt or grave
þ*e* powe*re* I grawnt ȝow here p*re*sent.

α *The ij knygt*ys *go with joseph to jh*esus and *stande* and *heldyn hym in þe face.*

ij^us^ miles

¶ Me thynkyth jh*es*u is sewr*e* anow
it is no ned his bonys to breke 1100
he is ded how þinkyth ȝow
he xal nevyr go nor speke.

i^us^ miles

¶ We wyl be sur*e* or þan we go
Of A thyng I am be-thowth [2]
ȝondyr is a blynd knyth I xal go to 1105
And sone A whyle here xal be wrowth.

α *here þe knyth goth to blynde longeys* and *seyth*

¶ heyl ser*e* longeys þ*ou* gentyl knyth
þe I prey now ryth hertyly

[1] *rest* first written as the first syllable and the *t* adapted to a *q*.
[2] The final *h* is blotted and seems to have been altered with darker ink.

þat þou wylt wend w*ith* me ful wyth
it xal be for *þi* prow veryly. 1110

longe*us*

¶ Sere at *3our* comauwndement w*ith* 3ow wyl I wende
in what place 3e wyl me haue
for I trost 3e be my fre*nd*
lede me forth sere · oure sabath *3ou* save.

i^us^ miles

¶ lo sere longeys here is a spere 1115
bothe long *and* brood *and* sharp a-now
heve it up fast *þat* it wore *þere*
for here is game show man show.

¶ here longeys showyth þe spere warly · and *þe blood comyth rennyng to his hand* and *he Auantorysly xal wype his eyne.*

longeys

Fo. 188 ¶ O good lord · how may *þis* be
þat I may se so bryth now 1120
þis thretty wyntyr I myth not se
and now I may se I wote nevyr how
but ho is *þis* *þat* hangyth here now
I trowe it be *þe* mayndonys[1] sone
and *þat* he is now · I knowe wel how 1125
þe jewys to hy*m* *þis* velany han don.

¶ here he ffallyth down on his knes.

¶ Now good lord fforgyf me that
þat I to *þe* now don have
for I dede I wyst not what
þe jewys of my*n* ignorans dede me rave 1130
M*er*cy m*er*cy m*er*cy I crye.

*¶ þan joseph doth set up þe leder*ys and *nychodem*us *comyth to help hy*m.

Nicodem*us*

¶ Joseph ab Aramathy · blyssyd *þou* be
ffor *þou* dost a fol good dede
I prey the · lete me help *þe*
þat I may be partenere of *þi* mede. 1135

Joseph

¶ Nychodemus welcome · indede
I pray 3ow 3e wole help *þer*-to

[1] So in MS.

he wole Aqwyte us ryth weyl our*e* mede
and I haue lysens for to do.

σ here joseph and *nychodem*us *takyn cryst of þe cros on on o ledyr* and *þe tother on An-other leddyr* and *qwan is had down joseph leyth hym in our*e *ladys lappe · seyng þe knyt*ys *turnyng heme* · and *joseph seyth*

Joseph

¶ Lo mary modyr good *and* trewe
her*e* is þ*i* son blody *and* bloo
ffor hym my*n* hert ful sor*e* doth rewe
kysse hym now onys · eer he go.

Maria *vir*go

Fo. 188v ¶ A Mercy Mercy my*n* owyn son so der*e*
þ*i* blody face now I must kysse
þ*i* face is pale w*ith*-owty*n* cher*e*
of meche joy now xal I mysse
þ*er* was nevyr modyr þ*at* sey this
so her*e* sone dyspoyled · w*ith* so gret wo
and my der*e* chylde nevyr dede A-mys
A mercy fadyr of hefne · it xulde be so.

Joseph

¶ Mary ȝ*our* sone ȝe take to me
in-to his grave it xal be browth.

Maria

joseph blyssyd evyr mot þ*ou* be
for þ*e* good dede þ*at* ȝe han wrowth.

σ here þei xal leyn cryst in his grave.

Joseph

¶ I gyf þe þis syndony þ*at* I haue bowth
to wynde þ*e* in whyl it is new.

Nichodem*us*

her*e* is An onyment þ*at* I haue browth
to Anoynt w*ith* all my*n* lord jhesu.

Joseph

¶ Now jhesu is w*ith*-inne his grave
wheche I ordeyn som tyme for me
on þe lord I vowche it save
I knowe my mede ful gret xal be.

Nichodem[*us*

¶ now lete us leyn on þis ston Ageyn
And jhes*u* in þis tombe stylle xal be

And we wyl walke hom ful pleyn
þe day passyth fast I se
Far*e* wel joseph *and* wel ȝe be
no lengere teryeng her*e* we make.

Joseph

Sere almythy god be w*ith* þe 1170
in-to his blysse he mote ȝ*ou* take.

Maria

Fo. 189 ¶ Ffar*e* wel ȝe jentyl p*rincys* kende
in joye evyr mote ȝe be
þe blysse of hefne w*ith*-owtyn ende
I knowe veryly þat ȝe xal se. 1175

℟ *here* þe *princys xal do reuerens to oure lady* and *gon here way* and
* *nota* *leve* þe *maryes at be sepulcre.* *Cayphas goth to Pylat seyng* þ*us*[1]
Incipit hic

Cayphas[2]

¶ herk ser*e* pylat lyst to me
I xal þe telle tydyng*ys* new
of o thyng · we must[3] war*e* be
er ellys her*e* after · we myth it rewe.

¶ þ*ou* wotyst weyl þat jhe*s*u 1180
he seyd to us w*ith* wordys pleyn
he seyd we xuld fynd it trew
þe thryd day he wold ryse agey[4]
yf þat hese dyscyplys come serteyn
and out of his g*r*aue stele hy*m* away[5] 1185
þei wyl go p*r*eche *and* pleyn seyn
þat he is reson þe thryd day.

¶ þis is þe cowncel þat I gyf her*e*
take me*n* *and* gyf he*m* charge þ*er*to
to weche þe grave w*ith* gret power 1190
tyl þe thryd day be go.

* These words *nota* and *Incipit hic* are written in the margin in another hand —apparently that of the scribe of Ff. 95, 96—and blacker ink. There are also some smudged indecipherable words (? scribblings) on the right-hand margin.

[1] These stage directions are written partly at the side of the text, and the ℟ stands before the last line only.

[2] *Cayphas* is in paler ink and may have been added.

[3] The *st* is blotted as if corrected from another letter. [4] So in MS.

[5] From line 1185 to line 1215 two lines are written in one.

Pylat

¶ Sere Cayphas · it xal be do
for as ȝe say þer is peryl in
And it happend þat it were so
it myth make our lawys for to blyn 1195
ȝe xal se ser er þat ȝe go
how I xal þis mater saue
And what I xal sey þer-to
and what charge þei xal haue.

hic ¶ Come foɪth ȝe ser Amorawnt 1200
N . . . *hic* *and* ser Arphaxat com ner also
Ser Cosdram *and* ser Affraunt
and here þe charge þat ȝe must do
Serys to Jhesu is grave ȝe xal go
tyl þat þe thryd day be gon 1205
and lete nother frend nor fo
in no wey to towche þe ston.

¶ yf ony of hese dyscipelys come þer
to fech̄ þe body fro ȝou a-way
bete hym down have ȝe no fere 1210
with shamful deth do hym day
in payn of ȝour godys *and* ȝour lyvys
þat ȝe lete hem nowth shape ȝou fro
and of ȝour chyldere *and* ȝour wyfys
for al ȝe lese *and* ȝe do so. 1215

i^us miles

o. 189v ¶ Sere pylat we xal not ses
we xal kepe it strong A-now.

ij^us miles

ȝa *and* An hunderyd put hem in pres
þei xal dey I make A vow.

iij^us miles

¶ And han honderyd fy on An C. *and* an C. þer-to 1220
þer is non of hem xal us with-stonde.

iiij^us miles

ȝa *and* þer com̄ An hunderyd thowsand *and* mo
I xal hem kylle with myn honde.

* This note in the margin, the first word of which is doubtful—probably Nota—is apparently in the same hand as that on p. 312.

Pylat

¶ wel ser*ys* þan ȝ*our* part ȝe do.
And to ȝ*our* charge loke ȝe take hede 1225
W*ith*-owtyn wordys ony mo *here þe knytys gon out of be place*
Wysly now þ*at* ȝe p*ro*cede.

¶ lo s*er* cayphas how thynkyth ȝow
is not þis wel browth Abowth.

Cayphas

in feyth s*er* it is sur*e* A-now 1230
hardely haue ȝe no dowth.

i*us* Arfaxat[1] ij

¶ let se ser amaraun*t* where wele ȝe be
wole ȝe kepe þ*e* feet or þ*e* hed.

ij*us* Ameraun*t*

At þ*e* hed so mote I the[2]
and ho so come her*e* he is but ded. 1235

i*us* Arfaxat ii

¶ *And* I wole kepe þe feet þis tyde
þow þ*er* come both jakke *and* gylle.

iij*us* cosdram iii

And I xal kepe þ*e* ryth syde
and ho so come I xal hym kylle.

iiij*us* Affraunt 4

And I wole on þe lefte hand ben 1240
and ho so come her*e* he xal nevyr then
fful sekyrly his bane xal I ben
* *nota* w*ith* dyntys of dowte ⁊ *syr pylat*[3]
Fo. 190 Syr pylat haue good day
(V quire) We xal kepyn þ*e* body in clay 1245
And we xal wakyn wele þ*e* way
and wayt*en* aꝫ abowte ⁊

Pylatus

Now jentyl ser*ys* wole ȝe vouch-saffe
to go w*ith* me *and* sele þ*e*[4] graffe
þ*at* he ne aryse out of þ*e* grave 1250
þ*at* is now ded

[1] These names were originally numbered to the right Ameraun*t* being 1, Arfaxat 2 — the order in which Pilate calls upon them. These numbers have been partly cut away with the margin, and the names have been renumbered to the left, Arfaxat the first speaker being 1, Ameraun*t* 2.

* Marginal note apparently in the same hand as those above.

[2] *So mote I the I wole be at þe h.* first written and crossed through.

[3] Catchword. [4] Some miswritten letter crossed through after *þe*.

Cayphas

We graunte wel lete us now go
Whan it is selyd *and* kepte Also
Than be we sekyr w*ith*-owtyn wo
And have of hy*m* no dred. 1255

Tunc ibunt ad sepulcrum pilatus Cayphas Annas et omnes milites et *dicit*

Annas

¶ Loo here is wax ful redy dyght
Sett on ȝ*our* sele anon ful ryght
þan be ȝe sekyr I ȝow plyght
he xal not rysyn agayn.

Pilatus

On þis corner my seal xal sytt 1260
And w*ith* þis wax I sele þis pytt
now dare I ley he xal nevyr flytt
out of þis grave serteayn.

Annas

¶ Here is more wax fful redy loo
aſſ þe corner*ys* ȝe sele Also 1265
And w*ith* a lokke loke it too
Than lete us gon oure way
And lete þese knyt*ys* abydyn þ*er* by
And yf hese dysciplys com p*re*uyly
to stele awey þis ded body 1270
to vs they hem brynge w*ith*-out delay.

Pilatas[1]

¶ On ev*er*y corner now is sett my seale
now is my*n* herte in welthe *and* wele
Fo. 190v This may no brybour a-wey now stele
þis body from vndyr stoñ 1275
Now syr buschoppe I pray to the
And Annas also com on w*ith* me
Evyn to-gedyr aſſ we thre
hom-ward þe wey we goñ.

¶ As wynde wrothe[2] 1280
knyght*ys* now goht

[1] A stroke in the *a* may indicate a correction of *as* to *us*.

[2] From line 1280 to line 1320 two lines are written in one. From line 1321 to line 1343 three lines are written in one.

clappyd in cloth
and kepyth hym well
loke ȝe be bolde
w*ith* me for to holde 1285
ȝe xul haue gold
And helme of stele.

Pylat Annas and *cayphas go to þer skaffaldys* · and *þe knyghtys sey*[1] . . [

Affraunt 4

¶ Now in þis grownnde[2]
he lyeth bounde
þ*at* tholyd wounde 1290
ffor he was ffals
þis lefft[3] cornere
I wyl kepe here
Armyd clere
bothe hed *and* hals. 1295

Cosdran̄ 3

¶ I wyl haue þis syde
what so betyde
If any man ryde
to stele þe cors
I xal hym chyde 1300
w*ith* woundys wyde
Amonge hem glyde
w*ith* fyne fors.

Ameraunt ij^{us} [4]

¶ The hed I take
here by to wake ⁄ 1305
A stele stake
I holde in honde
Maystryes to make
crownys I crake ⁄

[1] Fourth letter blotted, the rest cut off with margin.
[2] *tyde* first written and crossed through.
[3] *left* or *leftt* first written—corrected to *lefft*.
[4] The first of the two strokes is a later addition making *i*^{us} into *ij*^{us}.

Schafftys to shake 1310
And Schapyn schonde[1].

Arfaxat i^us[2]

¶ I xal not lete
to kepe þe fete
they ar ful w . . .[3]
walterid in blood 1315
He þat wyll stalke
be brook or balke
hedyr to walke
þo wrecchis be wood.

i^us miles

Fo. 191 Myn heed dullyth 1320
myn herte ffullyth
of sslepp
Seynt Mahownd
þis bereynge grownd
þou kepp. 1325

ij^us miles

I sey þe same
ffor Any blame
I falle
Mahownde whelpe
Aftyr þin helpe 1330
I calle.

3us miles

I am hevy as leed
ffor Any dred
I slepe
Mahownd of myght 1335
þis ston to nyght
þou kepe.

4us miles

I haue no foot
to stonde on root
by brynke 1340

[1] *sle fre* and *bonde* written above the last two words of this line in another ink and in the Ff. 95, 96 hand.

[2] An original 2 scratched out and i^us written in in darker ink.

[3] There has been some erasure and rewriting here by the later corrector; perhaps *wete* altered to *white*.

here I Aske
to go to taske
A wynke.

Tunc dormyent milites et *ueniet* Anima *Christi de inferno* cum *Adam et* Eua[1] · *Abraham johan baptista et Alijs.*

Anima Christi :

¶ Come forthe Adam *and* Eue w*ith* the
And all my fryndys þat here-in be 1345
to paradys come forthe w*ith* me
In blysse for to dwelle 35
þe fende of helle þat is ȝo*ur* ffoo
he xal be wrappyd *and* woundy*n* in woo
Ffro wo to welthe now xul ȝe go 1350
W*ith* myrthe evyr more to melle.

Adam

¶ I thanke þe lord of þ*i* grett grace
that now is for-ȝovyn my grett trespace
now xal we dwellyn in blysful place
In joye *and* endeles myrthe 1355
Thorwe my synne man was fforlorn
and man to saue þ*ou* wore all torn
and of a mayd in bedlem born
þat evyr blyssyd be þ*i* byrthe.

Eua

Fo. 191v ¶ Blyssyd be þ*ou* lord of lyff 1360
I am Eue Adam is wyff
þ*ou* hast soferyd strok *and* stryff
Ffor werkys þ*at* we wrought
þ*i* mylde mercy haht[2] All[3] for-ȝovyn
Dethis dentys on þe were drevyn 1365
now w*ith* þe lord we xul levyn
þ*i* bryght blood hath us bowth̄.

Johannes bap*tista*

¶ I am þ*i* cosyn my name is Joh*an*
þ*i* woundys hath betyn þe to þe bon
I baptyzid þe in flomjordon 1370
And ȝaff þ*i* body baptyze

[1] MS. Euã. [2] An *h* between *haht* and *All*e has been crossed through.
[3] This A seems to have been changed from a small to a larger form.

W*ith* þ*i* grace now xul we gon
Ffrom oure enmyes every-choñ
And fyndy*n* myrthis many on
In play of paradyse. 1375

Abraham

¶ I am Abraham fadyr trowe
þ*at* reyned after noes flowe
A sory synne Adam gan sowe
þ*at* clad us alł in care
A sone þ*at* maydenys mylk hath sokyn 1380
and w*ith* his blood oure bonde hath brokyn
helle logge lyth vnlokyn
Ffro fylth w*ith* frende we fare.*

anima christi

¶ Ffayre ffrendys now be ȝe wu*n*ne
on ȝow shyneth þ*e* sothfast sunne 1385
þ*e* gost þ*at* alł grevaunce hath gu*n*ne
Fo. 192 Fful harde I xal hym bynde
As wyckyd werme þ*ou* gu*n*ne Aper*e*
to tray my chylderyn þ*at* were so der*e*
þ*er*fore traytou*r* heu*er*-mo*re* here 1390
newe peynes þ*ou* xalt evyr ffynde.

¶ Thorwe blood I took of ma*n*nys kynde
Ffals devyl I here þ*e* bynde
In endles sorwe I þ*e* wynde
þ*er*-in evyr-mor*e* to dwelle 1395
now þ*ou* art bownde þ*ou* mayst not fle
Ffor þin envyous cruelte
In endeles dampnacion xalt þ*ou* be
And nevyr comy*n* out of helle.†

Beliał

¶ Alas herrow now am I bownde 1400
In helle gonge to ly on[1] grownde
In hendles sorwe now am I wounde

* The words *anima caym* are here written in the margin in the hand of Ff. 95, 96, and in the left-hand margin *Nota anima caym* followed by some erased words ending: *as folow fayere frendys*.

† A word or two, of which the first is *thowght*, is written here in the margin in the Fo. 95, &c., hand and smudged out.

[1] *on* is written (by the scribe) over the line above an erasure.

In car*e* evyr mor*e* to dwelle
In helle logge I lyȝ a-lone
now is my joye a-wey al gone 1405
Ffor all fendys xul be my fone
I xal nevyr com from helle.*

Anima Christi

¶ Now is ȝ*our* ffoo boundyn in helle
þ*at* evyr was besy ȝow for to qwelle
now wele I rysyn flesch *and* felle 1410
þ*at* rent was for ȝ*our* sake
my*n* owyn body þ*at* hynge on rode
And be þ*e* jewys nevyr so wode
It xal a-ryse both flesch and blode
my body now wyl I take. 1415

Tunc transiet anima christi ad resuscitandum corpus quo resuscitato dicat Jhesus

Fo. 192v (marked 190)

Jhesus

¶ harde gatys haue I gon
And peynes sofryd many on
Stomblyd at stake *and* at ston
nyȝ thre *and* thretty ȝer*e*
I lyght out of my fader*ys* trone 1420
ffor to Amende ma*n*nys mone
my flesch was betyn to þ*e* bon
my blood I bledde cler*e*.

¶ Ffor ma*n*nys loue I tholyd dede
and for ma*n*nys loue I am rysyn up rede[1] 1425
ffor man I haue mad my body in brede
his sowle for to fede
Man *and* þ*ou* lete me þus gone
and wylt not folwyn me a-none

* The hand of Fo. 95, &c., has interlined here and written down the left-hand margin:

'not*a* þ*e* devyll thowght many &c.
Thowght many begon I am glad &c. hens I wyll þ*e* bere.
þ*an* Cr*ist*.
and þ*an* cayme xall sey h*is* spech. *And* þ*an* cr*ist* xall sey now ys y*our* foo. etc.'

[1] The *r* is written above the line over an erased letter.

such a frende fyndyst *þou* nevyr none 1430
to help *þe* at *þi* nede.

¶ Salue sancta parens ! my modyr der*e*
All heyl modyr w*ith* glad cher*e*
Ffor now is A-resyn w*ith* body cler*e*
þi sone *þat* was dolve depe 1435
þis is *þe* thrydde day *þat* I ȝow tolde
I xuld a-rysyn out of *þe* cley so colde
now am I her*e* w*ith* brest ful bolde
*þer*fore no more ȝe wepe.

Maria

¶ Welcom my lord welcom my grace 1440
welcome my sone *and* my solace
I xal *þe* wurchep in every place
Welcom lord god of myght
mekel sorwe in hert I leed
whan *þou* were leyd in dethis beed 1445
but now my blysse is newly breed
All men may joye þis syght.

Jhesus

Fo. 193 ¶ All þis werlde *þat* was forlorn
Shal wurchepe *ȝou*[1] bothe evyn *and* morn
Ffor had I not of ȝow be born 1450
man had be lost in helle
I was deed *and* lyff I haue
A*n*d[2] thorwe my deth man do I saue
Ffor now I a*m*[3] resyn out of my graue
In hevyn man xal now dwelle. 1455

Maria

¶ A der*e* sone *þese* wurdys ben[4] goode
þou hast wel comfortyd my morny*ng* moode
blyssyd be *þi* precyous bloode
þat mankende *þus* doth saue.

Jhesus

Now der*e* modyr my leve I take 1460
joye in hert *and* myrth ȝe make

[1] Omitted and written above the line. [2] MS. Âd.
[3] MS. âresyn, with some mark under *a*, possibly indicating error.
[4] Some miswritten letter obliterated before *ben*.

Ffor deth̄ is deed *and* lyff doth̄ wake
now I am resyn fro my graue.

Maria

¶ Ffar*e* wel my sone far*e* wel my childe
Ffar*e* wel my lorde my god so mylde 1465
my*n* hert is wele þ*at* ffyrst was whylde
Ffare wel my*n* owyn der*e* love
Now all̄ ma*n*kynde beth̄ glad w*ith* gle
Ffor deth is deed as ȝe may se
and lyff is reysed endles to be 1470
In hevyn dwellynge Above.

¶ whan my sone was naylyd on tre
All̄ women myght rewe w*ith* me
Ffor grettere sorwe myght nevyr no*n* be
than I dede suffyr i-wys 1475
Fo. 193v but þis joy now passyth all̄ sorwe
þ*at* my childe suffryd in þ*at* hard morwe
Ffor now he is our*e* alderers borwe
to brynge us all̄ to blys.

Tunc evigilabunt milites sepulcri et *dicit primus miles*

i^us^ miles

¶ Awake Awake[1] 1480
hillis gyn qwake
And tres ben shake
ful ner*e* a-too
Stonys clevyd
wyttys ben revid[2] 1485
Erys ben devid
I am servid soo.

2us miles

¶ he is a-resyn þis is no nay
þ*at* was deed *and* colde in clay
now is resyn be-lyve þis day 1490
grett woundyr it is to me
He is resyn by his owyn myght
And fforth̄ he goth̄ his wey ful ryght

[1] From line 1480 to line 1487 two lines are written in one.
[2] *rewi* first written and crossed through.

how xul we now us qwytte
Whan Pylat doth us se. 1495

3us miles

¶ lete us now go[1]
pilat on-too
And ryght evyn so
as we han sayn
þe trewth̄ we sey 1500
þat out of clay
he is resyn þis day
þat jewys han slayn.

4us miles

¶ I holde it best
lete us nevyr rest 1505
but go we prest
þat it were done
Att̄ heyl pilatt
in þin A-stat[2]
he is resyn up latt 1510
þat þou gast dome.

Pilat

¶ What What What What
Out upon the why seyst þou þat
Ffy vpon the harlat
how darst þou so say 1515
Fo. 194 þou dost myn herte ryght grett greff
þou lyest vpon hym fals theff
howe xulde he rysyn ageyn to lyff
þat lay deed in clay:

1us miles[3]

¶ ȝa þow þou be nevyr so wroth̄ 1520
And of these tydandys nevyr so loth̄
ȝitt goodly on ground on lyve he goth̄
qwycke *and* levynge man
Iff þou haddyst a ben þer we ware[4]
in hert þou xuldyst han had gret care 1525

[1] From line 1496 to line 1511 two lines are written in one.
[2] Three red dots against this line.
[3] Some miswritten number or letter (? 2 or 3) crossed through between 1us and *miles*.
[4] The *a* is written as a correction over another (? *e*) letter.

And of blysse a ben ryght bar*e*
Of color*e* bothe pale *and* whan̄.

Pilatus

¶ Or ȝe com there[1]
ȝe dede aꝑ swer*e*
to fyght in fer*e* 1530
and bete *and* bynde
Aꝑ þis was trayn
ȝo*ur* wurd*ys* wor*e* vayn
þis is sertayn
ȝow fals I fynde. 1535

2us miles

Be þ*e* deth̄ þ*e* devyl deyd
we were of hym so sor*e* Atreyd
þ*at* ffor ffer we us down leyd
Ryght evyn vpon our*e* syde
whan we were leyd upon þ*e* grounde 1540
stylle we lay as we had be bounde
we durst not ryse for a thowsand pounde
ne not for aꝑ þis worlde so wyde.

Pilatus

¶ Now ffy upon ȝo*ur* grett bost
aꝑ ȝo*ur* wurchep is now lost 1545
in felde in town *and* in every cost
men may ȝ*ou* dyspravyn̄ :
Fo. 194^{v} now aꝑ ȝo*ur* wurchep it is lorn̄
And euery man may ȝow we[2] scorn
And bydde ȝow go syttyn in þ*e* corn 1550
And chare a-wey þ*e* ravyn :

3us miles

¶ ȝa it was hyȝ tyme to leyn our*e* bost
Ffor whan þ*e* body toke aȝen þ*e* gost
he wold a frayd many An ost
kynge knyght *and* knave 1555
ȝa whan he dede ryse out of his lake[3]
þan was þ*er* suche An erthe-quake
þ*at* aꝑ þ*e* worlde it gan to shake
þ*at* made us ffor to rave.

[1] From line 1528 to line 1535 two lines are written in one.
[2] So in MS. for *wel*. [3] Some letter (? *k*) crossed through before *lake*.

4us miles

¶ ȝa ȝa · herke ffelawys what I xal say 1560
late us not ses be nyght nor day
but telle þe trewth̄ ryght as it lay
In cou*n*tre where we goo
And than I dare ley my*n* heed
þat þei þat crystys lawys leed 1565
they wyl nevyr ses tyl they be deed
his deth̄ þat brought hym too.

i^{us} miles

¶ Be belyaḧ þis was now wele ment
to þis cownceḧ lete us consent
lett us go tellyn with on Assent 1570
he is resyn up þis day.

2us miles

I grawnt þer-to *and* þat forth̄ ryght
þat he is resyn by his owyn myght
Ffor þer cam non be day nor nyght
to helpe hym owte of clay. 1575

Pilat*us*

Fo. 195 ¶ Now jentyl serys I pray ȝow Aḧ
A-byde stylle a lytyl thraḧ
whyḧ þat I my*n* cowncel caḧ
And here of þer cownceḧ.

i^{us} miles

Syr att ȝour prayour we wyl abyde 1580
here in þis place a lytel tyde
but tary not to longe ffor we must ryde
we may not longe dwelle.

Pilat*us*

¶ Now jentyl serys I pray ȝow here
Sum good cowncel me to lere 1585
Ffor sertys serys without dwere
We stounde[1] in right grett dowte.

Cayphas

Now trewly sere I ȝow telle
þis matere is both ffers *and* felle
combros it is þerwith to melle 1590
And evyl to be browth a-bowte.

[1] Or *stonnde.*

Annas

¶ Syr pylat *þou* grett justyse
þow *þou* be of wittys wyse
ȝit herke fful sadly *with* good devyse
what *þat* *þou* xalt do 1595
I counsel *þe* be my reed
þis wundyrful tale p*ray* hem to hede
and upon þis ȝeve hem good mede
bothe golde *and* sylver also.

¶ *And* ser*e* I xall telle ȝow why 1600
in ȝour*e* erys p*rev*yly
be-tweyn us thre serteynly
now herk ser*ys* in ȝ*our* erys.

hic faciant pilatus cayphas et *Annas priuatim inter se consilium quo finito dicat*

Annas

Fo. 195v Ffor mede doth most in every qwest
and mede is mayst*er* bothe est *and* west 1605
now trewly ser*ys* I hold þis best
W*ith* mede men may bynde berys.

Cayphas

¶ Sekyr ser*e* þis counsell is good
pray þese knyht*ys* to chaunge *þer* mood
ȝeve them golde ffeste *and* ffood 1610
And *þat* may chaunge *þer* wytt.

Pylatt.

Ser*ys* ȝour*e* good councel I xall fulfylle
now jentyl knyht*ys* come hedyr me tylle
I pray ȝow ser*ys* of ȝ*our* good wylle
no ferther *þat* ȝe fflytt. 1615

¶ Jentyl knyht*ys* I ȝow p*ray*
A bettyr sawe *þat* ȝe say
Sey *þer* he was cawth away
w*ith* his dyscyplis be nyght
Sey he was w*ith* his dyscyplis ffett 1620
I wolde ȝe worn in ȝour*e* sadelys ssett
And haue her*e* gold in a purs knett
and to rome rydyth ryght.

4us miles

¶ Now syr pylatt[1]
we gon oure gatt 1625
We wyll not prate
no lengere now
now we haue golde
no talys xul be tolde
to whithtys on wolde 1630
we make þe A vow.

Pilatus

Now ȝe men of[2] myth
as ȝe han hyght
Euyn so forth ryght
ȝoure wurdys not falle 1635
And ȝe xul gon
with me A-non
All every-chon
in to myn halle.

i^us^ miles

Fo. 196 Now hens we go 1640
As lyth as ro
And ryght evyn so
As we han seyd
We xul kepe counsel
where so evyr we dwell 1645
We xul no talys tell
be not dysmayd.

*finem I^a die Nota** *hic uenient ad sepulcrum maria magdalene maria jacobi* et *maria Salome · et dicit maria magdalene*

Magdalen

¶ Swete systeryn I ȝow besech
heryght now my specyal speche **36**
Go we with salvys ffor to leche
Cryst þat tholyd wounde
he hath us wonnyn owt of wreche 5
the ryght wey god wyl us teche

[1] From line 1624 to line 1647 two lines are written in one, divided as before by double strokes to 1639, then by points.

[2] *men of* written twice and crossed through.

* This marginal direction is apparently in the Ff. 95, 96 hand. It seems to refer to the end of one day's performance.

ffor to seke my lorde my leche
his blood hath me vnbownde.

¶ vij develys in me were pyght
my loue my lord my god Almyght
A-wey he weryd þo ffyndys wight
with his wyse wurde
he droff fro me þe fendes lees
in[1] my swete sowle his chawmere I ches
In me be-levyth þe lord of pes
I go to his burryenge boorde.

Maria jacobi

¶ My systerys sone I woot he was
he lyth in here as sunne in glas
þe childe was born by oxe *and* Asse
Vp in a bestys stall
thow his body be gravyd vndyr gres:
Fo. 196v þe grete godhede is nevyr-þe-lasse[2]
þe lord xal rysyn *and* gon his pas
and comfortyn his ffrendys all.

Maria Salome

¶ My name is mary Salome[3]
his modyr *and* I systerys we be
Annys dowterys we be all thre
Jhesu we be þin Awntys
The naylis gun his lemys feyn
and þe spere gan punche *and* peyn
on þo woundys we wold haue eyn
þat grace now god graunt vs.

Maria Magdalene

¶ Now go we stylle[4]
with good wyll
þer he is leyd
he deyd on crowch
we wolde hym towch
as we han seyd.

Tunc respicit maria magdalene in sepulcro dicens

[1] Omitted and added against the line in the margin.
[2] *a* written as a correction over an *e*.
[3] Next line first written *systerys dowterys bothe* and crossed through.
[4] Lines 33 to 38 are written in two lines divided by points.

¶ Where is my lord þat was here
þat for me bledde bowndyn in brere 40
his body was beryed rygh[1] by þis mere
þat Ffor me gan deye
þe jewys ffekyll *and* ffals ffownde
where haue þei do þe body with wounde
he lyth not upon þis grownde 45
þe body is don A-weye.

Maria jacobi

¶ To my lorde my love my ffrende
Ffayn wolde I salve A spende
and I myght aught A-mende
his woundys depe *and* wyde 50
to my lord I owe lowlyte
both homage *and* fewte
I wolde with my dewte
a softyd hand *and* syde.

Maria Salome

Fo. 197 ¶ To myghtfful god omnypotent 55
I bere a boyst of oynement
I wold han softyd his sore dent
his sydys al a-bowte
Lombe of love with-owt loth
I ffynde þe not myn hert is wroth 60
in þe sepulcre þer lyth a cloth
And jentyl jhesu is owte.

Angelus

¶ wendyth fforth ȝe women thre
In-to þe strete of Galyle
ȝour savyour þer xul ȝe se 65
walkynge in þe waye
ȝour Ffleschly lorde now hath lyff
þat deyd on tre with strook *and* stryff
wende fforth þou wepynge wyff
and seke hym I þe saye. 70

¶ Now goth fforth ffast all thre
to his dyscyplys ffayr *and* fre
and to petyr þe trewth telle ȝe

[1] So in MS.

þer of haue ȝe no dreed
Spare ȝe not *þe* soth to say 75
he *þat* was deed *and* closyd in clay
he is resyn þis same day
And levyth *with* woundys reed.

M*aria* Magdalen

¶ A myrthe *and* joye in herte we haue
Ffor now is resyn out of his graue 80
he levyth now oure lyf to saue
þat dede lay in *þe* clay ⁝

M*aria* jacoby

Fo. 197v In hert I was ryght sore dysmayd
the Au*n*gel to us whan *þat* he sayd
þat cryst is resyn I was affrayd 85
þe Au*n*gel whan I say ⁝

M*aria* Salome

¶ Now lete us alł thre fulfylle
þe Angelys wurde *and* goddys wylle
lett us sey *with* voys wul shrylle
Cryst *þat* jewys dede sle 90
oure lord *þat* naylyd was on *þe* rode
And betyn out was his bodyes blode
he is a-resyn þoughh they ben wode
A lorde ȝitt wele *þou* be.

Maria magdalene ‖ *dicit petro et ceteris apostolis.*

¶ Bretheryn alł in herte be glad 95
bothe blythe And joyful in herte ful fayn
Ffor ryght good tydandys haue we had
þat oure lord is resyn Agayn [1]
An Aungel [2] us bad [2] ryght þus sertayn
to *þe* petyr *þat* we xulde telle 100
how cryst [2] is resyn *þe* whichh was slayn
A levynge man evyr more to dwelle.

[1] The scribe of Ff. 95, 96, 112 has written against this line: *and aperyd to us sertayne*, which he afterwards crossed out, writing longitudinally in the left-hand margin for insertion here:

† lyk as he dyede nakyd as he was borne
and commande us to go to pet*er and* john *and* hy*s* dyscipul*ys* alł
and tełł to yow he wolde apere in lyknes as he was befo[rn.

[2] The same hand has marked *aungel*, *bad*, and *cryst* with dots for deletion, and has written *bade* over *aungel* and *he* over *cryst*.

Maria jacobi

¶ To lyve is resyn a-geyn þ*at* lorde
the qwych judas to jewys solde
of þis I ber*e* ryght trewe[1] recorde 105
by wurdys þ*at* þ*e* Aungel tolde *
now myrth *and* joye to man on molde
Euery man now myrth may haue
he þ*at* was closyd in cley ful colde
Fo. 198 This day is resyn owt of his grave. 110

Petrus

¶ Sey me systeryn w*ith* wurdys blythe
may I troste to þ*at* ȝe say
Is cryst resyn ageyn to lyve
þ*at* was ded *and* colde in clay.

Maria Salome

ȝa trostyth us truly it is no nay 115
he is a-resyn it is no les
and so An Aungel us tolde þis day[2]
w*ith* opyn voys *and* speche expres.

Johannes

¶ ȝa þese[3] be tydyng*ys* of ryght gret blys
þ*at* oure mayst*er* resyn xulde be 120
I wyl go renne in hast i-wys
and loke my lord yf I may se.

[1] Some miswritten letter crossed through before *trewe*.

* The scribe of Ff. 95, 96 has written *for* in the margin just above *by* at the beginning of this line, and has marked for insertion here a passage written in the bottom margin:

for . . aperyd to us w*ith* hand*ys* fytte *and* hert borde
and . . . he schowyd us his wound*ys* fyve
both hand*ys* *and* fytte *and* þ*e* wound in h*is* syde
[?]
and þ*er*for be-leve us þ*at* he is man a-lyve.

In a darker ink but probably by the same hand, *for* in the first line has been crossed out and *All so* written before it, and *he* over the illegible second word; the last five words of the line have been crossed out, and *with body bolde* (or *belde*) written over them; the third line has been crossed out.

[2] An alternative line is written by the scribe of Fo. 95 against this longitudinally in the margin:

And so he badd us tell yow þ*is* daye.

[3] *be* ȝ*e* first written and crossed through.

Petrus

Ffor joye also I renne w*ith* the
my brothyr joh*an* as I þ*e* say
In hast anon evyn forth̄ go we 125
to his grave we renne our*e* way.

*hic currunt Joh*annes et *Petrus simul ad sepulcru*m et *Joh*annes *prius venit Ad monumentu*m *sed non intrat.*

Joh*ann*es

The same shete here I se
þ*at* crystys body was in wounde
but he is gon where so ever he be
he lyth not here up on þis grownde. 130

*Petrus intrat monumentu*m et *dicit Petrus*

Petrus

in þis cornere þ*e* shete is fownde
and here we fynde þ*e* sudary
Fo. 198v In þ*e* whiche his hed was wounde
whan he was take from calvary[1].

*hic intrat Joh*annes *monumentu*m *dicens*

Joh*ann*es

¶ The same sudary *and* þ*e* same shete 135
here w*ith* my syth I se both̄ tweyn
now may I wele knowe *and* wete
þ*at* he is rysyn to lyve ageyn̄
On to our*e* bretheryn lete us go seyn̄
þ*e* trewth̄ rygh̄t hevyn as it is 140
Our*e* mayst*er* lyvyth̄ þ*e* whech̄ was slayn̄
Alł myghty lorde *and* kynge of blys.

Petrus

¶ No lengere here wyłł we dwelle
to our*e* bretheryn þ*e* wey we take
the trewth̄ to them whan þ*at* we telle 145
grett joye in hert þan wul þei make.

*hic petrus loquitu*r *omnibus apostolis simul collectis*

Beth̄ mery bretheryn for Crystys sake
þ*at* man þ*at* is our*e* mayst*er* so good

[1] A *u* has been altered to *v* in this word, apparently by the scribe.

Ffrom deth to lyve he is A-wake
þat sore was rent up on þe rood. 150

Johannes

¶ As women seyd so haue we fownde
remevyd Awey we saw þe ston
he lyth no lengere vndyr þe grownde
out of his graue oure mayster is gon.

Omnes congregati[1] Thomas

We haue grett woundyr everychon 155
of þese wurdys þat ȝe do speke
A ston ful hevy lay hym up on
Ffrom vndyr þat ston how xulde he breke.

Petrus

Fo. 199 ¶ The trewth̄ to tellyn it passyth oure witt
Wethyr he be resyn thorwe his owyn myght 160
Or ellys stolyn out of his pitt
be sum man prevely be nyght
That he is gon we saw with syght
Ffor in his graue he is nowth
we can not tellyn in what plyght 165
Out of his graue þat he is browth.

*Maria magdal*ene *goth to þe graue* and *wepyth* and *seyth̄*

Maria Magdalen̄

¶ Ffor hertyly Sorwe myn herte doth breke
with wepynge terys I wasch̄ my face **37**
Alas ffor sorwe I may not speke
my lorde is gon þat here-inne wase
Myn owyn dere lorde *and* kyng of gras 5
þat vij develys ffro me dyd take
I kan nat se hym Alas Alas
he is stolyn awey owt of þis lake.

Aungelus

¶ Woman þat stondyst here Alone
Why dost þou wepe *and* morne *and* wepe so sore 10
What cawse hast þou to make such̄ mone
Why makyst þou such̄ sorwe *and* where-fore.

[1] MS. congregat̄—? so for *omnibus congregatis.*

Maria Magdal*ene*

I haue gret cawse to wepe evyr mor*e*
my lord is take out of his graue
Stolyn Awey *and* fro me lor*e*
I kan not wete wher*e* hy*m* to haue.

hic parum deambulet A sepulcro dicens

¶ Alas Alas what xal I do
my lord awey is fro me take
Fo. 199v A woful wrecche whedyr xal I go
My joye is gon owth of þis lake.

Jh*es*us

Woman suche mornynge why dost þ*ou* make
Why is þ*i* cher*e* so hevy *and* badde
Why dost þ*ou* sythe so sor*e* *and* qwake
Why dost þ*ou* wepe so sor*e* *and* sadde.

Maria Magdal*ene*

¶ A grettyr cawse had nevyr woman
Ffor to wepe both nyth *and* day
than I myself haue in serteyn
for[1] to sorwyn evyr *and* Ay
Alas ffor sorwe my*n* hert doth blede
my lorde is take fro me A-way
I muste nedys sor*e* wepe *and* grede
wher*e* he is put I kan not say.

¶ but jentyl gardener I pray to the
If þ*ou* hym took out of his graue
telle me qwer*e* I may hym se
þ*at* I may go my lorde to haue.

Jh*es*us *spectans*

M. A. R. I. A.

Maria Magdal*ene*

A mayst*er* *and* lorde to þ*e* I crave
As þ*ou* art lord *and* kynge of blys
Graunt me lord *and* þ*ou* vowchesave
thy*n* holy ffete þ*at* I may kys.

[1] *And* first written before *for* and crossed through.

Jhesus

¶ Towche me not As ȝett ⁝ Mary
Ffor to my fadyr I haue not Ascende
but to my bretheryn in hast þe hyȝ
with these gode wurdys here care Amende 45
Fo. 200 Sey to my bretheryn þat I intende
to stey to my fadyr *and* to ȝowre
to oure lord both god *and* frende
I wyl Ascende to hevyn towre.

¶ In hevyn to ordeyn ȝow A place 50
to my ffadyr now wyl I go
to merth *and* joye *and* grett solace
And endeles blys to brynge ȝow to
Ffor man I sufferyd both schame *and* wo
more spyteful deth nevyr man dyd take 55
ȝit wyl I ordeyn ffor all this lo
In hevyn an halle for mannys sake.

Maria Magdaly[n

¶ Gracyous lord at ȝ*our* byddyng[1]
to all my bretheryn I xal go telle
how þat ȝe be man levynge 60
quyk *and* qwethynge of flesch *and* ffelle
Now all hevynes I may expelle
And myrth *and* joy now take to me
my lord þat I haue louyd so wele
with opyn syght I dede hym se. 65

¶ Whan I sowght my lord in grave
I was fful sory *and* ryght sad
Ffor syght of hym I myght non haue
Ffor mornynge sore I was nere mad
Grettere sorwe ȝit nevyr whith had 70
Whan my lord A-wey was gon
Fo. 200v But now in herte I am so glad
so grett a joy nevyr wyff had non.

¶ how myght I more gretter joye haue
than se þat lorde *with* opyn syght 75

[1] The colour of the ink changes with this line from the darker colour which has prevailed since Fo. 193.

the whiche my sowle from synne to saue
From develys sefne he made me qwyght.

¶ There kan no tounge my joye expres
now I haue seyn my lorde on lyve
to my bretheryn I wyl me dresse 80
and telle to hem A-non ryght be-lyve
W*ith* opyn speche I xal me shryve
and telle to hem w*ith* wurdys pleyn
how þ*at* cryst ffrom deth to lyve
to endles blys is resyn Ageyn. 85

¶ Bretheryn alł bllyth̄ ȝe be
ffor joyful tydyng*ys* tellyn I kan
I saw our*e* lord cryst · lyste wel to me
of flesch̄ *and* bon̄ · quyk levynge man
beth glad *and* joyful as for than 90
Ffor trost me trewly it is ryght thus
Mowth̄ to mowth̄ þis sertayn[1]
I spak ryght now w*ith* cryst jh*e*sus.

Petrus

¶ A woundyrful tale for-sothe is this
ev*er* onowryd our*e* lorde mote be 95
we pray þ*e* lord *and* kynge of blys
Onys þ*i* presence þ*at* we may se
Fo. 201 Ere thu Ascende to thi mageste
Gracyous god if þ*at* ȝe plese
late us haue su*m* syght of the 100
our*e* careful hert*ys* to sett in ease Amen.

*Explic*it *apparic*io *Marie magdale*[*ne.*

[The remainder of Fo. 201 is filled up with what appears to be a roughly scribbled copy of part of Magdalene's speech on Fo. 200v.* Fo. 201v is blank.]

*[But now in herte I am so glad
So grete a jooy[2] nevyr wyff had non
how myght I mor*e* gretter haue than se
þ*at* lorde w*ith* opyn syght the wyche my soule

[1] So, with *is* omitted, in MS. [2] The first *o* blotted.

from synne to saue from develys sefne
he mad̃ me qwyght there kan no tounge
my joy expres now I haue seyn my lorde
on lyve to my brethryn I wyll me dresse
and thell to hem *with* wurdys pleyn hwow
þat cryst from deth to lyve to endles blys
ys resyn agayen̄[1] bretheryn alł bllyth̄
ye be for joyfull tydyng*ys* tellyn I kan I saw
Oure lord cryst lyste wel to me of flesch̄ *and* bon
quyk levyng man beth glad *and* Joyfull as for than
ffor trost me trewly it ys ryght thus
Mowth to mowth þis ys sartayne I[2] spak rgh̄t now
with cryst Jhesus.]

Fo. 202 hic incipit ap*a*ricio cleophe *et* luce.

¶ Cleoph*as*

MY brothir lucas I ȝow pray
plesynge to ȝow if þat it be
To þe castel of Emawus A lytyl way
þat ȝe vowche-saf to go *with* me.

38

lucas

Alł redy brother I walke *with* the
to ȝone castell̄ *with* ryght good chere
Euyn to-gedyr Anon go we
brother cleophas we to infere.

Cleophas

¶ A brother lucas I am sore mevyd
Whan cryst oure mayster comyth in my mynde
whan that I thynke how he was grevyd
Joye in myn herte kan I non fynde
he was so lowlye so good so kynde
holy of lyf *and* meke of mood
Alas þe jewys þei were to blynde
hym for to kylle þat was so good.

lucas

¶ Brothyr cleophas ȝe sey[3] ful soth̄
they were to cursyd *and* to cruell̄

[1] Some blotted letter after *agayen*.
[2] Some crossed out letters after *I*.
[3] The *e* is written over some erasure.

And judas þat traytour he was to loth̄
Ffor gold *and* sylvyr his Mayster to selle
the jewys were redy hym for to qwelle
With skorgys bete out att̄ his blood
Alas þei were to fers *and* ffelle
Shamfully þei henge hym on a rood.

Cleophas

¶ ȝa be-twen to Thevys Alas for shame
they henge hym up with body rent
Alas alas they were to blame
to cursyd *and* cruel was þer intent
Fo. 202v marked 200] Whan for thurste he was nere shent
Eyȝil *and* galle þei ȝovyn hym to drynke
Alas for ruthe his deth̄ thei bent
in a ffowle place of horryble stynke.

lucas

¶ ȝa And cawse in hym cowde they non fynde
Alas for sorwe what was here thought
And he dede helpe bothe lame *and* blynde
And att̄ seke men þat were hym browght
A-ȝens vice Alwey he wrought
synfull dede wold he nevyr do
ȝit hym to kylle þei sparyd nought
Alas Alas why dede they so.

Jhesus

¶ Well ovyr-take ȝe serys in same
to walke in felachep with ȝow I pray.

lucas

welcom̄ serys in goddys name
of good felachep we sey not nay.

Jhesus

Qwat is ȝour langage to me ȝe say
that ȝe haue to-gedyr ȝe to
sory *and* Evysum ȝe ben Alway
ȝour myrthe is gon why is it so.

Cleophas

¶ Sere me thynkyth þou art a pore pylgrym̄
here walkynge be þi selfe A-lone
and in þe cete of jerusalem̄
þou knowyst ryght lytyl what þer is done

Ffor pylgrymys comy*n* *and* gon ryth sone
Ryght lytyl whyle pylgrymes do dwelle
In all jerusalem as þ*ou* hast gone
I trowe no tydyng*ys* þ*at* þ*ou* canst telle.

Jh*es*us

Fo. 203 Why in Jerusalem what thynge is wrought
What tydyng*ys* fro thens brynge ȝe.

lucas

A ther haue they slayn a man for nought
gyltles he was as we telle the
An holy p*ro*phete w*ith* god was he
myghtyly in wurde *and* eke in dede
of god he had ryght grett pooste
Amonge þ*e* pepyl his name gan[1] sprede.

he hyght jh*es*u of nazareth
A man he was of ryght grett fame
the jewys hym kylde w*ith* cruel deth.
w*ith*-out trespas or Any blame
hym to scorne they had grett game
And naylid hym streyte on tyll a tre
Alas Alas me thynkyth grett shame
W*ith*-out cawse þ*at* this xulde be.

Cleophas

¶ ȝa sere *and* ryght grett trost in hy*m* we had
All Israel countre þ*at* he xulde saue
the thrydde day is this þ*at* he was clad
in coold cley *and* leyd in grave
ȝitt wou*n*dyrful tydyng*ys* of hym we haue
Of women þ*at* sought hym be-forn day lyth
wethyr they sey truthe or ellys do raue
We can not telle þ*e* trew verdyth.

¶ Whan cryst in grave þei cowde not se
they comy*n* to us And Evy*n* thus tolde
Fo. 203v How þ*at* An Aungell seyd to them thre
that he xuld leve w*ith* brest fful bolde
ȝitt petyr *and* johan p*re*ve this wolde
to Crystys graue they ran thei tweyne

[1] The *a* is written over some other letter (? *r*).

and whan they come to þe graue so coolde
They fownde þe women fful trewe sertayne.

Jhesus

¶ A ȝe Ffonnys And Slought of herte
Ffor to be-leve in holy scrypture
haue not prophetys with wurdys smerte
Spoke be tokenys in signifure
That Cryste xuld deye ffor ȝour valure
And syth entre his joye *and* blys
why be ȝe of herte so dure
And trust not in god þat myghtful is.

¶ Bothe Moyses and Aaron *and* othyr mo
in holy scrypture ȝe may rede it
of Crystis deth thei spak Also
And how he xuld ryse out of his pitt
Owt of ffeyth than why do ȝe fflitte
Whan holy prophetys ȝow teche so pleyne
turne ȝoure thought *and* chaunge ȝour witte
And truste wele þat cryst doth leve a-geyne.

lucas

¶ Leve Ageyn ? man be in pes
how xulde A ded man evyr A-ryse
I cownceil þe such wurdys to ses
Ffor dowte of pylat þat hyȝ justyce
Fo. 204 he was slayn At þe gre [1] A-syse
be cownceil of lordys many on
Of suche langage · take bettyr A-vise
In every company þer þou dost gon.

Christus

¶ Trewth dyd nevyr his maystyr shame
Why xulde I ses than trewth to say
Be Jonas þe prophete I preve þe same
Þat was in a Whallys body iij nyghtis *and* iij day
So longe Cryst in his grave lay
As Jonas was with-inne þe se
his grave is brokyn þat was of clay
to lyff Resyn Aȝen now is he.

[1] So in MS.

Cleophas

¶ Sey nott so man it may not be
thow thyn exaunple be su*m*dele good
Ffor jonas on lyve evyr mor*e* was he
And Cryst was slayn · vpon A rood
The Jewys on hym they were so wood 125
Þ*at* to his herte A spere they pyght
he bled owt Aħ his herte blood
how xulde he thanne ryse w*ith* myght.

Chr*ist*us

¶ Take hede at Aaron *and* his dede styk
Whiche was ded of his natur*e* 130
And ȝit he floryschyd w*ith* flowr*ys* ful thyk
and bare Almaundys of grett valur*e*
The dede styk was signifur*e*
how cryst þ*at* shamfully was deed *and* slayn
As þ*at* dede styk bar*e* frute ful pur*e* 135
So cryst xuld ryse to lyve a-geyn.

lucas

Fo. 204v ¶ That A deed styk ffrute xulde bere
I merveyle sor*e* þ*er* of i-wys
but ȝitt hym sylf ffro deth to rer*e*
And leve A-geyn mor*e* woundyr it is 140
That he doth leve I trost not[1] this
Ffor he hath bled his blood so[2] red
but ȝitt of myrthe evyr moor I mys
Whan I haue mende þ*at* he is ded.

Chr*istus*

¶ Why be ȝe so hard of truste 145
dede not Cryst reyse thorwe his owyn myght
lazar*e* þ*at* deed lay vndyr þ*e* duste
And stynkyd ryght foule as I ȝow plyght
To lyff Cryst reysid hym a-ȝen ful ryght
out of his graue þis is serteyn 150
why may nat Cryste hym self þus qwyght
and ryse from deth to lyve Ageyn.

Cleophas

¶ Now trewly sere[3] ȝo*ur* wurdys ben good
I haue in ȝow ryght grett delyght

[1] An *l* after *not* crossed through. [2] A *d* after *so* crossed through.
[3] The vowel is indistinct but should be *e*.

I pray ȝow sere *with* mylde mood
to dwelle *with* vs aſt̃ þis nyght.

Chr*ist*us

I must gon hens a-non ful ryght
Ffor grett massagys I haue to do
I wolde abyde yf þ*at* I myght
but at þis tyme I must hens go.

lucas

¶ ȝe xal not gon fro us þis nyght
It waxit aſt̃ derke gon is þ*e* day
þ*e* sonne is downe lorn is þ*e* lyght
ȝe xal not gon from vs A-way.

Chr*ist*us

Fo. 205 I may not dwelle As I ȝow say
I must þis nyght go to my ffrende
þ*er*fore good bretheryn I ȝow pray
lett me not my wey to wende.

Cleophas

¶ Trewly from vs ȝe xal not go
ȝe xal abyde *with* us her*e* stylle
ȝ*our* goodly dalyaunce plesyth us so
We may nevyr haue of ȝow our*e* fylle
We pray ȝow sere *with* herty wylle
Aſt̃ nyght *with* us abyde And dwelle
more goodly langage[1] to talkyn vs tylle
and of ȝ*our* good dalyaunce more ffor to telle.

lucas

¶ ȝa brothyr cleophas be my*n* Assent
lete us hym kepe *with* strenth̃ *and* myght
Sett on ȝoure hand *with* good entent
And pulle hym *with* us þ*e* wey weſt̃[2] ryght
The day is done sere *and* now it is nyght
Why wole ȝe hens now from us go
ȝe xal abyde as I ȝow plyght
ȝe xal not walke þis nyght vs ffro.

Cleophas

¶ This nyght fro us ȝe go not Away
we xal ȝow kepe be-twen us tweyne
to vs þ*er*fore ȝe sey not nay
but walke *with* us þ*e* wey is pleyne.

[1] The second *g* is written over another letter.
[2] Or woſt̃—the vowel is blotted.

Christus

Sythyn ȝe kepe me w*ith* myght *and* mayn̄
W*ith* hérty wyll I xal abyde. 190

lucas

of ȝ*our* abydyng we be fulfayn̄
no man more wel-kom · in þis werd wyde.

Cleophas

Fo. 205v ¶ Off our*e* maystyr Cryst J*he*su
Ffor ȝe do speke so mech̄ good
I loue ȝow hertyly trust me trew 195
he was bothe meke *and* mylde of mood
Of hym to speke is to me food
If ȝe had knowe hym I dare wel say
And in what plyght w*ith* hym it stood
ȝe wold haue thought on h*ym* many A day. 200

lucas

¶ Many A day ȝa · ȝa · i-wys
he was a man of holy levynge
Thow he had be þ*e* childe of god in blys
bothe wyse *and* woundyrfull was his werkynge
But Aftere ȝ*our* labo*ur* *and* feire walkynge 205
takyth þis loff *and* etyth su*m* bred
And than wyl we haue more talkynge
Of Cryst our*e* maystyr þ*at* is now ded.

Christus

¶ Peth̄ mery *and* glad w*ith* hert fful fre
ffor of cryst jh*e*su þ*at* was ȝ*our* ffrende 210
ȝe xal haue tydyng*ys* of game *and* gle
w*ith*-inne A whyle or ȝe hens wende
w*ith* myn hand þis bred I blys
And breke it here as ȝe do se
I ȝeve ȝow parte Also of þis 215
this bred to ete *and* blythe to be.

hic subito discedat christus ab oculis eorum.

[Cleophas]

¶ A mercy god what was our*e* happ*e*
Fo. 206 was not our*e* hert w*ith* loue brennynge
Whan cryst our*e* mayster so nere our*e* lapp*e*
Dede sitte *and* speke such̄ suete talkynge 220

[1] No name of speaker in the MS.

He is now quyk *and* man lyvenge
þat fyrst was slayn *and* put in grave
now may we chaunge all oure mornynge
Ffor oure lord is resyn his seruaunt*ys* to saue.

Lucas

¶ Alas for sorwe what hap was this
Whan he dyd walke w*ith* vs in way
he p*re*vyd by scripture ryght wel i-wys
þat he was resyn from vndyr clay
We trustyd hym not but evyr seyd nay
Alas for shame why seyd we so
he is Resyn to lyve þis day
out of his grave oure lord is go.

Cleophas

¶ latt us here no lengere dwelle
but to oure bretheryn þe wey we wende
W*ith* talys trewe to them we telle
that cryst doth leve oure mayst*er* *and* frende.

lucas

I graunt *þer*-to w*ith* hert ful hende
lete us go walke forthe in oure way
I am ful joyfull in hert *and* mende
þat oure lord levyth *þat* fyrst ded lay.

Cleophas

¶ Now was it not goodly don
of cryst jh*es*u oure mayster dere
he hath w*ith* us a large wey gon
And of his vprysyng he dede us l*ere*
Fo. 206v Whan he walkyd w*ith* us in fere
And we supposyd hym both dede *and* colde
þat he was A-resyn ffrom vndyr bere
be holy scripture þe trewth he tolde.

lucas

¶ Ryght lovyngely don for-sothe this was
What myght oure mayster tyl us do more
Than us to chere *þat* fforth dede pas
and ffor his deth we murnyd ful sore
Ffor loue of hym oure myrthe was lore
We were ffor hym ryght hevy in herte
but now oure myrth he doth restore
Ffor he is resyn bothe heyl *and* qwert.

Cleophas

¶ That he is þus resyn I haue grett woundyr
An hevy ston ovyr hym þer lay
how shulde he breke þe ston A-soundyr
þat was deed *and* colde in clay 260
Euery man þis mervayle may
And drede þat lord of mekyl myght
but ȝit of þis no man sey nay
Ffor we haue seyn hym[1] with opyn syght.

lucas

¶ That he doth leve I woot wel this 265
he is A-resyn with flesch *and* blood
A levynge man for-sothe he is
þat rewly was rent upon a rood
All heyl dere brothyr *and* chaunge ȝour mood
Ffor cryst doth levyn *and* hath his hele 270
We walkyd in wey with cryst so good
and spak with hym wurdys fele.

Cleophas[2]

Fo. 207 ¶ Evyn tyll Emawus þe grett castell
ffrom jerusalem with hym we went
Syxti ffurlonge as we ȝow telle 275
we went with hym evyn passent
he spak with us with good entent
þat Cryst xuld leve he tolde tyll us 278
And previd it be scripture verament *vade worlych*[3]
trust me trewe it is ryght thus. *not*[a] *worlych*

lucas

¶ ȝa *and* whan he had longe spokyn vs tylle
he wold ffrom vs agon his way
with strenght *and* myght we keptyn hym stylle
And bred we tokyn hym to etyn in fay
he brak þe loff As Evyn on tway 285
As ony sharpe knyff xuld kytt[4] breed
þer-by we knew þe trewth þat day
þat cryst dede leve *and* was not deed.

[1] The *y* is written over some other letter—? a half-written *e*.

[2] Under *Cleophas* in the bottom margin is written and smudged out *Vade Worlych*.

[3] These words seem to be written by the scribe of Ff. 95, 96.

[4] *be* crossed through after *kytt*.

Petrus[1]

¶ Now trewly serys I haue grett woundyr
of these grete merveylis *þat* ȝe vs telle 290
In brekynge of bred fful evyn A-soundyr
oure mayster ȝe knew · *and* lord ryght wełł
ȝe sey Cryst levith *þat* jewys dyd qwelle
tyłł us glad tydyng*ys* þis is serteyn
and *þat* oure mayst*er* w*ith* ȝow so longe dede dwelle 295
it doth wełł p*reve* *þat* he levith a-geyn.

¶ A brother Thomas we may be ryght glad
of these gode Novełł *þat* we now haue
þe grace of oure lorde god is ou*er* vs ałł sprad
Oure lord is resyn his seuauntys[2] to saue. 300

Thomas

Fo. 207v Be in pes petyr *þou* gynnyst to rave
thy wurdys be wantowne *and* ryght vnwyse
how xulde A deed man *þat* deed lay in grave
w*ith* qwyk fflesche *and* blood to lyve ageyn ryse.

Petrus

¶ ȝis Thomas dowte *þe* not oure maystyr is on lyve 305
Record of Mawdelyn *and* of here syster*ys* too
Cleophas *and* lucas *þe* trewthe ffor to contryve
Ffro jerusalem to Emaws w*ith* hym dede they go.

Thomas

I may nevyr in hert trust *þat* it is so
he was ded on cros *and* colde put in pitt 310
kept w*ith* knyht*ys* iiij. his grave sealyd Also
how xulde he levyn Ageyn *þat* so streyte was shitt.

Petrus

¶ Whan Mawdelyn dede tełł us *þat* cryst was a-resyn
I ran to his graue *and* Johan ran w*ith* me
In trewth *þer* we ffownde he lay not in p*re*syn[3] 315
gon out of his graue *and* on lyve þan was he
Therfore dere brother thomas I wole rede the
Stedfastly *þou* trust *þat* cryst is not deed

[1] *Nota Worlych* is again written under this name of speaker, and crossed through.

[2] So in MS.

[3] This *y* is written over an *o*.

Ffeythfully be-leve a qwyk man þat he be
A-resyn from his deth by myght of his godhed. 320

Thomas

¶ I may nevyr be-leve these wou*n*dyr merveles
tyl þat I haue syght of Euery grett wounde
and putt in my ffyngyr in place of þe nayles
I xal nevyr be-leve it ellys ffor no ma*n* on growunde
And tyll þat my*n* hand þe sperys pytt hath fowunde 325
Which dede cleve his hert *and* made hy*m* sprede his blood
I xal nevyr be-leve þat he is qwyk *and* sownde
In trewth whyl I knowe þat he was dede on rood.

Petrus

Fo. 208 ¶ Cryst be þi comforte *and* chawnge þi bad witt
Ffor ffeyth but þou haue þi sowle is but lorn 330
W*ith* stedfast beleve god enforme þe ȝitt
of A meke mayde As he was ffor us born.

Chr*ist*us

Pees be Amonge ȝow · be-holde how I am torn
take hede of my*n* handys my dere brothyr thom*as*.

Thom*as*

My god *and* my lorde nyght *and* every morn 335
I[1] Aske mercy lorde ffor my grett trespas.

Chr*ist*us

¶ Be-holde wele thom*as* my woundys so wyde
which I haue sufferyd ffor All mankynde
Put þin hool hand in to my ryght syde
And in my*n* hert blood þin hand þat þou wynde 340
So ffeythffull a ffrend were mayst þou fynde
be stedfast in feyth be-leve wel in me
be þou not dowtefful of me in þi mynde
but trust þat I leve þat deed was on A tre.

Thom*as*

¶ my lord *and* my god w*ith* syght do I se 345
þat þou art now quyk which henge deed on rode
more feythful þan I · ther may no man be
Ffor my*n* hand haue I wasch in þi p*re*cyous blode.

Chr*ist*us

Ffor þou hast me seyn þ*er*fore þi ffeyth is good
but blyssyd be tho of þis þat haue no syght 350

[1] *Ah* first written and crossed out after *I*.

And be-leve in me they ffor here meke mood
Shall com̄ in to hefne my blysse þat is so bryght.

Thomas

¶ As A ravaschyd man whos witt is all gon
grett mornynge I make ffor my dredfful dowte
Fo. 208v Alas I was dowteful þat cryst from vndyr ston 355
be his owyn grett myght no wyse myght gon owte
Alas what mevyd me thus in my thought
my dowtefful be-leve ryght sore me Avexit
the trewthe do I knowe þat god so hath wrought
Quod mortuus *et* sepultus· nu*n*c *re*surrexit[1]. 360

¶ He þat was bothe deed *and* colde put in grave
to lyve is A-resyn[2] by his owyn myght
In his dere herte blood my*n* ha*n*d wasch̄ I haue
Where þat þe spere poynt was peyn-fully pyght
I take me to feyth ffor-sakynge All vn-ryght 365
þe dowte þat I had fful sore me Avexit
Ffor now I haue seyn w*ith* ful opyn syght
Quod mortuus *et* sepultus nu*n*c resurrexit.

¶ I trustyd no talys þat were me tolde
tyll þat my*n* hand dede in his hert blood wade 370
My dowte doth̄ Ap*re*vyn cryst levynge fful bolde
And is a grett Argument in feyth us to glade
þ*ou* man þat seyst þis ffrom feyth nevyr þ*ou* ffade
my dowte xal evyr chere the || þat sore me Avexit
trust wele in cryst þat such̄ meracle hath made 375
Quod mortuus et sepultus nunc resurrexit.

¶ The p*re*chynge of petir myght not conu*er*te me
tyll I felyd þe wounde þat þe spere dyde cleve
I trustyd nevyr he levyd þat deed was on A tre
tyll þ*at* his herte blood dede renne in my sleve 380
Fo. 209 Thus be my grett dowte · our*e* feyth may we preve
be-hold my blody hand · to feyth þat me Avexit

[1] *Surrexit* apparently first written, and the initial *r* with a blotted over-written letter added. This line, in this stanza only, is written in slightly larger form.

[2] A small fragment of printed paper adhering to the MS. hides the last stroke of the *n*.

be syght of þis myrroure · ffrom feyth not remeve
Quod mortuus *et* sepultus nu*n*c resurrexit.

¶ Thow þat Mary Magdalyn in cryst dede sone be-leve 385
And I was longe dowteful ȝitt putt me in no blame
Ffor be my grett dowte oure ffeyth we may preve
Aȝens all þe Eretykys [1] þat speke of cryst shame.
Truste wel J*hes*u cryst · þe jewys kyllyd the same
The ffende hath he fferyd · oure feyth [2] þat evyr a-vexit 390
To hevyn ȝow brynge ·ꝛ *and* saue ȝow all in same
That mortuus *et* sepultus Iterum resurrexit. Amen.

[Remainder of Fo. 209—4½ inches—and Fo. 209ᵛ left blank.]

Scribble on Fo. 209. *That mortuus et se.*
Scribble on Fo. 209ᵛ. *hic Incipit Ascencio.*

Fo. 210 hic incipit Ascenc*io* do*m*ini no*st*ri · *cum maria* et *vndecim*
[W quire] *discipulis et duobus angelis sedentibus in albis et Jhesus dicit discipulis suis &c.*[3]

¶ Jhesus
39

PAx vobis [4] ·ꝛ Amonge ȝow pes
bothe love *and* reste *and* charyte
Amonge All vertues lete it not ses
Ffor Amonge All vertues prynspal his he
ȝe be to blame I may wel preve 5
Ffor I wyl vse to ȝow wordys pleyn
þat ȝe be so hard of herte to be-leve
þat from deth to lyve I am resyn Ageyn
Not-w*ith*-stondynge As ȝe knowe serteyn
to ȝow viij sythys · Aperyd haue I 10
be soundry tymes the trewth to seyn
and þis is þe ix · tyme sothly
Evy*n and* no mo
but Now sum mete
A-non doth gete 15
Ffor I wyl Ete
w*ith* ȝow And goo

[1] The *y* of the final syllable has been written over an *i*.
[2] Written *feyyth* with first *y* marked for deletion.
[3] The title and stage direction are written as one in the MS.
[4] *Pax vobis* is written in larger form.

¶ My dyscyplis here what I sey
And to my wourdys ȝevyth Attencion
From jerusalem loke ȝe go nott Awey
but mekely A-bydyth my fadyres promiscion
Off whiche be my mowth ȝe haue had informacion
whyll bodyly with ȝow I was dwellynge
Ffor johan Sothly ffor mannys Saluacion
Onlye in watyr was me baptysynge
but I ȝow be-hete:
With-inne ffewe days þat ȝe
Fo. 210v In þe holy goost xul baptyzid be
therfore rysyth up *and* ffolwyht me
On-to þe mownte of Olyvete.

Jacobus maior

¶ O lord vowche-saff vs for to telle
Iff þou wylt now with-owte more delay
restoryn þe kyngdam of israell
And ȝeve vs þe joye lord þat lestyth Ay.

Jhesus

Serys þe tymes *and* þe monthis knowe ȝe ne may
whiche my fadyr hath put in his owyn power
but ȝe xul tȧke with-in short[1] day
of þe holy goost þe vertu cler
thorwe whiche xul ȝe
In jerusalem And in jury
And more ovyr Also in samary
And to þe worldys ende vttyrly
My wyttnes only be.

¶ lovyth no wrath nor no wronge
but levyth in charyte with mylde stevyn
With myrthe *and* melody *and* Aungell songe
now I stey streyte ffrom ȝow to hevyn.

hic Ascendit ab oculis eorum et *in celo cantent etcetera.*

Angelus

¶ Returnyth ageyn to ȝour loggynge
to jerusalem ffor he wyl thus
his promys mekely þer Abydynge

[1] The *t* is blotted and written again—small—above the line.

Ffor dowteles þis forseyd jh*e*sus
Whiche from ȝow is take :/[1]
Fo. 211 In a clowde As ȝe hym seyn̄
Steyng vp so xal comy*n* A-geyn̄
Of Al mankynde þis is serteyn
jugement xal he make.[2]

¶[3]O ȝe bretheryn Attendyth to me
And takyth good hede what I xal seyn̄
it be-hovyth þ*e* scriptur*e* ffulfyllyd to be
þ*a*t of dauyd was seyd w*ith* wourdys pleyn
Of judas whiche was þ*e* gyde serteyn
Of hem þ*a*t cryst slow cruelly
Which̄ Aftyr ffrom deth ros vp ageyn̄
and hath abedyn in erthe · fful days fourty
and Aftyr Att̄ this
Before our*e* eye :/[4]
in A bryght skye
he dede up stye :/
to hevyn blys.

¶ This seyd judas was Among*es* us
noumbryd Apostylt̄ *and* had lych̄ dygnyte
but whan he be-trayd our*e* lord jh*e*sus
he hynge hy*m*-self vpon A tre
in whos sted muste nedys ordeyned be
A-nother our*e* noumbr*e* ffor to restor*e*
On of þ*o* whiche As weel knowe we
han be conu*er*saunt her*e* longe befor*e*
in our*e* co*m*pany
Whiche xal wyttnes :/[5]
berun expresse :/
to mor*e* *and* lesse :/
of crystys resurrexion̄ stedfastly.

[1] *Hic ascendit ab oculys* is scribbled at the bottom of the page.
[2] There is a larger space than usual between this and the next speech.
[3] This paragraph sign is slightly more elaborate.
[4] Lines 66 to 69 are written on two lines divided by the :/ stop.
[5] Lines 79 to 82 are similarly written.

hic statuent duos joseph justum[1] et *mathiam* etcetera.

Fo. 211v ¶ O sovereyn lorde whiche of Every man
The heitys dost knowe most inwardly
W*ith* all þe lowlyness we may or kan
to þe we prey fful benygnely
That þ*ou* vowche-saff thorwe þ*i* mercy
Vs hym to shewe · whiche in þis cas
þ*ou* lykyst to chesyn effectuously
To ocapye[2] þe lott of judas plas.

hic dabunt sortes et *cadet super Mathiam* etcetera.

Now gramercy lord
And to fulfylle[3]
þ*in* holy wylle
As it is skylle
We All Accorde.[4]

[Remainder of 211v—4⅝ inches—left blank.]

Fo. 212 ¶ M*odo* de die pentecost*es*. *Apostoli dicat*[5] *genuflect*entes *Spiritus sanctus descendat super eos* etcetera.

Petrus[6]	Andreas	jacob*us* major	40
¶ Honowr*e* ⁄	Wurchipp*e*	*and* reverens	
joh*ann*es[7]	Philippus	jacob*us* minor	
Glorye ⁄	Grace ⁄	And goodnes ⁄	
Thom*as*	Bartholome*us*	Symon	
Dygnyte ⁄	Vertu ⁄	*and* excellence ⁄	
matheus	Judas	Matheas	
Bewte ⁄	Blyssynge	*and* bryghtnes	

Be to th*at* lord heye wurthynes. Petrus

Whiche hath p*er*formyd þ*at* he vs hyght. Andreas

And vs enbawmyd w*ith* suche swetnes. Jacob*us* major

[1] MS. just.
[2] Written in larger form.
[3] Lines 96 to 100 are written in two lines divided by parallel strokes.
[4] Scribbled under last line *modo de die* and *hic dabunt so*.
[5] So in MS.
[6] These names are written in red, in larger but not more elaborate letters.
[7] The name *Johannes* was first written to the right of *Jacobus major*, but was erased; and the number *40* is written over it.

Joh*ann*es

Whiche to dyscrye ffer passyth oure myght.

Philippus

This we Aꝑ wel kenne.

Jacob*us* minor

Now gracyous lord jh*es*u. 10

Thom*as*

Conferme us in þ*i* vertu.

Bartholome*us*

And graunt us grace evyr it to sew.

Symoꝑ

Sey we Aꝑ togedyr Amen · Amen.

Et omnes osculant terram.

p*rim*us judeus

¶ Now ffelawys take hede ffor be my trewthe
ȝondyr syttyth A dronkyn ffelacheppe. 15

ij*us* judeus

To don hem good it were grett ruthe.

iij*us* judeus

ȝa I prey god ȝeve hem aꝑ shenscheppe.

i*us* judeus

Fo. 212v [marked 210] Muste in here brayn so sclyly doth creppe
þ*at* þei chateryn And chateryn As they jays were.

2*us* judeus

ȝa were they ony wel browth A-sclepe 20
it wore Almes to þe Revere hem to bere
there hem to baptyze.

i*us* judeus

þ*at* were as thynkyth me[1]
A jentyl sport to se
A bettyr game to be 25
Cowde no man devyse.

Petrus

¶ Serys Alas what do ȝe mene
why scorne ȝe now þus goddys grace
It is no thynge as ȝe do wene
þ*er* is no drunke man in þis place 30
Where-fore ryght grett is ȝowre trespace
but syr*ys* lyst what it doth sygnyfye
Ffulfyllyd is now to man*n*ys solace
Of Johel þe pregnau*n*t pro*p*hecye

[1] Lines 23 to 26 are written in two lines divided by parallel strokes.

In whiche *þat* he 35
That ȝe han seyn
In wourdys pleyn
Declaryth serteyn
Now blyssyd god be
Amen.

The remainder of 212ᵛ—2¾ inches—left blank. An interpolated quire—unmarked—begins with Fo. 213 (Ff. 213 to 222ᵛ). On Fo. 213 the words: '*The Lord be thanked for his g.*' are written in a sixteenth-century hand. And there are some scribbled single letters.

On Fo. 213ᵛ the words *Ad mea facta* are copied in a contemporary hand from the beginning of the next play.

Fo. 214[1] *Ad mea facta pater assit deus* et *sua mat*er.

Doctor

¶[2] RYht worchepful souereynes · liketh yow to here **41**
of the assu*m*pcioñ of the gloryous moder mary
that seynt Jhoñ the eu*a*ngelist · wrot and tauht as I lére
in a book clepid Apocriphu*m* · wyth-outyn dyswary
At fourten yer sche conseyved cryste in hire matere clere 5
and in the fiftene yer sche childyd · this avowe dar I
here lyvyng wyth that swete sone · thre and thretty yere
And after his deth in erthe · xii yer dede sche tary
Now acounte me thise[3] yeris wysely
and I sey the age was of this maide Marye 10
when sche assu*m*pte above the ierarchye
thre score yer · as scripture dothe specyfye
legenda *sanc*tor*um* autorysyth this trewely.

¶ She was inhabith in Jure by the mou*nt*e of syoñ
after the assencioñ of hir sone · conseyved in spoused 15
alle the holy placys in erthe · that criste duellyd[4] on
devouthly sche went hem · honouryng the godhed.
Fferste to the place there criste *cri*stenyd was clepid fflu*m* Jordoñ
there he fastyd and takyn was · by malicious falshed
there he beryed was and roos · vittoryously alon 20
there he assendid alle hevenys · god in his manhed

[1] This play is written in a different handwriting from the body of the MS.
[2] This paragraph sign is small and appears an afterthought.
[3] This *s* is written over some other letter.
[4] The *e* omitted and written above the line.

Thus was sche ocupyed I rede
and meche sche was in the temple preyand
now blissid mot sche be · we owe to be seyand
how sche was assu*m*pte · here men schul be pleyand 25
preyng you of audience · now ses and tak hede.

mi[. . .

Fo. 214v ¶ Pes now youre blaberyng in the develis name
what lousy begchis mow ye not se
owre worthy prynsis lo are gaderid in same
that are statis of this lond · hye men of degre
by there hye wisdam they schal now attayne 5
how alle Jure beste gou*er*nyd may be
and of this pillid prechouris · that oure[1] lawis defame
they schul ben slayn as they say · or fayn for to fle
Wherfore in pes be ye
and herkenyth ou to hem moste stille I[2] 10
ffor what boy bragge outh[3] · hym spilly I
as knave wyth this craggyd knad · hym kylle I
now herkenyth oure pryncis alle kneland on kne.

Ep*iscopus* leg*is*

¶ Now ye prysis[4] · I prest of the lave
of this demaunde responcyon̄ · I aske here anon 15
ys there ony renogat among vs · fer as ye knawe
or any that p*er*uertyth the pepil · wyth gay eloqu*e*ns alon
yif there be we muste on to hem set awe
for they feyne falsly oure feyth · hem preve I houre fon
Sweche schul ben bou*n*den vp be the beltys · til flyes he*m* blawe 20
and gnaggyd vp by the gomys · tyl the devyl doth he*m* grone
We may not won
to sweche harlotis settyn̄ reddure
that geynseyn oure lawe and oure scripture

[1] The *u* written over an erased *r*.

[2] *stilly* originally written, the *e* being written over the erased *y*. The rhyme line linking this with *spilly I* and *kylle I* passes through the following *I* which may therefore have been added.

[3] The *u* in this word appears a correction, and there are marks—possibly connected with the correction—under *o* and *t*.

[4] So in MS.

now let sere pryncis in purpure
In savynge of oure lawys now telle on̄.

primus[1] princeps

¶ Sere syn we slew hym · that clepid hym oure kyng[2]
and seyde he was goddis sone · lord ou*er* all
Syn his deth I herd of no maner rysyng
and lo yif he hadde levyd · he had mad vs his thrall.

Ep*iscopus*

Fo. 215 Therfore oure wysdam was to schortyn his endyng
Whoso clyme ou*er* hie · he hath a foule fall.

ij^{us} p*rin*ceps

ya yit of on thing I warne yow at the gynnyng
his dame is levyng mary that men call
Myche pepil halt hire wythall
wherfore in peyne of reprefe
yif we Suffre hyre thus to relefe
oure lawys sche schal make to myschefe
and meche schame don vs sche schall.

Ep*iscopus*

¶ A sere ye ben bolde i-now · art thou ferd of a wenche
What trowyste that[3] sche myht don vs agayn.

iij p*rin*ce[ps

Sere there are other in the contre that clenche
and prechyn he is levyng that we slewe they seyn
and yif they ben sufferyd thus this[4] will[5] bredyn a stench
for thorow here fayre speche oure lawys they steyn
and therfore devyse we now · vp on this pleyn bench
what is beste for to do · hem for to atteyn
we are but loste yif they reyn.

Ep*iscopus*

why let se than · sey me youre ententis.

P*rim*[us . . .

let vs preson hem til here myht schent is.

Se*cund*[us . . .

bettyr is to slen hem wyth dentis.

[1] The *m* is obliterated by a smear of later ink.

[2] In this and the following three lines the rhyme words are joined by red lines instead of black. This occurs only on Fo. 163 elsewhere in this MS.

[3] The word has been corrected and is not clear—possibly *that* corrected to *than* or *thou*.

[4] The *is* is written over an erasure.

[5] *will* omitted and written over the line.

Terc[ius . . .

nay best is to hang hem wyth peyn.

Ep[*iscopus*

¶ Nay seris[1] nowth so · youre bett*er* a-vyse
haue in syth before · what after may tide
yif we slewe[2] hem · it wolde cause the comownys to ryse 55
and rathere the devyl sle hym · than we schulde that abid[e
Fo. 215v But be that senstere ded mary that fise[3]
We shal brenne here body · and the aschis hide
and don here all the dispith · we can here devise
and than sle tho disciplis · that walkyn so wyde 60
and here bodyes devyde
halde ye not this beste as is sayde.

Prim*us*

Wyth youre wysdam sere we are wel payed.

Ep*iscopus*

than ye knyhtis I charge yow beth arayed[4]
and ye turmentouris · redy that tyde 65
α When mary is ded
and but she deye the sunere · the devyl smyte of here hed.

hic est maria in templo orans et dicens

Maria

¶ O hye wysdam in youre dygne deyte
youre infynyth lovnesse mad oure saluacyon
that it lyst you of me sympilest · to take here humanite 70
Wyth dew obeschyau*ns* · I make you gratulacyon
and gloryous lord and sone · yif it like youre benygnyte
nouth to ben displesid wyth my desideracyon
me longith to youre presense now co*n*ju*n*ct[5] to the vnyte
Wyth all myn herte and my sowle · be natures excitacyon 75
To youre domynacyon
Ffor all creaturis in you don affye
and myche more owe I · youre moder be alye
syn ye wern born god · and man of my bodye
to desyre yowre presens · that were oure ferste[6] formacyon. 80

[1] A smear of later ink over *se*.
[2] The *s* is incomplete and the *e* is written over an erasure.
[3] Some word has been erased after *fise*.
[4] The *yed* is written over an erasure.
[5] An extra stroke of the *u* before *c* is perhaps cancelled.
[6] The *s* is written over some correction.

Sap*ien*tia

¶ My suete moderis preyere on to me doth assende
here holy herte and here love · is only on me
Wherfore aungyl to here thou schalt now dyssende
seyinge here sche shal comyn to myn eternyte.

Fo. 216 ¶ myn habundau*n*t mercy · on here I extende 85
resseyuy*n*ge here to joye · from worldly p*er*plexite
and in tokyn ther of this palme now pretende
Seyinge here sche fere no man*er* [1] of diu*er*cyte.

Angelus i[

By youre myth I dissende to youre moder in v*ir*ginite.

Ang*el*us ij[

Ffor qwyche message injoyeth the hefnely co*n*sorcyte. 90

*hic discendet angelus ludent*ibus [2] *cithar*is [2] *et dicet marie*

p*rim*us A[ngelus

¶ Heyl excellent prynces · mary most pure
Heyl radyant sterre · the su*n*ne is [3] not so bryth
Heyl moder of m*er*cy and mayde most mure
the blessyng that god yaf Jacob vp on you now is [3] lyth.

Maria

Now wolcom bryth berde · goddis au*n*gel I seu*er* [4] 95
ye ben messager of all myhty · wolcom wyth my myhtis
I beseke you now say me · vp-on youre hie nortur
What is the very name · that to youre p*er*sone dith is.

Ang*el*us

α What nedith you lady my name ben desyrand.

Maria

A yis g*rac*yows au*n*gyl I beseke you requyrand. 100

Ang*el*us

¶ My name is gret and m*er*veylous · treuly you telland
the hye god youre sone abidyth you in blis
the thrydde day hens · ye schul ben expirand
and assende to the presence · there my god youre sone is.

Ma[ria

Mercy and grom*er*cy god now may I be seyand 105
thankyng you suete au*n*gyl for this message i-wys.

[1] MS. *ma*n̄.
[2] MS. *luden*ꝑ *citha*ꝑ.
[3] *is* written over the line; originally written after *bryth* and *lyth* respectively and erased.
[4] MS. *se*n̄ or *seu*er; cf. *Seuere*, Fo. 221, l. 411.

An[gelus

In tokenyng where-of lady I am here presentand
A braunce of a palme · owth of paradis com this
α before youre bere god biddith it be bore.

Mari[a

now thanke be to that lord · of his mercy euermore. 110

Angel[us

¶ yowre meknesse youre lovnesse and youre hie lore
is most acceptable in the trynite syth
youre sete ryall in hefne apparaled is thore
now dispose yow to deye · youre sone wyl thus rith.

Mar[ia

Fo. 216v I obbeye the commaundement · of my god here before 115
but on thyng I beseke · that lord of his myth
that my brether the appostelis · myht me be before
to se me and I hem · or I passe to that lyth
α But they ben so deseverid · me thynkyth it nyl be.

Angelus

A yis lady inpossible to god nothyng trowe ye. 120

¶ Ffor he that sent Abbacuc with mete to babylonye from Jure
In to the lake of lyonys to danyel the prophete
b Se the same myht god make may the appostolis here mete
a be an her of his hed lo · so myhty was he [1]

¶ And therfore abasche you not lady · in yowre holy mende. 125

Maria

no more I do glorious aungyl in kynde.

¶ also I beseke my sone · I se not the fende
What tyme outh of this word · I schal passe hens
his horible lok wold fere me so hende
ther is no thyng I dowte · but his dredfull presens. 130

Angelus

What nedith [2] it to fere you empres so hende
syn be the fruth of youre body · was convycte his vyolens
that horible serpent · dar not nyhyn youre kende
and yowre blosme · schal make hym recistens
that he schal not pretende 135
Desyre ye outh ellys now rythis.

[1] So written in reverse order and so corrected in MS.
[2] *yow lady* first written after *nedith* and crossed through.

Maria

nouth but blessyd be my god in his myhtys.

Angelus

to yow I recomaunde me than · moste excellent in sithis
and wyth this ageyn[1] to god I assende.

hic ascendit[2] *angelus.*

Maria

¶ Now lord thy swete holy name · wyth lovnesse I blysse 140
of qwyche hefne and erthe · eche tyme pshalmodyeth
that it lykyth youre m*er*cy[3] · me to you to wysse
my sympil sowle in serteyn · youre name magnefyeth
Now holy maydenys the s*er*uauntis of god as I gysse
I schal passe from this world · as the aungyl s*er*tefyeth 145
therfore to my sympil habitacyon̄ · I telle you now this
I p*ur*pose me to go · besekyng yow replyeth
Fo. 217 *a* And assedually wachith me be dayes and nythis.

P*ri*ma virgo

¶ We schal g*ra*cyous lady · wyth alle oure mythis
schal ye from vs passe swete sonne of socoure 150
that are oure sengler solas · radyant in youre lythis
youre peynful absence · schal make me doloure.

virgo s*ecun*da

Moste excellent princes in all̄ vertu that is[4] dith
alle hefne and erthe · lady you doth honure
We schal wachyn and wake as oure dewe *and*[5] ryth 155
In-to the tyme ye passe to that hye toure
w*ith*[6] [Ma[ria
God tha*n*ke you and so do I
now I wyl dispose me to this Jurne redy
so wolde god my brether were here me by 160
To bere my body · that bare jh*es*u oure savyoure.

hic subito apparet sanctus *Io*hannes *euangelista ante porta*m *marie.*

Joh*ann*es

¶ A myrable god meche is thy myth[7]
many wonderis thou werkyst evyn as thi wyll̄ is

[1] This *e* is corrected from an *a*.
[2] The *i* may have been written over an *e*.
[3] A miswritten letter crossed through before *mercy*.
[4] *is* written above the line, having been first written after *dith* and erased.
[5] *and* written above the line, an original *is* having been erased after *ryth*.
[6] Some word (? *honure*) rhyming with *savyoure* cut away here.
[7] Some word (? *now*) is erased after *myth*.

In pheso I was prechyng · a fer contre ryth [1]
and by a whyte clowde · I was rapt to these hyllys 165
here dwellyth cristis moder · I se wel in syth.[1]
Sum merveylous message is [2] comyn̄ that mayde tyll
I wyl go salusè that berde that in vertu is [2] moste brith
and of my sodeyn comyng · wete what is [2] the skele.

hic pulsabit super portam intrante domum marie sibi [3] *dicente*

α heyl moder mary mayden perpetuall. 170

Maria

¶ A wolcome mayde Johan · wyth all myn herte in specyall
Ffor Joye of youre presence myn herte gynnyth sweme
thynke ye not Johan how my child eternall
When he hynge on cros · sayd vs this teme
lo here thy sone woman · so bad he me you call
and you me moder · eche othir to queme 175
He betok you the gouernayl there of my body terestyall
on mayde to another as convenyens wold seme
Fo. 217v α and now that gracyows lord · hath sent me yow sone.

Johannes

¶ Now good fayr lady · what is ther to done
tellyth the cause why I am heder sent. 180

Maria

swete sone Johan so wyll I a-none
oure lord god sent to me an aungyl [4] that glent
and sayde I schulde pass hens · where thre were in one
tho I askyd the aungyl to haue you present.

Johannes

A [5] holy moder schul ye from [6] vs gone 185
My brether of this tydyngis sore wyl repent *þt ȝe schuld ben*
euer trybulacyon̄ lord meche þu vs sendyst *Absent* [7]

[1] Some word (? *now*) is erased after *ryth* and *syth*.

[2] The *is* on each of these lines is written as a correction above the lines, having been originally written at the end of the lines and erased.

[3] MS. *s* with overwritten *i*.

[4] An original *e* changed to *y*.

[5] A small *a* changed to a large.

[6] *fron* corrected to *from*.

[7] This, the missing ninth line of the stanza, is written in the right-hand margin in a different hand—apparently that of the scribe of the rest of the MS. The same words were written in the opposite margin by the scribe of this play, and crossed through in red ink—*ye schulde . . . bsen . . .* is legible.

b and now oure Ioye thy moder to take thou pretendist[1]
a thou oure mayster and oure comfort fron[2] us ascendist
thanne att oure comfort is from us detent 190
α but what seyde then[3] aungyl moder on-to you more.

Maria

¶ he brouth me this palme from my sone thore
qwyche I beseke as the aungyl me bad
that a-forn my bere · by you it be bore
saynge my dirige · devouthly and sad 195
α Ffor Johan I haue herde the Jewys · meche of me spelle.

Johannes

A good lady what likyth it you to telle.

Maria

¶ Secretly they ordeyne in here conseytis felle
When my sowle is paste where god is liste[4] is
to brenne my body · and schamly it quelle 200
For Jhesu was of me born · that they slew with here fistis
And therfore I beseke you Johan · both flech and felle
Helpe I be beryed · for yn yow my tryst is.

Johannes

Ffere yow not lady · for I schal wyth you duelle
wolde god my brether were here now and wyst this. 205

hic subito omnes apostoli congregentur ante portum[5] *mirates.*

Petrus[6]

¶ A holy brether wyth grace · be ye met here now
lord god what menyth · this sodeyne congregacyon
now swete brother powle wyl ye take this vp-on yow
preye to god for vs att · we may have relacyon.

Paulus

Fo. 218 Good brother[7] peter[8] how schuld I here pray now 210
that am lest and most vnworthy of this congregacyon[9]
I am not worthy to ben clepyd apostle sothly I say yow

[1] So written in reverse order and so corrected in MS.
[2] So in MS. for *from*. [3] So in MS.
[4] *Wytt* first written and crossed through in red ink; *liste* written above the line by the scribe.
[5] So in MS.
[6] *Petrus* written as name of speaker against preceding stage direction and crossed through.
[7] The *e* is written in between *h* and *r* as if a correction.
[8] *Powle* first written and crossed through. [9] MS. *congregacyn*.

Ffor as a woodman ageyn holy cherche I mad persecucyon
α but neuertheles I am the grace of god in that þat[1] I am lo.

Petrus

A gret is youre lownesse powle brother euer-mo. 215

Paulus

¶ the keyes of hevene peter · god hath you betake[2]
and also ye ben peler of lith · and prynce of vs all
it is most sittyng to you this preyere to make[2]
and I vnworthy wyth yow · preyen here schall.

Petrus

I take this vp-on me poule for youre sake[2] 220
now almythty god that sittiste · aboue cherubyn halle
In sygne of thyn holy cros · oure handis we make[2]
besekyng thy mercy · may vp-on vs falle
α And why we ben thus met yif it lyke vs lare.

Johannes

A holy brether alle welcom ye are 225

¶ Why ye be[3] met here I schal you declare
Ffor mary goddys moder by message is sent
that from this wrechid world · to blysse sche schal fare
and at here deying sche desyryth to haue vs present.

Petru[s

A brother Johan we may syhyn and care 230
yif it displese not god · for these tydyngis ment.

Paulus

Fforsothe so we may hevyin euermare
that oure moder and oure comfort schuld ben vs absent
α but neuertheles the wyl of god fulfyllid mot be.

Johannes

¶ that is wel seyd poule but her-of be-war ye 235
that non of you for here deth schewe hevy speche
Ffor a-non to the Jewys it schuld than notyd be
that we were ferd of deth and that is[4] a-geyn that we teche
Ffor we seyn all tho belevyn in the hol Trynyte
they schul ever leve and nouth deye this truly we preche 240
And yif we make hevynesse for here than wyl it seyd be

[1] The thorn letter þ is used here for the only time in the MS. except in two notes on Ff. 144ᵛ and 145; the form *y* is otherwise used.

[2] Some erasure (? *now*) has been made at the end of each of these lines before the rhyme lines were drawn; cf. ll. 360, 2.

[3] *sent* first written after *be* and crossed through.

[4] Omitted and written over the line.

lo yone *pre*chouris to deye · they fere hem ful meche
Fo. 218v α And therfore in god now beth glad eu*er*ychon̄.

Petrus

¶ We schal don as ye sey vs holy brother Joh*an*
now we beseke you let vs se · oure[1] moder marie.

Joh*ann*es

now in goddys name to here[2] tha*n* alł let vs gon
Sche wyl ben ful glad to se · this holy companye.

Petrus

Heyl moder and maydyn · so was neu*er* non
but only ye most blissid treulye.

Paulus

heyl incomp*ar*abil que*n* · goddis holy tron
of you spreng salvacyon̄ · and alł oure glorye
heyl mene for mankynde · and mendere of mys[3].

Maria

¶ A wyth alł myn hol herte brether ye are wolcom i-wys
I beseke you now to telle me · of youre sodeyne metyng.

Petrus

In dyu*er*is contreys we prechid · of youre sone *and* his blis
Diu*er*is clowdys eche of vs was · sodeynely curyng
w[4] in on were brouth before youre yate here i-wys
the cause why no man cowde telle of oure comyng.

Maria

now I thanke god of his m*er*cy · an hy merakle is this
now I wyl telle yow the cause · of my sonys werkyng
α I desyrid his bodily p*re*sence to se.

Joh*ann*es

no wonder lady · thow so dedo ye.

Maria

¶ Tho my sone jh*e*su · of his hye pete
sent to me an aungyl · and thus he sayd
that the thredde nyth I schuld assende to my sone in deite
thanne to haue youre p*re*sence brether · hertly I prayed
And thus at my request · god hath you sent me.

Petrus

Wys g*rac*yous lady · we are ryth wel payed.

Maria

blissid brethere I beseke you than tent me

[1] A capital *M* crossed through after *oure*.
[2] *let* first written and crossed through after *here*.
[3] A paragraph sign has been erased at the beginning of the line.
[4] So in MS. for *we*.

now wyl I rest me in this bed · that for me is rayed 270
ꝓ Wachith me besily wyth youre lau*m*pys and lithtis.

Paulus

We schal lady redy all thyng for you dith is.

Maria

¶ now sone schul ye se what god is myth is
my flech gynnyth feble be nature.
hic erit decenter ornatus in lecto.[1]

Petrus

Fo. 219 Brether eche of you a candele takyth now rithis 275
And lith hem in haste whil oure moder doth dure
and bisyli let vs wachyn in this v*ir*gyne sythis
That when oure lord comyth in his spoused pure
he may fynde vs wakyng · and redy wyth oure lithtis
for we knowe not the ho*ur* of his comyng now sure 280
ꝓ and yn clennesse alle · loke ye be redy.

Maria

¶ A swete sone Jhesu now mercy I cry
ou*er* alle synful thy mercy let sprede.
hic dissendet dominus cum omni celesti curia et dicet

Do*mi*nus

the voys of my moder me nyhith fulny
I am dyssend on to here of whom I dede sede. 285
hic cantabunt organa.

Maria

¶ A wolcom gracyous lord · Jh*es*u sone and god of m*er*cy
an aungyl wold a ssuffysed me hye kyng at this nede.

Do*mi*nus

In p*ro*pure p*er*sone moder I wyl ben here redy
Wyth the hefnely quer yowre dirige to rede
℣ Veni tu electa mea *et* ponam in te thronu*m* meu*m* 290
quia concupiuit rex speciem tua*m*.

Maria

℣ Paratu*m* cor meu*m* deus p*ar*atu*m* cor meu*m*
cantabo et psalmu*m* dicam d*omi*no.

Ap*ostol*i

℣ hec est que nesciuit thoru*m* in delictis
habebit requiem in respectu a*nima*ru*m* *sanct*aru*m*. 295

Maria

℣ Beatam me dicent om*n*es gèneracio*n*es
quia fecit michi magna qui potens est *et* s*anctu*m nome*n* eius.

[1] This stage direction is not underlined in red.

Dominus

℣ Veni de libano sponsa mea veni Coronaberis.

Ecce venio quia in capite libri scriptum est de me
vt facerem voluntatem tuam deus meus
quia exultauit spiritus meus in deo salutari meo[1].

hic exiet anima marie de corpore in sinu dei.

Dominus.

Fo. 219v ¶ Now com my swete soule in clennesse most pure
and reste in my bosom[2] brithtest of ble
alle ye myn apostelis · of this body taketh cure
In the valle of Josephat · there fynde schul ye
A grave new mad for maryes sepulture
there beryeth the body wyth alī youre solempnite
and bydyth me there stylle thre dayes severe[3]
and I schal pere ageyn to yow · to comfort your aduercyte
Wyth this swete soule now from you I assende.

Petrus

In oure tribulacyones[4] lord thou vs defende
We haue no comfort on[5] erthe · but of the alon
O swete soule of mary prey thy sone vs defende
haue mynde of thy pore brether · when thou comyst to þi tron.

Chorus martyrum

℣ Que est ista que assendit de deserto
deliciis affluens innixa super dilectum suum.

Ordo angelor[um

℣ Ista est speciosa inter filias Jerusalem sicut vidistis eam
plenam caritate *et* dilectione sic que in celum gaudens suscipitur
et a dextris filii in trono glorie collocatur.

hic cantabit omnis celestis curia.

Prima virgo *

¶ Now suster I beseke you · let vs do oure attendaunce
and wasche this gloryous body · that here in oure sith is
as is the vse among vs wyth-outyn ony varyaunce
now blessid be this persone that bar god of mythtis.

[1] MS. apparently *mes.*
[2] *brist* miswritten after *bosom* and crossed through.
[3] So in MS. for *seuere*? = *sure*, rhyming with *pure, cure, sepulture.*
[4] MS. *tribulacyoñs.* [5] Or *in.*
* *prima virgo* is written as name of speaker against the preceding stage direction and crossed through in red.

Secunda virgo

I am redy suster wyth all myn hol affyaunce
to wesche and worschepe · this body so brith is 325
alle creaturys therto owyn dew obeschaunce
ffor this body resseyved · the holy gostis slihtis.

et osculabunt corpus marie.

Johannes

¶ Now holy brother peter · I hertely you pray
to bere this holy palme · before this gloryous body
Ffor ye ben prince of apostelis · and hed of oure fay 330
therfore it semyth you best to do this offis treuly[1].

Petrus

Fo. 220 Sere and ye slept on cristis brest · seyng all celestly
ye are goddis clene[2] mayde · wyth-outyn ony nay
this observaunce is most like · you to do dewly
Wherfore tak it vpon you · brother we pray. 335
a and I schal helpe for to bere the bere.

Paulus[3]

¶ and I peter wyth oure brether in fere
this blessid body schal helpe to the ground
this holy cors now take we vp here
Seyng oure observaunce · wyth devouth sound. 340

hic portabunt corpus versus sepulturam cum eorum luminibus.

Petru[s

℣ Exiit israel de egipto · domus iacob de populo barbaro. Alleluia.

Apostoli

℣ Ffacta est iudea sanctificacio eius israel potestas eius. Alleluia.

hic angeli dulciter cantabunt in celo · Alleluia.

Episcopus

¶ herke sere princys · what noyse is all this
the erthe and the eyer · is ful of melodye
I herde neuer er · swyche a noyse now i-wys 345
con ye outh say · what they signefye.

primus princeps

I not be my god that of myht meche is
What-sumeuer they be · hougely they crye
I am aferd[4] there wyll be sum thyng a-mys
It is good prevely among vs we spye 350
wyth-owte.

[1] *tru* written before *treuly* and crossed out. [2] The *l* corrected from an *h*.
[3] Name not looped in red; perhaps added in a different ink.
[4] The *a* is squeezed in between the *m* of *am* and the *f* of *ferd*.

Secundus princeps

Now I haue levyd this thre skore yer
but sweche another[1] noyse · herd I neu*er* er
myn herte gynnyth ogyl · and quake for fer
there is su*m* newe sorwe · sprongyn I dowte. 355

Tercius princ[eps

¶ ya that there is sothly I say yow
the p*ro*phetis moder[2] mary is ded
the disciplis here beryn in gret aray now
and makyn alle this m*er*the in spyth of oure hed.

Episcopus

ffy on yon[3] lousy doggys · they were bett*er* nay[4] 360
outh harrow · the devyl is in myn hed
ye dodemvsyd prynces · faste yow aray[4]
or I make avow · to mahound youre bodyes schul blede
Fo. 220v Now that quene is ded
ye coward knytys in plate 365
and ye tormentours · thryfe schul ye late
Ffaste harlotys · go youre gate
and brynge me that bychyd body I red.

Primus princeps

¶ dowte you not sere byschop in peyne of repref
We schal don schame to that body · *and* to tho prechours. 370

Secundus

Sere I schal geyne tho glabereris or gramly hem gref
tho teynt tretouris schul tene · yif my loke on hem louris.

Tercius princeps

To hurle wyth tho harlotys · me is ful lef
I schal snarle[5] tho sneveleris · wyth rith scharp schowris.

Episcopus

hens than a devylis name · and take me that thef 375
and brnge[6] me that bygyd body · evyn to-fore these touris
and here disciplis ye slo
hye you hens harlotys at-onys
the devyl boyes mot breke youre bonys
Go stent me yone body wyth youre stonys 380
Outh harrow · al wod now I go.

[1] *another* first written and crossed through.
[2] *is* first written after *moder* and crossed through.
[3] Or *you*. [4] Some erasure at the end of these two lines; cf. ll. 216, 218, 220, 222.
[5] *snrle* first written and crossed through. [6] So in MS. ? bringe.

Secundus princeps

hic discendunt principes cum suis ministris vt feroci percutientes petras cum eorum capitibus.

¶ What devyl where is this mene
I here · here noyse · but I se ryth nouth
allas I haue clene lost my poste
I am ful wo · mad is my þowth. 385

Tercius princeps

I am so ferd I wold feyn fle
the devyl hym spede · hedyr me brouth
I renne I rappe · so wo is me
Wyndand wod wo hath me wrouth
To deye I ne routh. 390

Primus princeps

A cowardis vpon you now fy
are ye ferd of a ded body
I schal sterte ther-to manly
alle that company fere I ryth nouth.

Fo. 221 *hic saltat insanus af*[1] *feretrum Marie* et *pendet* per *manus.*

¶ Allas my body is ful of peyne 395
I am fastened sore to this bere
myn[2] handys are ser · bothe tweyne
o pet*er* now prey thy god for me here
In cayfas halle · when thou were seyne
And of the pet*er* a mayde · acusid there 400
I halpe the tho · now helpe me ageyne
that I were hol · outh of this fere
sum medycyn̄ me lere.

Petrus

I may not tend to the sere at this ho*ur*
Ffor ocupacyon̄ of this body of hono*ur* 405
but neu*er*theles · beleue in Jh*es*u criste oure save*your*
and that this was his moder that we bere on bere.

¶ I beleue in Jh*es*u mannys saluacyon̄.[3]

Petrus

In goddis name go dou*n* than and this body honure.

[1] So in MS. [2] A miswritten word (*hōdys?*) crossed through after *myn.*
[3] No name of speaker for the speech in MS.

Primus princeps

now m*er*cy god and grom*er*cy of this savacyon̄ 410
In Jh*es*u and his moder to beleve eu*er* I seuere.

Petrus

Than take yone holy palme · and go to þi nacyon̄
and bid hem beleve in god yif they wyl be pure
and towche hem ther wyth both hed · hand and facyon̄
and of her sekenesse · they schal have cure 415
and ellis in here peynys indure.

Primus princeps

Grom*er*cy holy fader peter
I schal do as ye me teche her
thankyng god eu*er* in my speche her
Wyth hye repentaunce · and herte most mure. 420

hic portabunt feretrum ad locum sepulture.

Petrus

¶ Now holy brether this body let vs take
and wyth alle the worschepe we may ley it in the g*ra*ue
kyssyng it alle atonys · for here sonys sake
now insence ye · and we schal put here in this cave.

hic ponent corpus in sepulcrum insensantes et *cantantes.*

Joh*ann*es

Fo. 221v ¶ De terra plasmasti me *et* carne induisti me 425
redemptor meus d*omi*ne resuscita me in novissimo die.
Now god blysse this body · and we oure synge make

hic vnanimiter benedicent corpus · In n*omi*ne pat*ri*s *et* filii *et* sp*iritus* s*anc*ti.

the fruth that it bar oure soules schal saue
Now reste we vs brether[1] vp-on this pleyn lake
tyl from oure god and oure lord · tydyngis we haue 430
Here must we belave.

Paulus

so muste we Joh*an* as ye say
thanne byde we here and pray
besekyng hym of comfort taht best may
restyng here abowtyn this g*ra*ue. 435

[1] Or *brother.*

Primus princeps

hic vadit princeps ad Iudeos cum palma.

¶ ye Jewys that langour in this gret Infyrmyte
belevyth in crist Jhesu and ye schal haue helthe
throw vertu of this holy palme · that com fro the trinyte
yowur sekenesse schal aswage · and restore yow to welthe.

Secundus princeps

I beleve in crist Jhesu · goddis sone in vnyte 440
and forsake my mavmentryes · fals in here felthe

hic tangat credentes cum palma et sanati sunt.

A I thanke the gracyous lord · and thy moder of pete
now are we hol of oure seknesse · and of oure foul belthe.

Tercius princeps

What harlotys forsake oure lawe.

Secundus princeps

So hald I best the do. 445

Tercius princeps

hens fro me in the develis name ye go
I deye outh outh harro
the wylde develys · mot me to drawe.

Primus demon

¶ herke belsabub and belyal sere sathan in the herne
vs fettyn oure servauntis to this presoñ 450
blow flamys of fer to make hem to brenne
mak redy ageyn we com to this demon.

Secundus demon

Ffaste for tho harlotis now let vs renne
to caste hem in this pet here that depe is adoñ
they schul brenne and boyle and chille in oure denne 455
go we now a dewelys name as faste as we moñ
Harrow harrow · we com to towñ.

Fo. 222 Primus demoñ

Drag we these harlotis in hye
In to the pet of helle for to lye.

ij^us demoñ

Gowe now helle houndis[1] ye crye 460
Sere sathan may heryn oure soñ.

[1] MS. *hoñdis.* The similar flourish on the rhyme words *presoñ, adoñ, moñ, soñ* in this stanza might also mark a contracted *u*, but it occurs also in *towñ* and not in *demon* the other rhyme words.

Dominus

¶ Now aungyl and alle this court celestyall
In to herthe now discendith wyth me
to reyse the body of my moder terestyall
and bryng we it to the blysse of my deyte 465
α assent ye here to now the vnyte.

Angeli

ya for yowre hye mercy lord · al hefne makyth melode.

hic discendit et venit ad apostolos dicens Dominus

¶ Pes be to yow alle · my postelis so dere
lo me here yowre lord · and youre god now rythtis.

Petrus

A wolcom criste oure comfort · in thy manhed clere 470
gret merveylous god · mekyl now thy myth is.

Dominus

What worschepe and grace · semyth you now here
that I do to this body · mary that hythtis.

Johannes

lord as thou rese from deth · and regnyst in thyn empere
so reyse thou this body · to thy blysse that lyth is 475
vs semyth this ryth is.

Mychael

ya gloryous god · lo the sowle here prest now
to this blissid body · likyth it you to fest now
hefne and erthe wold thynke this the best now
In as myche as sche bare you god in youre mythtis. 480

hic vadit anima in corpus marie.

Dominus

¶ Go thanne blyssid soule to that body ageyn[1]
arys now my dowe · my nehebour and my swete frende
tabernacle of Joye · vessel of lyf · hefnely temple to reyn
ye schal haue the blysse wyth me moder · that hath non ende
Ffor as ye were clene in erthe · of alle synnys greyn 485
so schul ye reyne in hefne clennest in mend.

Maria

A endles worchepe be to you Jhesu · relesere of peyn
I and alle erthe may blisse ye com of owre kend
Fo. 222v Lo me redy wyth you for to wend.

Dominus

A-bouen hefnys moder assende than we 490
In endles blysse for to be.

[1] This *e* has been corrected from an *a*.

Michael

Hefne and erthe now injoye may ye
Ffor god throw mary is mad mannys frend
Et hic assendent in celum cantantibus organis.
¶ Assumpta es maria in celum.

Dominus

Yow to worchepe moder · it likyth the hol trinyte 495
Wherfore I crowne you here · in this kyndam of glory
of alle my chosyn · thus schul ye clepyd be
qwen of hefne · and moder of mercy.

Michael

Now blysid[1] be youre namys we cry
ffor this holy assumpcyon̄ · alle hefne makyth melody. 500

¶ Deo gracias.

The remainder of Fo. 222v—4¾ inches—left blank and Fo. 223. On Fo. 222v *Deo gratias* enclosed in flourished lines is scribbled in another hand. On Fo. 223 are the scribbled words: 'lo me redy with wt yow to wend'. Cf. line 489.

¶ hic incipit dies iudicii *et Ihesu descendente cum Michaele* et
Fo. 223v[2] *Gabriele Archangelis* et *Michael dicet* etcetera[3]

¶ Michael

[marked 220] SVrgite! Att men Aryse **42**
venite Ad judicium
Ffor now is sett þe hyȝ justyce
And hath Assygnyd þe day of dom̄
Rape ȝow redyly to þis grett Assyse 5
bothe grett *and* smalt At An sum
And of ȝ*our* Answer*e* ȝow now Avyse
What ȝe xal sey · whan þ*at* ȝe cum
ȝowr*e* Ansuer*e* ffor to telle
Ffor whan þ*at* god xal ȝow appose 10
ther is non helpe of no glose
the trewth fful trewlye he wyl tose
And send ȝow to hevyn or helle.

Gabryell

¶ Bothe pope[4] prynce and prysste[5] w*ith* crowne
Kynge *and* caysere *and* knyht*ys* kene 15

[1] The third letter is a *y* written over an *s* or vice versa.
[2] Original handwriting resumed.
[3] Title and stage direction written as one in the MS.
[4] This word is written in larger form.
[5] *prynce* first written and corrected.

Rapely ȝe renne ȝ*our* resonys to rownc
Ffor this xal be þ*e* day of tene
Nowther por*e* ne ryche of grett renowne
ne aſt þ*e* develys in helle þ*at* bene
Ffrom þis day ȝow hyde not mowne 20
Ffor aſt ȝ*our* dedys her*e* || xal be sene
Opynly in Syght
Who þ*at* is fowndyn in deedly gylte
he wer*e* bettyr to ben hylte
in hendeles helle he xal be spylte 25
his dedys his deth xal dyght.

Fo. 224 *Omnes resurgentes subtus terram clamauit*[1] ha aa · ha aa · ha aa. ||
Deinde surgentes dicat[1] ha aa etcetera.

¶ Ha aa cleue a-sundyr ȝe clowdys of clay
A-sundyr ȝe breke *and* lete vs pas
now may our*e* songe be wele-Away
þ*at* evyr we synnyd in dedly trespas. 30
Omnes demones clamant[2]
Harrow *and* owt what xal we say
harrow · we crye owt And Alas
Alas harrow is þis þ*at* day
to endles peyne þ*at* vs must pas
Alas harrow *and* owt we crye 35
Omnes anime resurgentes dicant etcetera
A mercy lorde ffor our*e* mysdede
And lett þ*i* mercy sprynge *and* sprede
but alas we byden in drede
It is to late to Aske mercye.
Deus
¶ Venite benedicti[3] || my bretheryn aſt 40
Patris mei || ȝe childeryn der*e*
Come hedyr to me to my*n* hyȝ haſt
Aſt þ*o* my*n* suterys *and* servaunt*ys* be[4]
Aſt þo ffowle wyrmys ffrom ȝow ffalle
W*ith* my ryght hand I blysse ȝow her*e* 45
my blyssynge burnyschith ȝow as bryght as beraſt

[1] So in MS. [2] MS. *clamant*. [3] Or *benedicti*.
[4] So in MS. for *were*.

As crystall clene it clensyth ȝow clere
All ffylth ffrom ȝow ffade
Petyr to hevyn ȝatys þou wende *and* goo
þe lokkys þou losyn *and* hem vndo 50
my blyssyd childeryn þou brynge me to
here hertys for to glade.

Fo. 224v

Petrus

¶ The ȝatys of hevyn I opyn þis tyde
Now welcome dere bretheryn to hevyn i-wys
Comȝ on *and* sytt on goddys ryght syde 55
Where myrthe *and* melody nevyr may mys.

Omnes saluati

On kne we crepe we gon we glyde
to wurchepp oure lorde þat mercyfful is
Ffor thorwe his woundys þat be so wyde
he hath brought us to his blys 60
holy lorde we wurcheppe þe.

Deus

Welcome ȝe be in hevyn to sitt
wel cum fro me xul ȝe nevyr flitt
so sekyr of blys ȝe xul be ȝitt
to myrth *and* joye welcum ȝe be. 65

Anime dampnandum[1]

¶ Ha · Ha · mercy mercy we crye *and* crave
A mercy lorde for[2] oure mysdede
A mercy mercy we rubbe we rave
A help us good lord in þis nede.

Deus

How wolde ȝe wrecchis Any mercy haue 70
Why Aske ȝe mercy now in þis nede
What haue ȝe wrought ȝour sowle to saue
to whom haue ȝe donȝ Any mercyful dede
Mercy for to wynne.

Primus diabolus

Mercy nay nay they xul haue wrake 75
And þat on here fforehed[3] wyttnes I take
Ffor þer is wretyn with letteris blake
Opynly all here synne.

[1] MS. *dāpnādu*, the final *u* with circumflex and dot.
[2] ore miswritten and crossed through after *for*.
[3] The *r* is blotted.

Deus

Fo. 225 ¶ To hungry *and* thrusty þat Askyd in my name
[The number is written on a patch in the paper.]
mete *and* drynke wolde ȝe ȝeve noñ 80
Of nakyd men had ȝe no shame
ȝe wold nott vesyte men in no *pre*son
ȝe had no pete on seke nor lame
Dede of mercy wold ȝe nevyr don
Vn-herborwed men ȝe servyd þe same 85
to bery the deed por*e* man ? wold ȝe not gon
These dedys doth ȝow spylle
ffor ȝour*e* love was I rent on rode
and for ȝour*e* sake I shed my blode
Whan I was so mercyfull *and* so gode 90
Why haue ȝe wrought aȝens my wylle.

2us diabolus

¶ I ffynde here wretyn in þi*n* fforheed
þ*ou* wor*e* so stowte *and* sett in pryde
þ*ou* woldyst not ȝeve a por*e* man breed
but ffrom þ*i* dor*e* þ*ou* woldyst hy*m* chyde. 95

3us diabolus

And in þ*i* face her*e* do I rede
þ*at* if a thrysty man com any tyde
Ffor thrust þow he xulde be deed
drynk ffrom hy*m* þ*ou* woldyst evyr hyde
On covetyse was all thy thought. 100

1us diabol*us*

In wratth þ*i* neybor*e* to bakbyte
them for to hangere was þ*i* delyte
þ*ou* were evyr redy them to endyte
On þ*e* seke man rewyst þ*ou* nought.

2us diabolus

Fo. 225v ¶ Evyr-mor on Envye was All þi mende 105
þ[*ou*][1] woldyst nevyr vesyte no *pre*soner
to All þ*i* neybor*ys* þ*ou* were vnkende
þ*ou* woldyst nevyr helpe man in daunger.

3us diabol*us*

The synne of slauth þ*i* sowle xal shende
masse nor*e* mateynes woldyst þ*ou* non here 110
to bery þ*e* deed man þ*ou* woldyst not wende.

[1] The over-written letter is blurred by a stain, of which there are many on this page.

þ*er*fore þ*ou* x̄alt to endles ffe*re*
to slowth̄ þ*ou* we*re* ful prest.

i^us diabol*us*

Thou haddyst rejoyse in glotonye
In dronkesheppe *and* in rebawdye 115
Vnherborwyd w*ith* velonye
þ*ou* puttyst from her*e* rest.

2us diabol*us*

¶ Sybile sclutte[1] þ*ou* ssalte sewe
All̄ ȝ*our* lyff was leccherous lay
to all̄ ȝ*our* neybor*ys* ȝe wor*e* a shrewe 120
all̄ ȝ*our* plesauns was leccherous play
Goddys men ȝe lovyd but fewe
nakyd men *and* ffebyl of array[2]
ȝe wolde nott socowr*e* w*ith* a lytel drewe
nott w*ith* A thred þ*e* soth̄ to say 125
Whan they Askyd in godys name.

O*mn*es da*m*pnandi

A mercy lord mekyl of myght
w*e* Aske þ*i* mercy *and* not þ*i* ryght
not after our*e* dede so us quyth̄[3]
we haue synnyd we be to blame. 130

Deus

The rest is missing, this page forming the last of the W quire.

[1] The *l* in this word is faint, only just legible.
[2] *ray* first written and *ar* added above the line. [3] The final *h* is blotted.

Scribbled on right-hand margin in (?) contemporary hand—1. at*que*. 2. I must go to þ*e* most.

HOMERTON
COLLEGE
LIBRARY

LIST OF PROPER NAMES OCCURRING IN THE *LUDUS COVENTRIÆ*

Aaron, 340.
Abacuch, 60. Habakkuk.
Abbacuc, 359.
Abdias, 60. Obadiah.
Abel (Abelle, Abeel, Abeell), 2.
Abias, 59. Abia (Matt. i. 7).
Abraham, 3.
Abyacar, 4. Abiathar or Abiacar pontifex (*De Nativitate S. M.*).
Abysakar, 83. Isachar reading of some MSS. for Abiathar (*De Nat. S. M.*).
Achas, 61. Achaz (Matt. i. 9).
Ada, 29 (note). (Gen. iv. 19.)
Adam, 1.
Affraunt, 313. Pilate's 4th knight.
Aggeus, 61. Haggai.
Almonye, 198. Germany (Almayne in *C. of P.* list).
Alpheus, 62 (note). (Matt. x. 3.)
Amon, 61. (Matt. i. 10.)
Amorawnt (Ameraunt), 313. Pilate's 1st knight (cf. Poem on the Resurrection, MS. Ashmole 61).
Amos, 147.
Andreas, 250.
Andrewe, 269.
Anna, prophetissa, 164.
Annas, 230.
Anne, 63 (note).
Apocriphum, 354.
Aragon, 198.
Aran, 37 (note). Haran (Gen. xi).
Archage, 153, 198. Realm of Jasper and country seen from the Mount (cf. Artage apparently for Arcadia in *Parl. of Three Ages*, l. 347. Achaia and Arcadia occur together in geographical lists. 'Ibi est Achaia, ibi est Arcadia.' Honorius Augustodunensis *De imagine mundi*; cf. *Rel. Ant.* i, p. 272, and MS. Arundel 123).
Archas, 198. Country seen from the Mount (see last note).
Arfaxat, 37 (note). Arphaxad (Gen. xi).
Arfaxat (Arphaxat), 313. Pilate's 2nd knight (MS. Ashmole 61, Arfax).
Arfexe, 231. Messenger of Annas.
Artyse, 292. Pilate's servant.
Asa, 59. (Matt. i.)
Asmaria, 62 (note). Mother of Joachim—no authority has been found for this name. Esmeria is the name of Joachim's sister-in-law in *Legenda aurea*.

Babolony, 198.
Babylony, 74.
Bakbytere, 125. Also a character in *Castle of Perseverance*.
Balaam, 147.
Baltazare, 152. 1st king (of Saba).
Barpanter, 62 (note). Father of Joachim (see *Legenda aurea*, cap. 131).
Barrabas, 12.
Bartholomeus, 350.
Baruk, 61. Baruch.
Bedleem, 136. Bethlehem.
Bedlem jude, 283.
Belsabub, 193. Beelzebub.
Belyall (Beliall), 194. Belial.
Belyard, 193. Name of a devil (mistake for Belyall ?).
Bertylmew, 270. Bartholomew.
Bertylmewe þe bochere, 123.
Betany, 261.
Bette þe bakere, 123.
Bettrys Belle, 123.
Boosdras, 148, l. 62 (? Is. lxiii: 'Who is this that cometh with dyed garments from Bozrah').
Boosras, 146. 1st shepherd.
Boutyng þe brewstere, 123.

Calsydon, 244. (?)
Caluerye (Kalvarye), 294.
Caton, 178. Catonis Disticha.
Cavdas, 269. Candace.
Caym, 2.
Caynan, 31 (note). (Gen. v.)
Cayphas (Cayfas), 10.
Cenacle, 15. Upper Chamber (Luke xxii. 12, Vulgate 'Coenaculum').
Cephas, 269.
Cham, 37. Ham.

Cleophas, 62 (note). Second husband of Anne (*Legenda aurea*).
Cok crane, 123.
Colett Crane, 123.
Coleyn, 7. Cologne.
Colle Crakecrust, 123.
Contemplacio, 62.
Cosdram (Cosdran). Pilate's 3rd knight (cf. Poem on the Resurrection, MS. Ashmole 61).

Danyel, 59.
Dauyd, 5.
Davy Drydust, 123.
Diabolus, 176.
Doctrynal, 178. Name of part of the *Speculum* of V. de Beauvais.
Dysmas, 293. The penitent thief.

Ebrew, 10.
Egypth, 7.
Egythp (Egypt), 254.
Elyud, 62 (note). Son of Emeria and brother of Elizabeth, mother of John (*Legenda aurea*).
Elyzabeth, 105.
Emanuel, 58.
Emawus (Emaws), 337.
Emeria, 62 (note). Sister of Anne, mother of Elizabeth and Elyud; generally found as Esmeria or Ismaria (*Legenda aurea*).
Eminem, 62 (note). Daughter of Elyud, mother of St. Servasius.
Enoch, 29 (note). (Gen. iv.)
Enok, 31 (note). (Gen. v.)
Enos, 31 (note) (Gen. v.)
Erlonde, 199. Ireland.
Eve (Eua), 2.
Ezechiel, 59.
Ezechias, 61.

Ffraunce, 198.
Flomjordon, 188.

Gabryell, 5.
Galelye, 169.
Galys, 199. (Cf. *Castle of Perseverance* and *Croxton* play.) ? Galicia.
Gamalyel, 246.
Geffrey Gile, 123.
Geruasius, 62 (note). For Seruasius.
Grek, 15. Greek.
Grw, 169. Greek.
Gryscysme, 178. (?)
Gylle Fetyse, 123.

Heber, 37 (note). Eber (Gen. xi).
Herownde (Herodes, Herowde, Herowdys, Herowndys), 7.
Hierusalem, 185.

Jabel, 29 (note). (Gen. iv.)
Jacob, 147.
Jacobus major, 62 (note).
Jacobus minor, 62 (note).
Jak at the Style, 123.
Jamys, 269.
Jamys the lesser, 269.
Jane, 123.
Januense, 298. Genoa? (Jenyse and Genewaye occur in the *Croxton* list).
Japhet, 37.
Jared, 31 (note). (Gen. v.)
Jasper, 153. 3rd king (of Ypotan and Archage).
Jeremye (Jeremias), 58.
Jesmas 293; Jestes, 299. The impenitent thief. *Legenda aurea*, cap. liii, Gesmas; *Gesta Pilati*, Gestas.
Jesse, 4.
Jewry, 234.
Jewys (Juwys), 10.
Jherosolyman, 269.
Jherusalem, 8.
Joachym, 4. Father of the Virgin.
Joathan, 60. (Matt. i.)
Joel, 60.
Jhon, 8.
Jon, 300.
Johan Jurdon, 123.
Johannes apostolus, 239.
Johannes baptista, 62 (note).
Johannes Evangelista, 62 (note).
Jonas, 59.
Joras, 60. Joram (Matt. i).
Jordon, 8.
Joseph, 5.
Joseph ab Aramathy, 310.
Joseph of Baramathie, 308.
Josephat (Valley of), 366.
Josophat, 60. (Matt. i.)
Iradh, 29 (note). (Gen. iv.)
Isaye, Ysaias, 57.
Israel, 169.
Itayl, 198. Italy.
Juda, 62 (note). Jude.
Judas, 10.
Jude, 283. Judea.
Jurye (Jure, Jury, Jewry), 216.

Kate Kelle, 123.
Kytt Cakelere, 123.

Lamech, 29 (note). (Gen. iv.)
Lameth, 39.

Latyn, 15.
Lazarus (Lazare), 10.
Letyce lytyl trust, 123.
Leyon (Leon, Lyon) 231. Jewish judge.
Longeus (Longeys), 12. Longinus.
Luce lyere, 123.
Lucyfer (Locyfere), 1.
Luke, 14. (Lucas, 337.)
Lumbardye, 198.

Mabyle, 123.
Mahound (Mahownde), 154.
Malachel, 31 (note). Mahalaleel (Gen. v).
Malcheus (Malcus), 266.
Malkyn mylkedoke, 123.
Manasses, 61. (Matt. i.)
Martha, 210.
Mary (Maria), 6.
Maria jacobi, 327.
Maria magdalene, 327.
Mary Mawdelyn, 14.
Maria Salome, 327.
Matheus, 250.
Mathew, 270.
Mathias, 352.
Matussahel, 29 (note). Methusael (Gen. iv).
Matussalem, 31 (note). Methuselah (Gen. v).
Maunfras, 146. 2nd shepherd. Mauferas occurs as name of a devil in two French mysteries: *Le Martyre de S. Pierre et de S. Paul* and *Les Miracles de Ste Geneviève* (ed. Jubinal).
Mawdelyn, Magdalyn, 210.
Megge Mery weder, 123.
Melchizar, 152. 2nd king (of Tarys).
Melophat, 63 (note). Alternative name for the mother of Anne. (No source has been found for this name.)
Messy, 59. Messiah.
Miles the miller, 123.
Montana, 116. The hill country (Luke i. 39, Vulgate 'in montana').
Mors, 174.
Moyse, 147. 3rd shepherd.
Moyses, 3. Moses.
Mychael (Michael), 372.
Mycheas, 59. Micah.

Nachor, 37 (note). Nahor (Gen. xi. 22).
Nacor, 37 (note). Nahor (Gen. xi. 26).
Nasaphat, 63 (note). Alternative name for mother of Anne. ('Est tuus Anna pater Jozafath, Nazafath tua mater', *Rel. Ant.* i, p. 287. Cf. also Pedigree of our Saviour, Harl. 6148, fo. 114ᵛ: 'Ex vetusto libro manuscripto de vaticiniis et genealogis', which gives Isakar and Nazaphat as parents of Anne.)
Naverne, 198. Navarre (cf. *Croxton Play of the Sacrament*).
Nazareth, 96.
Neptalym, 198. Naphtali—seen from the Mount.
Noe, 3.
Normandye, 198.
Nycodemus (Nichodemus), 13.

Olyvet, 262.
Osyas (Ozyas), 61. Hosea.
Ozias (Ozyas), 60. Ozias (Matt. i).

Parys, 199.
Pernal Prane, 123.
Petyr (Petir, Peter), 11.
Petrus, 224.
Peyrs Pottere, 123.
Phaleg, 37 (note). Peleg (Gen. xi).
Pharao, 255.
Pharasy, 232. Land of Pharisees.
Phariseus, 202.
Pharysewys, 9.
Phelypp, 269. (Philippus, 250.)
Phelypp þe good flecchere, 123.
Pheso, 361. Ephesus.
Poperynge, 199. (Cf. *Sir Thopas.*)
Portyngale, 199. Portugal.
Poul, 270. Paul.
Powle, Paulus, 363.
Powle Pewterere, 123.
Pownteys, 199. ? Pontoise.
Pycardye, 199.
Pylat (Pilatus), 11.

Rachel, 68.
Raguel, 63 (note). (Tobit vi.)
Rebecca, 93. One of the Virgin's handmaidens. (Cf. Pseudo-Matthew.)
Reu, 37 (note). (Gen. xi.)
Rewfyn (Rufyne), 231. Jewish judge.
Reyse sclaundyr, 125.
Roboas, 59. Roboam (Matt. i).
Robyn rede, 123.
Romaynes, 245.
Rome, 198.

Saba, 152. Realm of Baltazar (Ps. lxxii. 10).

Sabyn Sprynge, 123.
Salamon, 59. (Matt. i.)
Sale, 37 (note). Salah (Gen. xi).
Salmana, 198. Country seen from Mount (cf. Ps. lxxxiii. 11).
Salome, 62 (note). 3rd husband of Anne.
Salome (Salomee), 140. The doubting midwife.
Samary, 350. Samaria.
Samaryan, 269. (?)
Sampson, 68.
Samuel, 63 (note), 68.
Sara, 68.
Sarazyn, 230.
Sarug, 37 (note). Serug (Gen. xi).
Sathan, 9.
Sathanas, 25.
Sawdyr Sadelere, 123.
Scottlonde, 199.
Scriba, 201.
Sella, 29 (note). Zillah (Gen. iv).
Sem, 36 (note).
Sephore, 93. One of the Virgin's handmaidens (cf. Pseudo-Matthew).
Seraphyn, 2.
Seruasius, 62 (note). Geruasius by mistake, grandson of Elyud brother of Elizabeth mother of John. (*Legenda aurea.*)
Sesar, 231.
Seth, 29 (note). (Gen. iv.)
Shem (Seem), 36.
Sophonye (Sophosas), 61. Zephaniah.
Spayn, 198.
Stevyn Sturdy, 123.
Susanne, 93. One of the Virgin's handmaidens (cf. Pseudo-Matthew).
Sybile Sclutte, 377.
Sybyly Slynge, 123.
Sym Somnore, 127.
Symeon Justus, 162.
Symeon Zelotes, 270.
Symme Smalfeyth, 123.
Symon, 62 (note), 250.
Symon leprows, 243.
Tarys, 152. Realm of Melchizar Tharsis (Ps. lxxii. 10).
Thadeus, 251.
Thare, 37 (note). Terah (Gen. xi).
Thomas of Ynde, 15.
Thom Tynkere, 123.
Tobie, 63 (note). Tobias for Tobit (Tobit i).
Tyffany Twynkelere, 123.

Veronica, 296.

Walys, 199. Wales.
Whatt at þe Welle, 123.
Wyttsunday, 15.

Ynde, 15.
Ypotan, 153. Realm of Jasper (source of name not found).
Ysaac (Isaac), 3.
Ysaias (Ysaie), 57.
Ysakar, 65. (Isakar, 73.) High priest.
Ysakar, 63 (note). Father of Anne ('Achar' name given in Pseudo-Matthew. Isakar and Nazaphat parents of Anne in pedigree.—Harl. 6148).

Zabulon, 198. Country seen from the Mount.
Zabulon 'þe devylle of helle', 58. 'Terra Zabulon et terra Nephthalem . . . populus qui sedebat in tenebris vidit lucem magnam.' In the 'Gospel of Nicodemus', cap. xiii. 9, the text in this form is applied to the Harrowing of Hell. Zabulon is thus associated with the 'regio umbrae mortis.' Zabulus occurs with the meaning 'devil' in *De Nugis Curialium.*
Zakarye, 116.
Zebedeus, 63 (note).
Zebee, 198. Country seen from the Mount (cf. Ps. lxxxiii. 11).
Zelomy, 140. The believing midwife.

GLOSSARY

This glossary cites, in most cases, the first occurrence only of the word in the text, and it includes only those variations in spelling which seem significant with regard to the date or the locality of the composition. Initial *ff* is given as *f*; ȝ follows *g*; *i* and *j*, and *u* and *v* are taken together.

a, 50, *have*; abene, 175, *have been*; abrowth, 281, *have brought*; affaylid, 40, *have failed*; ago, 186, *have gone*; asavyd, 176, *have saved*.
abey, 260, *obey*.
abhomynabyl, 100.
abought, 117, *about*; aboute, 39; abowth, 20.
abyl (v.), 72, *enable*.
accende (v.), 201, *kindle*.
acorde (v.), 58, *agree*.
ad, 14, l. 471, ? for *as*.
adon, 371, *adown*.
adrad, 68, *afraid*.
advowtrye, 203; adultrye, 9.
affendyn, 37, *offend*.
afferde, 196, *afraid*.
afflyght, 208, *afflicted*.
affye (v.), 357, *trust*.
aglottys, 227, *ornamental tags*.
agryse (v.), 36, *horrify*; agreseth, 307, *terrifieth*.
aȝen, 11, *again*; agayn; ageyn, 6.
aȝens, 2, *against*; ageyns, 21.
alkende, 184, ageyn nature and alkende, *kind*.
allether, 12, oure allether, *of all of us*; alderers, 322; althere, 217; altheris, 191.
almyght (adj.), 328, *almighty*. *See* myht, 359, l. 123, *mighty*.
alow, 253, *approve, accept*; alowyht, 33; alowe, 56, *permit*.
alwyse, 292, in alwyse, *in every way, inevitably*.
alyaunce, 211, syb of alyaunce, *related in blood*.
alye, 136, *kindred*.
alythe (v.), 297, *alight*. *See* lyght.
amat, 274, make ȝou amat, *dismayed*.
an, 76, *have*.
anameryd, 125, *enamoured*.
angelle (pl.), 1, *angels*; aungelle (pl.), 197.
anodyr, 64, *another*.
anow, 21, *enough*; inow, 253.
antecer, 80, *ancestor*.
anvemynyd, 67, l. 124, an vemynyd, *have poisoned*.
apayed, 59, *pleased*.
apparens, 255, to ȝour apparens, *in appearance to you*.
apperyd, 182, l. 122, ? *apportioned*.
applyande, 229, *applying*.
applyed, 239, *apply it*.
appose, 2, *confront, examine*; apposyn, 195.
aqwhyte, 49, *pay, acquit*; aqwhyte (pp.), 239.
arde, 196, *hard*.
are, 40, *hare*.
arere, 124, *raise*.
aresyn, 14, *arisen*.
arn, 30, *are*; ar, 95. *See* be.
arneys, 264, *harness, armour*.
arryn, 294, *torment, irritate*.
arwe, 40, *arrow*.
aseyth, 100, *satisfaction*.
ashyrlyng (adj.), 170, ? *a shrilling*. *See* shyrle, 113, *shrill*.
askuse, 2, *excuse*.
askyght, 144, *asketh*.
aslake, 224, oure deth ȝe may aslake, *cause to cease*. *See* slake.
asmatryk, 178, *arithmetic*.
asoyle, 34, *solve*.
assedually, 360, *assiduously*.
assumpte (? pt.), 354, ? *rose to heaven*; (pp.), 355.
ast, 50, *hast*.
astat, 11, *state*.
asyse, 53, þe last asyse, *judgement*.
atast (v.), 27, *taste*.
atent (sb.), 3, *entent*.
athreste, 179, *athirst*.
atreyd, 324, *vexed, afflicted*.
atwynne, 197, *in twain*.
auantorysly, 310, *by chance*.
aventure, 75, per aventure, *chance*.
auerte (sb.), 80, ? *averter*.

avexit, 348, *vexeth.*
augrym, 178, *arithmetic.*
aunge, 80, 115, *angel.*
avoyd, 123, *stand aside, depart.*
autere, 65, *altar.*
avyse (sb.), 47, *advice, counsel.*
avysement, 247, *advice.*
awe, 9, 203, vndyr þer awe, *power.*
awyse (v.), 47, *advise.*

baftys, 170, *behind.*
bake, 48, *prepare, make ready.*
balke, 317, be brook or balke, *mound.*
bane, 314, *destroyer.*
baptyme, 188, *baptism*; baptyze, 318, l. 1371.
bargany, 253, *bargain*; cf. gracy, 75, towaly, 260.
barne, 147, *child*; baron, 154, 157, 172.
barrany, 65, *barren.*
barynes, 68, *barrenness.*
bat, 11, *debate.*
batte, 277, *stroke, blow.*
bawmys, 58, *of balm.*
baye, 170, be bestys baye, *stall.*
bayle, 273, *jurisdiction.*
bayn, 163, *ready.*
be, 3, *by.*
be, 2, *to be*; bene, 2; beyn, 228; byn, 218.
be, 14, *are*; byn, 24; beth, 43. *See* arn.
be, 165, *been.*
bede (pt.), 33, *offered.*
bede (pp.), 167, *awaited.*
bedellys, 226, *heralds.*
bedene, 2, *forthwith.*
beetys, 19, *beet plants.*
begchis, 355, *bitches.*
beglete, 155, l. 125, ? *for* begete, *begotten*; cf. vnbegete, 258. *See* glete.
begownne, 110, *begun.*
behete (v.), 350, *promise.*
behovyth, 135, *belongs.*
belde, 17, *build.*
beleve, 170, *remain*; belave, 370; belevyth, 328.
belthe, 371, *evil.* *See* note in Dr. Greg's *Assumption of the Virgin.*
belyff, 171, *quickly*; blyff, 2; belyve, 192.
bemys, 107, *beams.*
bende (sb.), 153, *bond*; 169, *band.*
berde, 247, 361, *woman, lady.*
bere, 22, *bear*; berun, 351; bar (pt.), 70; bore (pp.), 69; beryght, 137, *beareth.*
berst, 181, *breast*; bryst, 12; brest, 14.
beryed, 10, *buried*; burry, 215.
beryels, 16, *graves.*
berynt, 294, 296, *for* beryn it. *See* levynt, reysynt.
beschop, 87; bushop, 4.
beseche, 65; beseke, 121, 358.
beseyn, 230, *beseen, arrayed.*
bestad, 69, 150, 305, *appointed, placed.*
betake, 65, *commit*; beteche, 63; betake (pp.), 363.
bete (sb.), 23, *bite.*
bete, 170, on bete, *unbeaten.*
bethwen, 92, *between.*
bett, 40, *better.*
bety, 268, *beat.*
bey, 228, *buy*; bye, 58.
bey, 227, *boy.*
blaberyn, 154. *chatter.*
blad (pp.), 172, *bled.*
blake, 2, 19, blewe and blake (colour of flowers), *yellow*; cf. blayke, *Pearl*, l. 27.
ble, 169, *complexion, colour*; bryth as ble, 17.
blewe, 2, *blue*; bloo, 311, *livid.*
blomyght, 136, *bloometh.*
blyff, 12. *See* belyff.
blyn, 313. *cease.*
blysse (v.), 44, *bless*; blyssyd, 4.
bobbyd, 307, *buffeted.*
bone, 37; boun, 137, *boon, prayer, request*; 25, *command.*
bone (adj.), 27, þi byddyng bone, ? *good, gracious.*
bonet, 228, *cap.*
boorde, 328, his burryenge boorde, ? *his burial feast.*
boot (pt.), 26, *bit.*
boot (sb.), 26, *remedy*; bote, 4.
borwe (v.), 37, *save.*
borwe (sb.), 68, *pledge.*
borys, 297, *bored holes.*
bothers, 25; þour bothers, 25; here botherys, 71; *of both*
bountevous, 225, *bounteous.*
bourgh, 135, bourgh and cety, *town.*
bow, 21, *bough.*
bowne, 157, *ready, willing*; 248, *destined.*
bowth, 11, *bought.*
boyst, 329, *box.*
brace (sb.), 248, *embrace.*
brake, 19, *fern.*
brast, 15, *burst*; breste, 68.
braunce, 359, *branch.*
brayde, 218, *short space of time.*
brede, 40, *breadth.*
breganderys, 264, *body armour for foot-soldiers.*
breng, 258, *bring.*

brent, 2, *burnt*; brenne, 248, *burn*; brynnyng, 188.
brest, 321, 339, brest ful bolde, *dauntless breast.*
brethellys, 287, *abandoned wretches.*
brethere, 364, *brothers*; bretheryn, 77.
brewe, 38, *bring about.*
bronde, 47, *brand, torch*; 173, *sword.*
bronston, 287, *brimstone.*
brook, 36, 197, brennyng brook, *hell.*
brothel, 205, *abandoned woman.*
brouth, 10, *brought*; brout, 10; brougth, 15.
brybory, 228, *bribery.*
brybour, 173, *scoundrel.*
brymmys, 152, *brinks*, by brymmys browne.
bryth, 1, *bright*; bryght, 33.
buske, 148, *set out, go.*
buxom, 46, *obedient*; buxhum, 94.
byche, 205, byche clowte. *See* bychyd.
bychyd, 368, *cursed, execrable*; bygyd, 368. *See* bicched, N. E. D.
bylle, 36, *list.*
bynne, 150, *manger.*
bysmare, 132, *shameless creature.*
byth (v.), 22, *bite*; boot (pt.), 26.
byware, 54, *beware*; bewhare, 54.

cadace, 227, *cotton wool.*
calabere, 228, *a kind of fur.*
call, 178, l. 24, be call, ? *by repute.* Cf. callyng, *proclamation, Cleanness*, l. 1362.
calsydon, 244, *chalcedony* (Rev. xxi. 19). *See* Note at end of Glossary.
cammaka, 154, *a fine fabric.*
care (v.), 363, *lament.*
careyn, 43, *carcass, corpse.*
carpynge, 147, *talking.*
cast (sb.), 121, *intention, purpose.*
cast (v.), 116, *intend, purpose.*
catel, 176, *property.*
celestly, 96, 367, *celestially.*
cenacle, 15 *See* List of Proper Names.
cevyle, 179, canon and cevyle lawe, *civil*; sevyle, 228.
chalys, 259, þese chalys, *chalice.*
chare, 303, 324, *drive away.*
chargyng (sb.), 89, *charge, command.*
charle, 131, *churl.*
chase, 20 (r. w. place), ? *for* chaste (v.), *discipline, correct.*
chastement, 228, *chastisement.*
chaumbyr, 39, *chamber*; chawmere, 107.
chauncel, 64, *chancel.*
chavyl bone, 33, *jawbone.*
chef, 228, l. 114, ? *chief.*
chene, 269, *chain.*
cheryse, 213, *cherish.*
chese, 48, *choose*; ches, 100, *chose*; chose (pp.), 133.
cheselys, 50, *pebbles.*
chete, 214, *sheet* (*corrected to* schete).
cheve, 151, *fare, succeed*; cheuith, 300, *befalls.*
cheverelle, 227, *kid leather.*
chevesauns, 228, *resource, shift.*
childely, 76, *childlike.*
chownys, 170. *See* schelchownys.
chyldere, 26, *children*; childyr, 74; childeryn, 7.
chyse (adj.), 170, *choice.*
clappyd, 316, clappyd in cloth; *for* clyppyd, *clasped, wrapped. See* N. E. D.
claryfieth, 96, *illumines*
clenche, 356, *affirm, insist.*
clepe, 63, *call.*
clergyse, 178, *learning.*
cleue (v.), 48, 143, *stick.*
cleue (v.), 156, *split*; clovyn (pp.), 40.
clos, 119, l. 98, ? in clos.
clowdys, 374, *clods.*
clymbyn, 170, ? A barn . . . wolde clymbyn kynge and knytys.
clyne, 106, *incline.*
clynge, 48, my hert doth clynge and cleue as clay, *shrink*
com, 29, *come*; cum, 63; comyn, 15; cam (pt.), 8; com (pp.), 65.
comawnde (v.), 278, *commend.*
comberaunce, 211, *trouble.*
combros, 325, *difficult.*
comeryd, 98, *cumbered, overwhelmed.*
comfortation, 108, *comfort.*
contownys, 236, 298, 357, *commons, people.*
compellyd (pp.), 42, coupellyd (*coupled*). Emended by J. M. Manly on authority of Kittredge.
con, 163 (r. w. son), *can*; kan . . . thanke, 249; cone . . . thanke, 274, *thank.*
conclave, 13, *locked chamber*; his dede conclaue, 13, *tomb*; my conclave, 97.
concludyd, 275, *overcome in argument.*
conseyve, 246, l. 442, ? *for* constreyn. *See* p. 262, l. 913.
consorcyte, 358, *fellowship* (*consortium*); N. E. D. consorce, earliest quotation 1512.
contenawns, 254, 265, be contenawns, *by gesture, signs.*
contewnyng, 244, *continuing*; contenue, 232.

convycte, 9, *prove in error*; convycte (pp.), 359, *overcome.*
coors, 214, *corse.*
corde, 241, of corde, *accord.*
cordewan, 227, *Cordovan leather.*
cornys, 32, *corn.*
coryous, 173, *elaborately wrought.*
cost, 173, helle cost, *region, country.*
cost, 32, croft and cost, ?*field and hill-side.*
costyous, 227, *costly.*
cote, 88, *cottage.*
coverte, 132, *covering.*
covetyse, 198, *covetousness.*
counawnt, 279, *covenant.*
countyrfe, 226, *contrive.*
cownde, 288, *could.*
cowthe, 96, *known.*
craggyd, 355, *rough.*
credyl, 129, *cradle*; cradyl, 179.
crenseyn, 227, *crimson.*
creppe, 36, 353, *creep*; crepe, 37.
cressetys, 254, *a form of lantern.*
crowch, 328, *cross.*
croyse, 147, *cross.*
cunnyng, 2, *knowledge.*
curat, 64, *pastor.*
curryd, 154, cruel and curryd, *for* cursyd, *curst, evil-tempered.*
curteys, 152, *courteous.*
curyng, 228, 364, *covering.*
cus, 81, *kiss.*
custom, 108, a custom ocupacion, *customary.*

dalyawnce, 127, *dalliance*; dalyauns, 203, *trifling*; 342, *talk.*
dampne, 9, *condemn*; dempt (pp., r. w. ment), 293.
datys, 21, *dates* (fruit).
daungere, 26, *danger*; 233, *power.*
dawe, 272, 275, bryng o (a) dawe, *slay.*
dawe, 279, the day dawe, *dawn.*
day, 236 (r. w. say), 313 (r. w. away), *die*; deye, 25; deyd (pt.), 222, *altered to* dyyd.
declinande, 229, *declining.*
dede (v.), 1, *did*; dyd, 7; dude, 129.
dede, 320, *death.*
defame (sb.), 127, *ill report.*
defendyd, 300, *for* offendyd. *See* defense *for* offence, N. E. D.
defens (v.), 58, *defend*; dyffende, 74.
defens (sb.), 121, without defens, *opposition.*
delacion, 234, *delay.*
delayde, 144, *mitigated.*
dele, 213, every dele, *bit.*
delve, 29, *dig*; dolve (pp.), 321, *buried.*
delyre, 193, *act madly.*
dene, 289, *din.*
dent, 153, *blow.*
depart (v.), 65, 204, *part, go asunder*; deth me xulde departe, 211, *dismiss.*
deppe, 36 (r. w. scleppe, creppe), *deep.*
deprave, 189, *decry*; dyspravyn, 324.
dere (v.), 54, *injure.*
derth, 139, *dearness, costliness*; dyrthe, 175.
derwurthy, 193, *precious.*
desideracyon, 357, *desire.*
dete, 96, *ditty, song.*
detent (pp.), 362, *kept away.*
deve, 156, *stun*; devid, 322.
devele (pl.), 139 (*see* aungelle), dewelys, 371, *devils.*
devoyde, 229, *withdraw*; devoydyng, 240, *casting out.*
devyr, 112, *duty.*
devys, 8, *mechanical contrivance*; devyse, 132, bettyr it is to telle þe trewth devyse, *for* þe trewe devyse. *device, trick.*
diffynicion, 100, *limitation.*
diplois, 204, in deploydo, *doublet.* In a statute of 1433 of the Cistercian Order ecclesiastics are bidden to abstain 'a gestatione diploidorum'—among other secular garments.
displesaunte, 45, to þi displesaunte, ?*for* to þe d., *displeasing.*
dissponsacion, 81, *betrothal.*
do, 169, *cause*; 226, I gan þer do play, *I did cause there to be played*; 275, we han þe do sowth, *we have caused thee to be sought.*
dobbelet, 227, *doublet.*
dodemvsyd, 368, ? *amazed.*
dompnesse, 117, *dumbness.*
domys, 64, *judgements*; 178, *prescriptions.*
dowcet, 21, *sweet.*
dowe, 3, 166, *dove*; dove, 43.
dowm, 118, *dumb*; dum, 193.
dowse, 83, *sweet.*
dowth (sb.), 77, *doubt*; doute, 9; dowhte, 85; dowt, 173, *fear*; no dowth of, 77, 88, *without doubt.*
doyl, 42, *dole, sorrow.*
doyst, 261, *dost.*
dragonys, 272, 287, dragons.
drawe, 40, *draw*; drowe (pt.), 225; drawe (pp.), 246.
dredynge, 26, deth dredynge, *fear of death.*
drepe (v.), 161 (r. w. slepe), *drop.*
dresse, 61, *prepare.*

dreve (pp.), 297, *driven*; drevyn, 60.
drewe, 32, 377, *morsel.*
drey (adj.), 159, *dry*; dreye (v.), 216.
dreynt (pp.), 39, *drowned.*
dronkesheppe, 377, *drunkenness.*
drynchyng (sb.), 38, *drowning.*
dur (inf.), 261, *dare.*
dure, 340, *hard.*
dwelle, 101, *abide.*
dwere (sb.), 15, *fear.*
dyet, 178, *diet, regimen.*
dyght, 40, *bring about, make ready*; dyth, 16; dyht (pp.), 31; dyth (pp.), 238; dyght (pp.), 162, *put, removed.*
dygne (sb.), 173, for dygne of my dygnyte þei haue of me dowt, ? *worth.*
dylexcion, 78, *love.*
dyng, 27, *strike, knock.*
dyngne, 154, *worthy.*
dyrknes, 164, *darkness.*
dyscres, 10, 211, *fall off, decrease.*
dysprave, 267 (r. w. rave), *disprove*; dysprevyd, 293; dyspravyn, 324. *See* deprave.
dyssend, 365, *descended.*
dystrye, 235, *destroy*; dystroyt, 231, *destroyeth.*
dyswary, 354, *doubt.*
dyvide, 241, *divided.*

echon, 19, *each one.*
edyfy (v. intr.), 237, *grow, prosper*, Mannys sowle in blys now xal edyfy; edyfied (pp.), 76, *built.*
eerly, 240, *earthly.*
effectuously, 352, *effectually.*
efne (sb.), 261, *heaven.*
egal, 260, *equal*; 255, eqwall.
elde (sb.), 65; (adj.), 255, *old.*
empere, 190 (r. w. manere), *emperor.*
empere, 307, *empire.*
empres, 108, *empress.*
enbrace, 87, 208, *embrace.*
encheson, 257; incheson, 108, *cause, reason.*
enclosyd, 227, ? hosyn enclosyd.
ende, 262, an ende to make, *agreement, compact.*
ende (adj.), 165, *for* hende, my savyour ende.
enform, 36, *inform*; 85, *explain.*
engynes, 226, *contrivances.*
enjonyd, 108, l. 340, *enjoined*; *or see* injouyid.
enjoyd, 117, *was joyful, rejoiced.*
ensure, 82, 112, *assure.*
entent, 5, *intention.*
enterly, 254, *entirely.*
er, 277, *before, ere.*
erbe, 19, erbys, 11, *herb.*
erde, 281, *heard.*
erde, 265, *earth.* *See* herd.
erdon, 263, *errand.*
erraunt, 192, *wandering, departing.*
erste, 98, *sooner.*
erte, 191 (r. w. smerte), *for* arte.
erthelech, 19; erthly, 27, *earthly.*
erthqwave, 307, *earthquake.*
ete, 20, *eat*; ete (pp.), 254 (r. w. fete), *eaten*; etyht, 251, *eateth.*
eternalyte, 269 (r. w. paraclyte), *eternity.*
ethe, 136, *easy.*
eve song, 120, *evensong.*
everich, 3; every, 13, *every*; everychon, 10, *every one.*
eueryth, 228, *or* eneryth, ? *inherit.*
evy, 46, *heavy.*
evyn-crysten, 77, *fellow Christian.*
evysum, 338, *heavy.*
exorte, 65, 241, *request, entreat.*
exort, 99, *issue, spring forth.*
expowne, 14; expownd, 85, *expound.*
exys, 254, *axes.*
ey (pl. ?), 75; eyn, 13, *eyes.*
eyd (sb.), 303, *heed.*
eye, 114, sen at eye; cf. *at hand.*
eyȝil, 302, *vinegar.*
eylight, 211, *aileth*; heylith, 131.
eylsum, 82, *wholesome.*
eynes, 98 (r. w. peynes), *eyes.*
eyre, 230, *heir.*
eyte, 122, *eight*; eyted, 76, *eighth.*

facyon, 370, *face.*
fad (pp.), 21, *fed*; fade, 159 (r. w. glade).
fade, 24, 51, *fade*; ffrom feyth nevyr þou ffade, 348, *grow weak, fall away.*
falfage, 34 (*for* falsage), *falsehood.*
falle (pp.), 98, *fallen.*
falle, 195, 327, ? *cause to fall*; felle (v.), 58, *fell, cast down.*
falsed, 9, *falsehood.*
fame (v.), 131, *defame.*
famyt, 97, *famished.*
fast (pt.), 226, *fasted.*
fawe, 274 (r. w. lawe), *fain, glad.*
fawte, 64, *fault*; for fawth of, 259, *default.*
fay, 14, in fay; feye, 299, *faith.*
fe, 173, ffryth and ffe, *property.*
fekyll, 329, *treacherous.*
felas, 67, 275; felawus, 228, *fellows, equals.*
fele, 21, 50, 345, *many*; in fele degre, 9.
felle, 177, 320, *skin.*

fenaunce, 210, *end.*
fend, 58; fynde, 9, *fiend.*
fende (v.), 248, *protect.*
fenne, 157, 248, *fen, marsh.*
ferder, 231, *further.*
fere, 84, *companion*; in fere, 46, *together.*
fere (sb.), 19; fer, 23, *fear*; fere (v.), 64, *to fear*; ferying, 68 (*gerund*).
fere, 7, *far.*
fere, 15, 68; feyr, 2; fyer, 18; fyre, 19, *fire.*
ferforthe, 118, so ferforthe ffor þat (*ecce enim ex hoc, for behold from henceforth*), *for* lo forsooth; cf. Wycliffite version, Luke i. 49.
ferly (adj.), 15, *marvellous.*
ferne (r. w. sterne), 158, be feld and ferne, ? *fern.*
fers, 76 *fierce.*
ferth, 161, *forth, forwards.*
ferthe, 19, *fourth.*
feryth, 96, *causeth to fear, aweth.*
fest (v.), 372, *fasten.*
fest (sb.), 15, 63, *feast*; ffeste and food, 326; *feast* or *fast*, 303, l. 894.
fett (v. inf.), 215, *fetch*; fettyn (3 pl. pres.), 371.
fewte, 329, *fealty.*
feyn, 231, *feign, make a pretence.*
feyn, 328, The naylis gun his lemys feyn, ? *make faint.* *See* feynnesse.
feyn, 227, *fine.*
feynnesse, 39, *faintness, weakness.*
feyt, 304, *feet.*
fise, 357, ? *term of abuse.*
flamke, 160, *flank.*
flem (v.), 262, *drive away.*
flescly, 24; flesly, 24; fleschly, 76, *fleshly.*
flokkys, 227, *stuffing for doublet.*
flomjordon, 8, *river Jordan.*
florens, 157, *florins.*
flowe (sb.), 38, *flow, current*; flowe (sb.), 319, *flood.*
flyth (sb.), 2, *flight.*
flytt (v. inf.), 176, *save, cause to flee*; flytt (pp.), 248; flytt (inf.), 315; flitte, 340, *flee, depart.*
fode, 305, now xal wepynge me fode and fede, *supply food to.*
fol, 310; fole, 295, *full.*
fole, 284 (r. w. to þe), *folly.*
folwe, 11, 36, *to follow*; folwygh, 202, *followeth.*
fon, 10, *foes.*
fond (v.), 22, *try, attempt*; fonde, 28, l. 386, leve spowse now þou fonde, *findest, provest.*
fonge (v.), 36, *begin, undertake.*
fonnyng, 284, *foolishness, madness.*
fonnys, 340, *fools.*
fonnyst, 32, *art mad.*
food, 140, *person, creature*; 185, *child, offspring.*
footmayd, 64, *waiting-maid.*
fop, 276, *fool.*
fordere, 226, *further.*
forfare (v.), 42, *be lost, come to destruction.*
form faderys, 256, *forefathers.*
forme, 202, *form of a hare.*
formere, 49, *former, creator.*
fors, 316, with fyne fors, *by main force*; 32, I ȝeve no fors, *I attach no importance to.*
forsake, 2; forsakyn (inf.), 36; forsake (pp.), 264, *avoid, forsake, deny.*
fortyfye, 226, his deth to fortyfye, *secure, ensure*; 292 (v. intr.), *resist, stand firm.*
foulyng, 286, *wretch.*
fourtene, 77, *fourteenth.*
fowth, 76, *fought.*
foyson, 58, *abundance.*
fraught, 129, *freight, burden.*
fre (adj.), 3, *noble.*
freke, 27, *man, fellow*; 303, þe fend þat freke.
frelnes, 101, *frailty.*
frelte, 195, *frailty.*
frensche, 110, *French.*
fryght, 149 (r. w. kyth), *for* fryth.
fryth, 138, *wood, forest.*
fullich (adv.), 58; fulleche, 263, *fully.*
furryd, 228, furryd with ermyn, calabere or satan, *lined.*
fyftene, 77, *fifteenth.*
fynyaly, 102, *finally.*
fynyte, 82, for *infinite.*
fyth, 228 (r. w. delyte); fytyn (inf.), 298, *fight.*
fytt, 176, *tune, song.*

gadere (3 pl.), 9; gaderyn (inf.), 21; gaderyd (pp.), 15, *gather.*
galaunt (sb.), 125, *gallant.*
gan, 2; gonne, 147; gun, 10; gunne, 319 (aux. v.), *did.* *See* gynne.
gast (2 sg.), 323, *gavest.* *See* gyff.
gate, 26; ȝatys, 28, *gate(s).*
gate, 46; ȝatt, 327, *way*; al gatys, 291, *at all costs.*
geawnt, 13, *giant.*
gef (inf.), 231, *give.* *See* gyff.
geneologye, 61; genealogye, 62, *lineage, pedigree.*
gent (adj.), 127, *pretty.*

gentyl, 4, *noble*; jentylle, 59.
gentyllys, 1, *gentles.*
gerlys, 171 (knave gerlys), *children, young people.*
gerth (r. w. byrth), 175, on ground nere on gerth, ?*garth.*
geste, 155, *event, tidings.*
geste, 161, *guest.*
gesyne, 141, *childbed.*
gett, 302, *fashion, manner.*
geyne, 368, *meet, oppose.*
glabereris, 368, ?globaris, *gluttons; or for* glavereris, *deceitful talkers. See* Dr. Greg's *Assumption of the Virgin.*
glathe, 161, 222, *welcome, gladden.*
glent (pt.), 361, *shone.*
glete, 155, *mud, slimy matter. See* Note at end of Glossary.
gleterynge, 152, *glittering.*
gleyvis, 254, *glaives, weapons.*
glose (sb.), 8, *gloss, commentary*; 373, *explanation.*
glyde, 26, *crawl*; 155, *go, pass.*
gnaggyd, 355, *knotted, tied up.*
godhyd, 81, *godhead, deity.*
godnysse, 94, *goodness.*
gomys, 169, 355, *men, fellows.*
gonge, 319, in helle gonge, *pit, latrina.*
gouernayl, 361, *government, direction.*
grace (sb.), 9; gracy, 75; gracys, 79; gras, 82, 94, *grace.*
graffe, 314 (r. w. saffe), *grave.*
grame (v. inf.), 24, 202; gramyd (pp.), 133, *make angry, enrage.*
grame (sb.), 2, *wrath, anger.*
gramercy, 30; gromercy, 242, *thanks.*
gramly, 368, *fiercely.*
gratulacyon, 357, *thanks.*
graunt (pp. r. w.), 209, *granted.*
graue (pp.), 214, 223, *buried.*
grecys, 77, *steps, stairs* (Psalms 119–133, Wyclif. Version, The Song of Grees *or* Greces).
grede, 171, 334, *cry out.*
grees, 74, *steps. See* grecys.
greff (r. w. sheff), 32; greve (r. w. preve), 143, 151, *grief*; at greff, 253, *amiss.*
gres, 26; gresse, 19, *grass.*
gret (pt.), 115, *greeted, saluted.*
grote (sb.), 88, *groat.*
ground, 14; grounde, 34; grownd, 32; grownnde *or* growunde, 316, 347, *ground.*
grownde (pp.), 139, *grounded.*
grugge (v.), 215, *grudge.*
grw, 161, *Greek.*
grylle (adj.), 217, *fierce.*
gryse, 149, grysly as a gryse, ?*pig, boar.*
grysly, 149, *horribly.*
gryth, 7, *peace.*
gyff, 98; gyf, 226; gef, 221, *give*; geve, 252; gevyth, 77; gast (2 sg.), 323.
gyldyn, 69, *golden.*
gynne (1 pl.), 16; gynnyth, 1; gun (pt.), 10, *begin.*
gynne (sb.), 39, *gin, trap, plan of destruction.*
gynny, 156, ?*scribal error; altered by reviser to* wonyt.
gyse (sb.), 11, *way, manner of life.*
gysse (v.), 360 (r. w. wysse), *guess.*

ȝa, 38, 67, *yea, yes.*
ȝardys, 86; ȝerde, 5, *rod(s).*
ȝede (v. pt.), 182, *went.*
ȝelde (v.), 91, *yield, give.*
ȝemanry, 1, *yeomanry.*
ȝenge, 139, *young*; ȝonge, 182.
ȝer (pl.), 8; ȝere (pl.), 4; ȝerys, 72, *years*; yer, 374.
ȝerd, 278, *garden.*
ȝevene, 165; ȝevyn, 167, *even.*
ȝewys *for* jewys, 307.
ȝiff, 102; ȝyff, 64; ȝyf, 252; ȝef, 272, *if.*
ȝit, 13; ȝitt, 30; ȝyt, 10, *yet.*
ȝonge, 44, *youth*; ȝough, 76; ȝoughȝ, 183.
ȝyf (v. 1 sg.), 29, *give*; ȝevyn (inf.), 32; ȝeve (inf.), 54; ȝevyn (pp.), 21; ȝovyn (pp.), 77; ȝove (pp.), 73; ȝaff (pt.), 103; ȝove (pt.), 144; ȝovyn (pt. pl.) 338.
ȝyft, 31, *gift.*
ȝyng, 71; ȝynge, 59, *young*; cf. ȝenge.
ȝys, 276, *yes.*

[Words having an adventitious initial aspirate, not otherwise needing explanation, are not included.]

ha, 34, *have*; han, 33; haht, 318, *hath.*
hald, 293, *hold*; halt, 293, *holdeth.*
halpe (v.), 301, *help.*
halse, 300, *embrace.*
halwe, 54, *keepholy, hallow*; halwyd, 96.
hand-maydeȝe, 118, *handmaiden.*
haras, 138, hous of haras, *stable for a stud of horses.*
hardely, 314, *certainly.*
harrow, 199, *the devil's cry.*
harwere, 150, *harrower.*
hat (sb.), 11, *hate.*
hattyht, 147, *is called. See* hyght.
hay, 154, *net for catching wild animals*, or *hedge.*

hede, 10, *hide*; heyde, 41; hede (pp.), 159, *hidden*.
hedyr, 67, *hither*; hider, 361.
hefly, 240, *heavenly*.
hele, 211, *health*.
helpyngys, 116; cf. Wyclif. Version, 1 Cor. xii. 28, aftirward graces of heelingis, helpyngis, . . . kyndis of langagis.
hem, 2, *them*; þem, 2.
hem, 159, 228, *him*.
hende, 5, *gracious, gentle*; hynde, 57, 113; hende, 358, *near at hand*; ter and hynde, 113, *far and near*; hendyr, 81, *nearer*.
hens partyng, 31, *death*.
hent, 87, *take, hold*; hent (pp.), 11.
herborwe (sb.), 137, *shelter*; (v.), 227; herboryd (pp.), 138.
herd, 264, *earth*.
herde, 147, *herdsman*.
here, 2, *their*.
here, 4. *her*; hire, 71.
here, 227, *hair*.
herne, 371, sere sathan in the herne, *corner*.
hert, 40, *hart*.
hertely, 36, 122, *hearty*; hertyly, 31, *heartily*.
heryght, 327, *heareth*.
hese, 1, *his*; *more usual form* his.
hest (v.), 40, *promise*; hestyd (pp.), 85.
hete (sb.) (r. w. mete), 175, *hit, blow*.
hevy (v.), 93, *make heavy*; hevyin, 363, *grow heavy*.
hey, 44, *high*; heyȝ, 131; hieȝ, 135; hy, 11; high, 58.
heyd (sb.) (r. w. dede), 49, *heed*.
heyn, 169, *villain, wretch*; cf. Chaucer, C. Y. Tale, 1319.
heyned, 76, *lifted*.
ho, 68, *who*; hoo, 52.
ho, 100, byddyth him ho, ? *bid him cease*.
hoberd, 169, 302, *a term of reproach*, ? *clown*
holond, 227, *holland cloth*.
hond, 22 (r. w. husbond), *corrected from* hand.
honge, 303, *hang*; hyng, 12; henge (pt.), 338; hynge (pt.), 320; hangyn (pp.), 236.
hont (v.), 154, *hunt*.
hool, 16, *whole*.
hors, 246, 272, *horses*.
howeth, 89, *oweth, ought*.
howlott, 169, *owl* (*here used for woman*).
howte, 172, *hoot*; howtyn, 169.
hurle, 368, *strive, contend*.
hy (v.), 6, *hast*; hyȝ, 27.
hyde, 303, *skin*.
hyght, 6; hyth, 15; hygth, 69; *be called*; hyth, 116, *was called*. *See* hattyht.
hyȝht, 7, 87, 165, in hyȝht, *in haste*; on hyght, 107, ? *in excelsis*.
hylle (adj.), 248, *ill*.
hylle (v.), 33, 238, *cover*; hylte, 374, *hidden*.
hyn, 298, *him*.
hynde. *See* hende.
hyth, 25, *promised*.
iche, 86, *each*; ilke, 170.
ichon, 135, *each one*.
jematrye, 178, *geometry*.
jentylle, 59, *noble*; 109, *gentle*.
ierarchie, 98, *hierarchy* (*of angels*).
jewus, 301 (r. w. þus); jewys, 301 (r. w. þis), *Jews*.
i-fownde (pp.), 149, *found*.
i-knowe (pp.), 132, *known*.
ilke, 170, *each*.
inbassett, 70, *embassy*.
incheson, 108, *occasion, cause*; encheson, 251.
indeployde, 204 (Latin st. dir.). *See* diplois.
indute, 193, *clothed*.
inhabith (pp.), 354, *settled, established*.
injouyid, 118, *rejoiced* (*exultavit*); enjoyd, 117; injoyeth, 358; injoye, 373. *See* enjonyd.
inow, 253, *enough*; 21, anow.
inportable, 272, *unbearable*.
i-nym (pp.), 149, *taken*.
irke (v.), 168, *grow weary*.
juge, 116, hese juge, ? *See* Intro., p. xliii.
i-wys, 7, *certainly*; i-wus (r. w. þus), 122; wys *for* i-wys, 73, l. 32.
kachyd, 274, *caught*.
kage, 153, 156, *scaffold, pageant, seat of honour* (*Prompt. Parv.* cage, catasta); cf. myn hyȝ cage, *Hymns to the Virgin and Christ*, ed. Furnivall, E. E. T. S; castell and cage, *Castle of Perseverance*.
kan, 140, þat kan ȝour good, *know what to do*.
kayser, 173, *emperor*; caysere, 190.
ken (v.), 117, *acknowledge*.
kend (sb.), 148, *kind, nature*; kendely, 30, *natural*; kende, 80, *race*. *See* kynde.
kepe (sb.), 171, *care*.
kepyng knyght, 34, *keeper, protector*.

kerchere, 49; kerchy, 296, *kerchief.*
kest (pt.), 278, *cast.*
knad, 355, craggyd knad (Halliwell, *knife*), ? *for* knag, *stump*; hence *bludgeon.*
knawe (r. w. awe), 79, 355, *know*; knawe (r. w lawe), 284, *known*; knove, 267 (r. w. lowe), *know.*
knawe, 177, 272, *gnaw.*
knelande, 355, *kneeling*; knelende, 67.
knett (pp.), 326, *knotted, fastened*; knyt, 17, *joined.*
knowlych (v.), 60, *acknowledge*; knowlage, 115.
kure (v.), 49, *cover. See* curyng.
kynde, 119, *generation* (*progenies*); 39, *species.*
kynde (adj.), 149, *natural, native.*
kyngdham, 229, 350, *kingdom*; kyngham, 282.
kynnys, 110, in no kynnys wyse, *by no means.*
kyrke, 168, *church*; cherch, 55; chirch, 121.
kyth (sb.), 149, kynde in our kyth, *kindred.*
kythe (v.), 170, *make known, show*; kydde (pp.), 45; kyd (pp.), 125, *well known.*
kytt (v.), 345, *cut.*

lacche (v.), 26, *gain, receive.*
lak, 129, withowtyn lak, *without fail.*
lake, 288, in lake, *lake of Hell*; 324, 333, 334, 370, *pit, grave*; 359, lake of lyonys, *pit.*
lappe, 343, so nere our lappe, *so near our persons.*
lappyd, 117, *closed, bound.*
las (r. w. was, pas), 14, *less.*
las (sb.), 25, *net, snare.*
lasse, 19, *lest.*
lat (imp. s.), 47; late (imp. p.), 65, *let.*
latyng, 243; lettyng, 4, *hindrance.*
laue (r. w. haue, raue, saue), 91, whyte as laue. *See* Note at end of Glossary.
lave (r. w. knawe), 355, *law*; lawe, 4; lay, 55, 377 (lede no lecherous lay), *rule of life, conduct.*
lawhs, 125; lawgh, 141, *laugh*; lowh (pt.), 21.
lay, 154, the lesse lay, *laity, ignorant people.*
lech (v.), 179, 192, *heal, cure.*
lechory, 42; lychery, 228, *lechery.*
ledys, 173, *peoples.*
lees, 328; les, 331, *falsehood, deceit.*
lef (v.), 32; leff, 164, *leave.*
lefful, 82, *lawful.*
leke (adj.), 289, *like*; lycke, 274.
lely, 90, lely whyte, *lily.*
lemys, 163, *limbs*; lyme, 160.
lenage, 183, *lineage, pedigree*; lynage, 58.
lende (v.), 160, *land, arrive.*
lenger, 81, *longer.*
lenyall, 60, *lineal.*
lere (v.), 24, *learn.*
lese (inf.), 48, *lose*; lore (pp.), 56; lorn (pp.), 42.
lesse than (conj.), *unless.*
lest (v.), 20 (*corrected from* last, r. w. best); leste, 68; lestyght (3 sg.), 92, *last, endure.*
lest, 126, 179, *list, listen*; leste (imp.), 84; lyst (imp.), 45.
lesyng, 4, *deception.*
lete (v.), 5, *cause*; lete, 28, late, lett, 72, *let.*
lete (v. pt.), 16, *leave.*
leve (v.), 233, *believe.*
leve (inf.), 26, *live*; levyn (inf.), 26; lyff (inf.), 30; levyth (3 sg.), 47; levyd (pp.), 71.
leve (adj.), 28, *dear*; levyr, 30, *rather.*
levers, 106, *livers.*
levyn, 146, *lightning.*
levyng, 29, 31, *living, life.*
levynt, 258, i. e. levyn it, *leave it*; cf. reysynt.
levyth (pp.), 256, *left.*
lewd, 62, *ignorant*; lewdness, 144, *ignorance.*
leyke (v.), 148, *go quickly.*
leyn (inf.), 276, *lay, pledge*; leyn (inf.), 324, *lay, cast down.*
leysere, 184, *leisure*; 298, leysere seyng, *slowly, mournfully* (N. E. D. *leisurely*).
lofsummere (adj.), 151, *more lovesome.*
lofte, 76, on lofte, *aloft, on high.*
logge, helle logge, 26, 176, 305, 319, *prison of hell.*
lokygh, 154, *looketh.*
lokyn (pp.), 17; loky *for* lokyn, 26, *locked.*
longe (v.), 36, *belong*; longyht (3 sg.), 57; longygh (3 sg.), 199.
loof, 195, *loaf*; 345, loff; ? 91, laue.
losel, 33, *scoundrel.*
loth (sb.), 329, *evil*; loth (adj.), 338, *wicked.*
loveday, 103, *day of agreement.*
lovnesse, 357, 359, *lowness, humility.*
lowlyte, 329, ? *for* lewte, *loyalty* (r. w. fewte).

lowte (v. 3 pl.), 53; lowth (inf.), 156, *bow.*
loyn (pp.), 13, 97, *lain*; loyn (1 pl. pt.), 274, *lay.*
lullay, 129, *lullaby.*
lullyd (v. int. pt.), *slept peacefully.*
lurdeyn, 41, *lazy person.*
lyberary, 80, we xal lerne you þe lyberary of oure lordys lawe, *the whole of, the canon.*
lyce (v. 3 sg.), 85 (r. w. servise), *lies. See* lyggyst.
lyche, 154 (r. w. dich); lycke, 274, *like. See* leke.
lyfflad, 1, *life.*
lyfte (pp.), 75, *lifted.*
lyfte (adj.), 294, *left (hand).*
lyggyst (2 sg.), 150, *liest*; lyght (3 sg.), 140; lyce (3 sg.), 85; lyth (3 sg.), 7, *lies*; lyne (inf.), 214 (3 pl.), 172.
lyght (v. pt.), 320, I lyght out of, *lit, alit*; lyth (pp.), 5; (inf.), 241; lyhtyn (inf.), 157.
lympe, 160, *lump.*
lymyd, 56, handys lymyd; cf. *later expression* lime-fingered, *given to pilfering.*
lynacion, 178, *measuring.*
lyste (sb.), 161, 309 (r. w. resquest), *desire, liking*; lyst (v.), 67.
lyst, 235, *lest.*
lytenyth (3 sg.), 96, *lighteneth, illumines.*
lyth (sb.), 17; (adj.), 68, *light.*
lyty *for* lytyl, 147.

maculacion, 133, *spot, stain,* fig. *defilement* (1st example N. E. D.).
make (sb.), 19, *mate.*
make (inf.), 1; makyn (inf.), 3, *make*; made (pt.), 1; mad (pt.), 58; makyght (3 sg.), 172.
males, 9, *malice.*
mankende, 53, 60, *mankind, men*; 163, 302, *human nature.*
mankynne (r. w. begynne), 148, *mankind.*
mansclawth, 290, *manslaughter, murder.*
many, 87, *for* man.
marryn (inf.), 153, *mar, destroy.*
massage, 43, *messenger.*
massage, 108, *message.*
mast (v.), 219, *mayest.*
mavmentryes, 371, *idolatries.*
mayn (sb.), 22, *power.*
maystrye, 179, *mastery*; maystryes, 199, 226, *cunning tricks.*
me, 220, *men.*
meche, 68, *great (much).*
mede (sb.), 16, *reward*; mede (v. inf.), 50.
medele (v.), 34, *have to do with*; 148, *mingle.*
medylerth, 26, *earth, world.*
meke (v.), 97, *make meek, temper.*
mekell, 159, *great*; mekyl, 18, 125; mykyl, 172.
melle (v.), 184, *declare, utter.*
melle (v.), 18, *mingle.*
melyon, 266, *million.*
men (v. inf.), 87, *mean.*
mende (sb.), 7, *mind* (haue mende = *know*); mendys, 226, *thoughts, memories.*
mene, 108, l. 327, *mine, my.*
mene (sb.), 166, 364, *mediator.*
mene (sb.), 369; meny, 254, *company, gang.*
mercyabyl, 100, *merciful.*
mere (sb.), 162, 329, *boundary,* hence *way, place.*
merke (v.), 195, *darken.*
merveyl (adj.), 90; mervaylle, 224, *marvellous.*
mete (sb.), 8, *food, meal.*
meth, 147 (r. w. breth, deth), *moderation, gentleness.*
methowut, 274, *methought.*
meve (v.), 192 (r. w. greve); meef (r. w. breff), 229, *move.*
mo (adv.), 26; moo (adj.), 109, *move.*
mokador, 179, *bib, napkin.*
molde (sb.), 1, *earth, world.*
mon (sb.), 110, *moan.*
mon (aux.), 277, *may, shall.*
monyth, 106, *month.*
moote (sb.), 4, *dispute.*
morny (adj.), 97, *mournful.*
mote (aux.), 45, *may*; moty *for* mot I, 29.
mothalle, 278, *moot hall, judgement hall.*
mow (aux.), 99, *may*; mowne (pl.), 374, *may, can.*
mowe, 302, *grimace.*
mownth, 294, *mount.*
mullinge, 151, *term of endearment* (1st example N. E. D.).
mure, 358, 370, *demure, modest.*
muse (v.), 131, *think, consider.*
muste (sb.), 353, *new wine.*
mut *for* mot, 65, 118, *may.*
myche, 153, on a myche? *See* Note.
myche, 356, *much.*
myght (sb.), 35; myghtys, 73; myth, 1, 226, *might.*

myht (v.), 31; myth, 2, *might.*
myht, 359, *mighty.* *See* almyght.
mynstrelle (pl.), 152; mynstralle (pl.), 174; menstrelle (pl.), 176, *minstrels.*
myrable, 360, *wonderful.*
myrke (adj.), 161, *dark*; myrkenes (sb.), 217, *darkness.*
mysse (sb.), 38, *wrong, injury.*
mys (v.), 7; mysse, 44, *miss.*
myscheve (v.), 127, *to come to grief, go astray*; myschevyd (pp.), 100, *in misery.*
myschyf (sb.), 18, *evil plight*; myschevys, 98, *miseries.*
mystyr, 132, *kind.*

nale, 54, at þe nale, *at the ale (house).*
nat, 25; natt, 56, *not.* *See* noth.
ne, 20, *nor.* *See* nere.
negramauncye, 178; nigramansye, 266, *witchcraft, necromancy.*
nemene (v.), 164 (r. w. evene), *name*; nempne (r. w. hevynne), 47.
nere, 40, 228, *nor.*
nesch, 28, *soft, tender, weak.*
nest, 107, *next.*
neyhand (adj.), 162, *near at hand.*
neyth (v. inf.), 226, *draw near.*
neyther (. . . nor), 32; neythyr (. . . nore), 53, *neither.* *See* nowther.
nome (pp.), 88, *taken.* *See* i-nvm.
norche (inf.), 196, *nourish*; norchyth, (3 sg.), 77.
norsshere, 225, *nourisher.*
norture, 29, 358, *nurture.*
noth, 11; nott, 67; notht, 109; notwh, 299; notȝ, 305; nought, 24; nowght, 41; nowth, 11, 97 (r. w. wrought); notwth, 233, 267, *not.*
novelle, 346, *news.*
nought, 75; nowht, 38; nowth, 20, l. 126; not, 68, *nought.*
nowther (. . . ne), 56; nother (. . . ne), 87; noyther, 253, *neither.* *See* neyther.
nowthty, 34, *worthless.*
noyis, 76, *noise, sound.*
noyous, 76, *grievous.*
nyghe (v.), 156; nyhyn, 359, *draw near.* *See* neyth.
nyn . . . nyn, 62; nyn, 117, *nor.*

o, 2, *one*; oo, 17; on, 21.
obecyon, 281, non obecyon of errour, *obstacle* (Fr. *obicion.* *See* Godefroy, *Dict. de l'anc. langue française*). Not in N. E. D.
obeschyauns, 357, *obedience*; obeschaunce, 367; obedyens, 57, 75.
oble, 255, *sacramental wafer.*
oblocucyon, 62, *bad delivery* (N. E. D. in this case only); or *interruption* (Halliwell, Latin obloqui, *to interrupt a speaker*).
ocapye, 352, *occupy*; occapyed, 81.
odyr, 264, *other.*
ogyl, 368, *shudder.*
olyff, 186, *in life, alive*; on lyve, 323.
onethys, 137, *with difficulty.*
onpossible, 180, *impossible.*
ony, 96, *honey.*
onyth, 227, *at night.*
or (conj.), 48, *before.*
ordenaryes, 79, *ecclesiastical officials.*
ore, 71, *grace.*
ore, 299, *for* oure, *our.*
ostage, 137, *hostelry.*
outh, 359, 367, *aught*; owght, 81; owgth, 86; owught, 115.
outrage (v. int.), 127, *commit excess, transgress.*
ouyr (v.), 8, *hover.*
ovyrdon (adv.), 21, *exceedingly.*
ouyrlede, 245, *domineer over* or *lead astray.*
ovyth, 46, *behoveth.*
owe (adj.), 25, *own.*
owth, 79, *out*; owughte, 131 (r. w. aboute).
owtrage (adj.), 35, *violent.*
owyght, 92, *oweth, should.* *See* howeth.
oyle, 262, oyle of mercy.

pace (v.), 12, *pass.*
pad, 154, *toad*; padde, 175.
page, 13, *slip for* pagent.
page, 174, *fellow, knave.*
paphawk, 154, 169, *suckling* (only instances given in N. E. D.); Halliwell *parrot* (? *for* papjay, popinjay).
paramoure (sb.), 44, *darling,* (Abraham, of Isaac)
parayl, 231, *equipment, ways and means*; 253, *apparel.*
parlement (of hefne), 82, 264, *Divine Council.*
parochonerys, 64, *parishioners.*
partabyl, 259, *capable of sharing.*
pas (sb.), 74, *pace, step.*
pasche, 244, *Easter.*
passage, 106, 108, *period of pregnancy.*
passent, 345, we went with hym evyn passent, *journeying.*
pawsacion, 81, *pause.*

pay, 43, *pleasure, satisfaction.*
payd, 72, *pleased.*
pelle, 158, 232, *fur, furred cloak.*
pende, 169, prevyn pychyn and topende; 302, with stronge peyn yt gynnyth to peynde, *press, pinch,* (N. E. D. dial.).
pepyr, 20, *pepper.*
perchyn, 225, *pierce*; perysch, 181.
perdure, 239, *continue.*
pere (v.), 123, *appear.*
pere (sb.), 20, *pear.*
persevere, 106, *for* perservere, *preserver.*
pertly, 1, *briskly, readily.*
perverte, 232, *perverteth.*
perysch, 181, *pierce.*
pet, 371, *pit.*
pete (sb.), 9, *pity.*
pete (v.), 25 (r. w. ete), *put*; 155 (r. w. wete), *put* or *pity.*
peusawns, 245, peusawns of pepyl, *crowd, number of.*
peys, 223, *weight.*
pillid, 355, *bald.*
plasmacion, 180, *fashioning, creation.*
playn, 207, telle us þe playn, *the plain case*; 237, *openly.*
pleand, 62, *acting*; pleyand, 355.
plentevously, 35, *plenteously*; plentevous, 227.
plesande, 229, *pleasing.*
plesawns, 17, *pleasure.*
pleson, 101 (r. w. reson), *pleasure* (not in N. E. D.).
plete (v.), 175, *plead.*
pleyn (v.), 197, *act.*
pleyne (adj.), 157, poundys pleyne, *full*; 356, 370.
pleynge fere, 107, *playfellow.*
pleynyn, 112, *complain.*
ply, 232, *work at, carry out.*
plyght (v.), 144, *fold.*
plyth (v.), 2, *promise*; plyghtys (1 sg.), 170; plyght (pp.), 56.
plyth (sb.), 12, *plight, state*; plyght, 37.
poer (sb.), 282, *power*; pore, 119.
poer (adj.), 245, 249, *poor*; pore, 24; power, 144, 249; porys, 99, *of the poor.*
popetys, 169, *dolls, puppets.*
portature, 30, *figure, form.* (*See* N. E. D. portraiture.)
portys, 50, *gates.*
postelis, 372, *apostles.*
povert, 64, *poverty.*
pouste, 17, *power*; pooste, 339.
powndys, 157, *ponds for fish.*
praty, 96, *pretty.*
pray, 300, to delyuere man fro þe develys pray, *action of preying, capture* (earliest instance in N. E. D. 1523); pray, 204, *prey.*
prayr, 89 (r. w. her), *prayer.*
prayt, 233, *prayeth*; preyand, 355, *praying.*
precepte (pp.), 255, *prescribed* (earliest instance in N. E. D. 1534).
pref (v.), 289 (r. w. belef), *prove.*
prendyd, 175, ?*taken*; Halliwell, *pricked* (?*for* prened).
prerogatyff, 107, *prerogative, peculiar right.*
present, 235, *now, at this place.*
prest, 11, *ready.*
presyn, 346, *prison* (*altered from* preson *to rhyme with* aresyn).
pretende, 74, *aim, direct*; 358, *hold out, offer*; 359, ?*make an attempt*; 362, pretendist, *intendest.*
prevyde (v.), 102, *look* (meaning not given in N. E. D.).
prevydens, 81, ?*for* provydens, *provision.*
prevyn, 154, 169, *put to the proof, test.*
prise, 37, 149, of prise, *excellent, precious.*
promyssyon, 58, *promise*; promiscion, 350.
pronunciation, 62, *delivery, declamation.*
properyd, 230, I am properyd, *mine is the property, the right*; N. E. D. proper (v.), *to make master of.*
prose, 231, *story, narrative.*
provaylys (v. 3 pl.), 226, *for* prevaylys, *avail, benefit.*
prow, 111, 308, *advantage.*
prune, 154, *trim, preen.*
prynsesse, 252, *princes*; prysis, 355.
psalmus, 96; psalmys, 74, *Psalms.*
pshalmodyeth, 260, *sing psalms.*
punchement, 100, *punishment*; punchyth, 68, *punish.*
punche, 328, *pierce.*
purvyauns, 228, *provision.*
pyan, 20, *peony*; pepyr, pyan; cf. peper and piones . . . hote spices, *P. Pl.* B. v. 312.
pychen, 169, *stab.* *See* pyth.
pygth, 96 (r. w. with), *pith.*
pyke (v.), 91, *pick.*
pylis, 199, *towers.*
pylle, 277, whele and pylle, *a game.*
pyn (sb.), 142, *suffering*; 211, pyne.
pynde (pp.), 28, to pynde, *wasted.*

pynne (sb), 129, *centre of the target*; cf. prycke, 40; pynne, 197, *peak.*
pynne (v.), 24, 251, *shut up, confine.*
pyth, 12, *thrust*; pyth (pp.), 1, *placed, fixed*; pyght (pp.), 6, 144, 196; pyht (pp.), 18.

qwalle, 60, *whale*; whallys (gen.), 340
qwan, 241, *when*; quan, 109.
qwart (sb.), 211, *health*; qwerte, 191, 344 (adj.), *whole, sound.*
qweche, 259, *which*; qwhich, 216; qwyche, 331. *See* weche.
qwed (adj.), 13, *evil, wicked.*
qwedyr (v.), 114, *quiver.*
qweke (v.), 27 (r. w. freke), *quake*; qwake, 15.
qwelle (v.), 12, *kill.*
qwelp (sb.), 45, helle qwelp, *whelp.*
qweme (v.), 101, *unite, reconcile*; queme, 361, *please, be agreeable.*
qwenys, 171, *women, queans.*
qwens, 75, *whence.*
qwere, 269; quer, 365, *choir.*
qwere, 102, *where*; qwher, 278.
qwethynge, 335, quyk and qwethynge, *speaking* (earliest quotation of this phrase N. E. D. 1529).
qwhethe, 204, *bequeath, consign*; qwethe, 213.
qwhyl (conj.), *while.*
qwy, 112, *why.*
qwyght (adj.), 336 (r. w. syght), *free, clear*; (v. inf.), 16, 39, *requite, reward.*
qwyke, 96; qweke, 227; quyk, 335, *quick, alive.*
qwyle (sb.), 65, *time.*
qwyppys, 294, *whips.*
qwyte (adj.), 19, *white.*

race (v.), 185, *break asunder.*
raftys, 170, *beam, pole, shaft* (*of a spear*). *See* rakynge.
rake (v.), 173, rake þem on rought, *rake, sweep.*
rakyl (adj.), 21, *hasty, unstable.*
rakynge, 170, with rakynge raftys, *dashing, violently moving.*
ransake, 143, *search carefully.*
rape (sb.), 169, *violence.*
rape (sb.), 218, *haste*; (v.), 369, I renne I rappe; (imp.), 373.
rapely, 374, *swiftly, quickly.*
rappe (v.), 173, *smite, hit.*
ravaschyd, 348, *transported.*
ray (sb.), 170, a reed ray, *aphetic for* array *or* ray, *dance* (dance of blood).
ray (adj.), 233, ray tabardys . . . ray hodys, *made of ray*, i. e. *striped cloth.*
rebate (v.), 69, *abate.*
reclyne (v.), 132 (r. w.), *to return, to earlier condition* (quoted in this sense N. E. D.).
record (sb.), 40, record of my boy (absol. phrase), *my boy bearing witness.*
recure (v.), 82, *win, obtain*; recuryd (pp.), 226, *recovered, restored*; recuryn (inf.), 281, *restore to life.*
reddure, 355, *rigour, severity*; redrure, 239, ? *for* reddure.
rede (sb.), 5, *counsel*; reed, 23.
rede (v.), *counsel, advise.*
redolent, 80, *sweet-smelling*; redolens (sb.), 241, *perfume.*
reducyd, 263, ? *brought back again.*
reed, 48 (r. w. god), *for* rod.
reed (adj.), 170, *red, blood-stained.*
regyon, 80, Regina of regyon, *rule, government*; 102, *region.*
rejoyse (v.), 56, *enjoy*; (sb.), 377, *joy, pleasure.*
relacion, 82, *regard, consideration*; 38, *report.*
relefe (v.), 356, *remain.* *See* relevys.
relese (v.), 195, *relieve, alleviate.*
releve (v.), 220, *relieve, assuage.*
relevys (sb.), 81, *remains.*
rem (r. w. bedleem), 146, *realm.*
remeffe, 229, *remove.*
renne, 10; ronne, 10, *run.*
replye (v.), 256; replyeth (imp. pl.), 360, replyeth and assedually wachith, *apply.*
repreff, 32; repreve, 127, *reproof.*
reprevable, 232, *conducing to the reproof of.*
rere (v.), 56, *raise up.*
resch (sb.), 161, resch and root, *rush*; rosch, 28.
rescu (sb.), 106, *rescue.*
rese (v.), 372, *rose*; reson (pp.), 312, *risen.*
responcyon, 355, *answer, response* (earliest example N. E. D. 1502).
respyt, 113, *delay.*
rest (v.), 228, *remain.*
resydens, 182, *seat, position.*
reve (v.), 156, *spoil, plunder*; 302, *take* (*away*); 322 (pp.), wyttys ben revid.
revere, 353, *river.*
revyfe, 231, *revives.*
rewe (sb.), 10, *row*; rowe, 35.
rewelerys, 236, *rulers.*

rewly (adj.), 10, *rueful, piteous*; ruly, 12.
rewthe, 99, *matter for regret.*
reynenge, 16, *sovereignty.*
reynes, 227. *cloth of Raines, linen made at Rennes.*
reysynt, 246, 299, *for* reysyn it, *raise it. See* levynt.
roberych, 260, *rubric, stage direction.*
rochand, 287, I reyne as a rochand, ? *ruler* (Halliwell), *or cognate with* roch (= *rough*).
rought (sb.), 173 (r. w. dowt, abought), *rout.*
rowel, 169, *spur* (I ryde on my rowel = *I ride spurring*).
rownceys, 170, *horses, hackneys.*
rowne (v.), 374, *utter.*
rowse (v.), 83, *proclaim, make public*; 97, *utter.*
rowte (sb.), 35, *company, assembly*; rowthte, 176.
rowth (v.), 167, 281 (impers.), 369, *reck, care* (*past tense form with present meaning*).
rubbe (v.), 375, ?.
rustynes, 42, rustynes of synne, *corruption.*
ruyne, 104, *fall, overthrow.*
ryff, 3, his chylde þer offered xuld be upon an hylle full ryff; þe busshop toke here iii maydonys ryff; 92, I wedde here ryff; (adv.), *readily.*
ryght, 34; ryte, 2; ryth, 1, *right.*
ryghtwysnes, 100, *righteousness.*
ryme, 227, *match.*
rys, 20, gentyl rys, *rice* (*choice rice*).

sa, 119, ? *so.*
sacryd (pp.), 255, *consecrated.*
sad, 1, *serious*; 154, *firmly fixed* 223, *heavy*; sadly, 41, *resolutely.*
sage (adj.), 8, 105, *wise*; 117 (sb.), *wise woman.*
sage, 93 (r. w. age), in old sage, *saying*, cf. *Pearl*, l. 226, saghe; sawys, 80, *sayings, speech.*
salver, 166, *healer.*
saluse (v.), 361, *salute.*
same, 19, in same, *together.*
satan, 228, *satin.*
sawe, 267 (r. w. have), *save.*
sawtere book, 95, *Psalter.*
sayll (v.), 40, *assail.*
scappys, 197, *slips, mistakes.*
schadu (v.), 105, *shadow.*
schal. *See* shalle.
schamly, 362, *shamefully.*
schape (v.), 132, *escape*; shape, 313.
schapman, 252, *chapman, trader.*
schapyn, 317, *cause, bring about.*
scharlys, 171 (r. w. gerlys), *churls.*
schedyn (v.), 156, *shed.*
schelchownys, 170, MS. schel chownys *altered to* schel chowthys, ? *for* selcouthys, *marvels*; possibly in the first case a mistake for scheltrownys, *troops.*
schende, 46, *injure, destroy*; schent (pp.), 110, *disgraced.*
schep, 139, *cheapness, abundance.*
schet (pp.), 28, *shut*; shytt (pp.), 248; shytte (pr. t.), 215.
schet (pt.), 40, *shot. See* shete.
schon, 52, *shoes*; sho, 25, *shoe.*
schonde (sb.), 317, *shame, disgrace.*
schrowde, 25, *garment.*
scle, 46, *slay*; scloo, 46, 171; slo, 33; slen, 147; slewe (pt.), 34; scle (pt.), 41; sclow (pt.), 28; slawe (pp.), 9; slayn (pp.), 3.
scleppe (v.), 36, *slip.*
sclepyr, 93, *slippery.*
sclyde, 216, our sorwe doth slake and sclyde, *slip away, be forgotten.*
scorn, 220, *shorn, rent.*
scowte, 128, *a term of abuse.* 205/145
se (v.), 2, *see*; sene, 4; seyn, 52; saughe (pt.), 162; sey (pt.), 277, 311; say (pt.), 330; sowe (pt.), 253; seyn (pt.), 351: sene (pp.), 2; sayn (pp.), 34, 323; seyn (pp.), 81; seyne (pp.), ? 152.
se (sb.), 18, *seat.*
se, 359, l. 123, *for* so *or for* be = *by.*
se, 103, l. 193, ?.
seche, 246, *such. See* soch, suych.
seche (r. w. leche), 152; seke, 153, *seek*; south (pt.), 7; sowth (pt.), 8; sought (pp.), 30; sowte (pp.), 102; sought, 303, *attacked, visited*; han de sowth, 275 (*see* do).
seke, 65, *sick.*
sekyr, 38, *safe, sure.*
sel, 258, 276, *time*; seyl, 274.
selkowth, 146, *strange.*
selph, 93, *self*
semlant, 153, *appearance.*
sen, 298, *saint.*
sen (conj.), *or* son, 66, *since. See* syn *and* sythyn.
sende, 103 (r. w. kende), *sent.*
sengler (adj.), 360, *singular.*
senstere, 357, ? *seamstress* (disparaging term for women).
senues, 297, *sinews.*

sequens, 65, *sequence, chant.*
ser, 369, *withered, sere.*
serge (sb.), 273, *search.*
serteyn (sb.), 241, a serteyn of chylderyn, *a number.*
seryattly, 357, *one by one, seriatim.*
sese, 220, *cause to cease.*
sesyd, 112, *seized, in possession, established.*
sethe (adv.), 118, *since.*
settynt, 275, settyn it, *set it.* *See* levynt, reysynt.
seuer (v.), 358, 370, *promise, assure, declare.*
severe (v.), 232, *sever.*
severe (adj.), 366 (r. w. cure), ? *sure*; sewre, 242.
sevyle, 228, *civil law.* *See* cevyle.
sew (v.), 13, *proceed, go with speed*; 108, 247, *petition, sue*; *follow*, 289; suenge, 29; sewyng, 256, *following.*
sewe (sb.), 377, *sow* (?).
sewte, 182, *suit, cause, action*; hold the sewte, *defend the action.*
sey, 25, *say*; seyn, 40, 52, l. 22 (or *see*); seyt, 299; sythe, 235, *saith*; seyand, 355, *saying*; ȝe haue me herd seyd, 30.
seyd, 125, *seed.*
seyn, 228, *sign.*
seyne (r. w. tweyne, regne), 152, ? *for* sheyne, *bright.* *See* shene.
seynge, 76, l. 123; 311, st. dir., ? *seeing.*
seyse, 228, seyse nere sessyon, *assize* (form not given in N. E. D.). *See* syse.
shadyr, 148, *shudder.*
shaftys, 154, *spears.*
shalle, 58, 202, 207, 348; shal, 204; schal, 229, 262; sal, 41; shalt, 50; schulle, 4; shulle, 8, 30; shul, 195; schulde, 48; schuld, 256; and throughout the *Assumption* Play *for usual* xal, xul, xulde.
shamfastnes, 105, *shyness.*
shede, 266, *sheath.*
sheff, 32, *sheaf.*
shendynge (sb.), 162, *harm.*
shene (adj.), 146, *bright.*
shenshipp, 45, *shame.*
shete (v.), 40, *shoot*; schet (pt.), 40; shote (pt.), 40; shet (pp.), 212.
shetyng (sb.), 41, *shooting.*
shewyght, 147, *sheweth*; shewyth (3 pl.), 80, l. 235, ? *cause to appear.*
shove (v.), 229, *show.*
showe (v.), 33 (r. w. loue), *shove, thrust*; 310 (r. w. anow).
shray (s.), 170 (r. w. clay) ? *clamour.* *See* N. E. D. scry. Cf. day *for* die.
shrevyn (pp.), 193, *shriven.*
shrewe (sb.), 125, *wretch*; brede a shrewe, 194, *something troublesome, vexatious, mischief.*
shrewe (v.), 124, *curse.*
shrynkyd (as a snayle), 198, *shrank, drew back.*
shulderyn, 172, *shoulders.*
shynand, 167, *shining.*
shyne (sb.), 146, *radiance* (earliest instance in N. E. D. 1529).
shyrle (r. w. fylle), 113, *shrill.*
skore (v.), 120, *scour.*
skorgyd, 165, *scourged.*
skye, 147, out of Jacob xulde shyne a skye, ?; in a bryght skye, 351, *cloud.*
skylle, 32, *reason*; skele, 361; it is skyl, 92, *reasonable, proper.*
slake, 216, *grow less, come to an end*, cf. *Pearl*, l. 942.
slauth, 376, *sloth.*
slawdyr, 124; *for* slawndyr (64), *slander*; sclaundyr, 125.
slawe (r. w. lawe), 236, *slow.*
sle, 7, *slay.* *See* scle.
sleytys, 146, sleytys slyȝ, *skilful devices*; slithtis, 367; 200, *artifices, wiles.*
slought, 340, *slothful.*
slye, 9, *cunning*; slyȝ, 146, 156, *clever.*
smyght, 48, *smite*; smyth, 253; smyth, 266, *smiteth*; smet (pt.), 277; smete (pp.), 266.
snarle, 368, *catch in a snare, strangle.*
snelle, 113, *quick, ready* (i. e. to help mankind).
sneveleris, 368, *snivellers, whiners.*
soch, 307, *such.* *See* seche, suych.
socurraunce, 207, *succour* (not in N. E. D.).
soferauns, 76, *endurance*; 271, *sanction.*
sofreynes, 81, *masters*, term applied to audience by Contemplacio and in the *Assumption* Play. *See* sovereynes.
sofron, 66, *suffer, allow*; soferyd (pp.), 42, *endured.*
sokelyng, 254, *clover.*
solempne (adj.), 15, *solemn.*
solennyte, 293, *solemnity.*
somowne, 123, *summon.*

son *for* sen, 66, *since.*
sond, 22 *send.*
sonde, 28, 46, 109, 132, *message, dispensation, ordinance*; 87, 169, *messenger.*
sondyr, 40, on sondyr, *asunder, in pieces.*
sool, 227, *sole of the foot.*
sor (sb.), 144, *injury, ailment.*
sore (adv.), 297, *violently*; 5, *greatly.*
sorwatorie, 306, ? (Halliwell, *place of sorrow*).
sorwe (sb.), 8, *sorrow, trouble*; sorwyn, 154, (? pl.).
sorwyth, 68, *sorroweth.*
soserye, 284, *sorcery.*
sote, 4, *sweet.*
sotely, sotylly, 253, *subtly, stealthily.*
sottys, 153, *fools.*
sotyl (adj.), 195, *cunning.*
sotylte, 9, *trick, crafty device.*
sovereynes, 71, *masters, sirs*; soueyreynes, 354. *See* sofreynes.
sought, 303. *See* seche.
sowe (*or* sewe), 229, *follow.*
sowe, 253 (r. w. knowe), *saw.*
sowkyn (v.), 7, *suck*; sokyn (pp.), 24.
sownd (adj.), 32, *reasonable*; 34, *sound*; 134, on sownd, *unsound, ill-founded.*
sownde, 35, Thy seruauntys saue lord fro synful sownde, *swoon, torpor.*
sowre (adj.), 154, *sour.*
sowyght, 67, *seweth.*
spede (sb.), 16, 71, *help, helper.*
spede (v.), 30, *assist, cause to succeed*; sped (pr. t.), 20, *hasten*; spedly sped (pp.), 148, *accomplished.*
spedful, 82, *helpful.*
spedly, 148, *successfully, prosperously.*
spekyn, 30, *speak*; all maner langage hem spak, 15 (refl.); cf. bot than hym spake Gallerone to Gawayne, *Anturs of Arth.*, N. E. D.
spelle (sb.), 115, *language.*
spelle (v.), 147, *relate, preach.*
spere (v.), 27, *shut off, bar*; sperd (pp.), 59, 288, *fastened.*
spetously, 268, *despitefully.*
splayed, 228, *spread out.*
spoused, 354, *wedlock*; 365, *espousal.*
spowsage, 83, *espousal, marriage.*
spowsyng, 85, *marriage.*
sprede, 35, *spread*; sprad (pp.), 21.
sprynge (v.), 4, *spring, arise*; spreng (pt.), 364; spronge (pp.), 60.
spryte, 191, *spirit*; sprytt, 60.
spylle, 12, *destroy*; spilly (1 sg.), 355; spylt (pp.), 41, *shed.*
stage, 35, strete and stage, *raised platform for spectators*; 93, xul dwelle with ȝow in stage, ? *now, at this time, straightway*; cf. N. E. D. stage, *a period of time. See* stownde.
starkly, 116, *stoutly.*
statis, 355, *dignitaries.*
sted (sb.), 23, with-inne a sted; þis steed, 134; as if used for point in time instead of point in space, cf. *on the spot*; stede and place, 43, *situation.*
steke (v.), 154, *stab, kill.*
stent (v.), 368, *stop, bring to a stand.*
steracle, 197, *spectacle, show.*
sterre, 7, *star.*
steryd (pp.), 25, *excited, instigated*; styrth, 138, *stirreth.*
stevene, 146, *sound*; 350, *speech*; stewyn, 90, *voice, speech.*
stey, 335, *arise, ascend*; stye, 351.
stody (sb.), 75, *study, acquisition of learning*; in a cold stodye, 207, *state of abstraction.*
stomachere, 227, *waistcoat.*
stomble, 28, *stumble*; stumbyll, 161; stomele, 197.
stondyn (v.), 10, *stand*; stonde, 52; stound, 87; stant, 138, *standeth*; stonde (pp.), 40.
stondynge (that), 179, 205, 214, *it being the case that.* (This absol. use not given in N. E. D.)
stotte, 205, *term of contempt for a woman.*
stow (v.), 204, *stop.*
stownde, 12, *time*; þis stownde, *now*; in stownde, 156, *at this moment, now.*
strekyn, 157, *passed, come forth. See* stryke.
strengere, 229, *stronger.*
strenght (sb.), 75, *strength*; strenghthis, 78; strenthis, 163, *powers.*
strengthe (v.), 217, *strengthen.*
streyte, 26, *straight, directly*; streyth, 140.
stryke, 153, *go, make one's way.*
sty, 26, helle sty, *pigsty.*
stye, 160, ouer þis stye, *path* (cf. AS. stīg).
stynkygh, 222, *stinketh.*
styrte (v.) (r. w. herte), 231, *go.*
styward, 7, *steward.*
submyt (pp.), 260, *submissive.*
sudary, 332, *napkin.*
suerd, 3, *sword*; swerd, 2.
sumdele, 341, *somewhat, to some extent.*
sumtyme, 26, *formerly.*

supportacion, 122, *countenance, support.*
sustyr, 107, *sister.*
sute, 193, folwyth þe fowle sute of þe devyl, *train.*
sutere, 190, *follower.*
suych, 8, *such*; suech, 76; swyche, 68; sweche, 239; seche, 246; soch, 307.
swage (v.), 35, *abate*; 170, *cease.*
swap (sb.), 7, *blow.*
swappynge (adj.), 172, *slashing.*
swelle, 75, *swallow, devour.*
swem (sb.), 65, *grief*; 101, *a pity.*
sweme (v.), 361, *faint, be overcome.*
swemful, 64, *full of grief.*
swemynge (sb.), 74, *grief, emotion.*
swete (v.), 32, *sweat.*
swetyng, 151, *darling.*
swoot (sb.) 26, *sweat.*
swowne (v.), 12 (r.w. stownde), *swoon*; swownde, 139; swonge, 299 (st. dir.); swuonyng, 298.
swynk (sb.), 26, *toil.*
swynke (v.), 32, *toil.*
swythe, 38, *quickly.*
sybb (adj.), 48, sybb blood, *kindred*; sybbest blood, 213; sybbe, syb, 126, *akin.*
syeng (sb.), 35, *sighing.*
sygnes, 80, *signs*; syne, 77, 146; synge, 370.
sygnyfure, 188, *signification, sign*; signifure, 340.
syhyn (v.), 863, *sigh.* *See* sythe.
sympyl, 193, *of little value.*
syn (prep.), 118; (conj.), 165, *since.* *See* sen, sythe, *and* sythyn.
syndony, 311, *fine linen* (sindon, word in Vulgate for linen cloth used by Joseph of Arimathea).
syne (sb.), 77, 146, *sign.* *See* sygnes.
syre, 61, a grym syre, *lord, master.*
syse, 27, *sighs.*
syse, 11, *assize*; in syse, 161, *in state, in proper fashion.* *See* seyse.
syt, 101, *is fitting, seemly*; syt, 26, *sitteth.* *See* syttyn.
syth, 2, *sight*; syght, 33; syht, 39; syte, 52; sygth, 223; sythte, 273; syghtys, 76, *for* fyghtys.
sythe (v.), 334, *sigh.*
sythe, 235, *for* seythe, *saith.*
sythe (adv.), 62, *afterwards*; (prep.), 165, *since*; syth (conj.), 240, *since, as*; sythe þat, 110, *since.*
sythym (adv.), 260 *for* sythyn, *afterwards, then.*
sythyn (conj.), 258, *since, as*; sythyn þat, 130, *since.* *See* syn *and* sen.
sythys, 349, *times.*
syttyn (v.), 18, *sit.* *See* syt.
syttynge, 101, *suitable, fitting.*

take (imp.), 39, *take*; take (pp.), 41; tan (inf.), 38; tan (pp.), 13; takyght (imp.), 171; take, 92, l. 308, *give*; toke (pt.), 2; tokyn (pt. pl.), 345, *gave.*
takke (v.), 288, *tack, fasten*; takkyd (pp.), 297, *nailed.*
takylle, 40, *weapon, bow.*
taske, 318 (r. w. aske), to taske a wynk, *tax, levy*, i. e. *seize, take.*
tast (v.), 142, *touch, examine.*
taught, 162; tauht, 354, *taught.*
tee (v. inf.), 29, *go.*
tekele (v.), 227, *tickle*; tekyl (adj.), 126, *ticklish.*
teme (sb.), 361, *theme*, sayd us this teme, *spoke to us to this effect.*
tempte (v. pt.), 226, *tempted.*
tende (v.), 43, *intend*; 369, *attend*; tent (imp.), 364, *take care of.*
tendyrly, 83, *carefully*, ? *for* tentily.
tene (sb.), 6, *malice, anger.* *See* trey.
tene (v.), 85, *rex*; 368, *suffer vexation.*
tent (sb.), 86, take tent, *take heed.*
terme, 9, *space of time*; 92, terme, tyme and tyde.
termynable, 232, To seuere ryth and wrong in me is termynable; 272, In ȝow alle jewgement is termynabyle. Hence 'to be terminable in' = to be derived from, to have its ultimate source in. (Not in this sense in N. E. D.)
terrewth, 102 (*for* trewth), *truth.*
teryeng, 72, *tarrying, delaying*; tary (imp.), 87; tery (subj.), 278.
testyficacion, 61, *testimony, witness.*
teyl (v.), 26, *till, obtain by tillage.*
teynt, 368, *attainted, corrupt, guilty.*
the (v.), 29, *in phrase* so moty (= mot I) the, *prosper, thrive*; then, 87.
thedom, 131, evyl thedom, *bad luck.* *See* the.
thedyr, 65, *thither.*
therkeness, 96; thyrknes, 270, *darkness.*
therlys, 171, *thralls, serfs.*
thes (r.w. pres, *press*), 61, *these*; theys, 57.
thes, 231, l. 24, *this*; þese, 259. *See* chalys.
tholyn (v. inf.), 153, *suffer*; tholyd (pp.), 172.

thore (r. w. more), 74, *there.*
thorw, 17; thorwe, 9; thour, 108; thourghe, 57; throwh, 31, *through.*
thouht (sb.), 35; thouth, 11; thought, 28; thowth, 57, *thought.*
thralle (sb.) (r. w. alle), 325 = thraw, *period, space of time.*
thrawe (pp.), 233, *thrown*; throwe (pp.), 24.
thredde, 364, *third*; thryd, 2; thrydde, 75.
threttene, 76, *thirteenth.*
threwth, 100, *for* trewth, *truth.*
throwyht (3 sg.), 67, *throweth.* *See* thrawe.
throwys, 139, *throes.*
thrylle (v.), 8, *pierce*; thyrlyng, 15; thyrlyd (pp.), 268.
thryste (sb.), 98; thrust, 302, *thirst.*
thy, 112, what for thy, *that.*
thyk, 25, þi fals fablis þei be ful thyk, *plentiful*, '*thick*'.
thynke, me thynke (r. w.), thynkyth, thynkeht (v. impers.), 32; thynkyht, 40; thynkygh, 173, *methinks.*
thynkyth (3 pl.), 106, *think*; thynkys (imp.), 120.
thyrknes, 270. *See* therkeness.
to, 111, l. 72, *two.*
to-breke, 48, *break open.*
tokenynge, 146, *sign, betokening.*
too (sb.), 126, *toe*; ton (pl.), 131.
toost (sb.), 173, (not) wurth a toost, *a piece of toast.*
to-pende (v. inf.), 169. *See* pende.
to-pynde, 28. *See* pynde.
tormentry (sb.), 184, *tormenting.*
tose (v. inf.), 373, '*tease out*', *search out.*
tother, 53, þe tother, *other.*
to-torn (pp.), 27, *torn in pieces.*
towaly, 260, *towel.*
trace (sb.), 50, *track*; returnyth his trace, *comes back.*
tras (v.), 13, *trace, follow a course*; 249, on of ȝow my treson xal tras, *contrive, work out.*
trast (v.), 84, *trust.*
tray (v.), 319, *betray.*
trayn, 324, *deceit.*
trekyl, 64, *trickle.*
trenyte, 17, *Trinity.*
trepett, 174, I ȝeve hym . . . a trepett, *I trip him up.*
tretable, 201, *tractable, yielding.*
trewth, 2, *truth*; trowth, 56, l. 160.
trey, 6, 97, *anger, spite.* *See* tene.
tribus, 63, *tribes.*
tron, 8, *throne.*
trow (v.), 33, *think, consider.*
trowth, 56, l. 145, *troth.*
trus (v.), 171, *truss, pack.*
tryne, 81, *threefold, triune.*
tryste (v. 3 sg., r. w. myste), 161, *trusteth.*
tundyr, 40, *tinder.*
turtelys, 64, *doves.*
twey, 99; tway, 345, *two.*
tweyners, 117, here tweyners (g. pl.), *of them two.*
twynne, 303, *part in two, cleave asunder.*
tyde (sb.), 1, *time.*
tydandys, 289, 323, 330, *tidings*; tydyng, 3; tydyngys, 43.
tyght, 222, *readily, promptly*; tyth, 16; as tyght, 34, *as quickly* (*as possible*), i. e. *at once.*
tymbre wryth, 136, *carpenter*
tythyng, 3, *tithing, paying tithes.*
tyxt, 226, *text.*

þar (v.), 257, *need.*
þare, 66 (r. w. bare), *there.*
þenge, 2, *thing*; thyng, 19.
þenke (v.), 1, *think, purpose*; thynk, 7.
þese, 259, þese chalys, *this.*
þo, 1, *those.*
þurowe, 1, *through.*

valure, 340, for your valure, *for your worth*, i. e. *to redeem you*; 341, *value.*
varyable, 232, *contrary to.*
varyauns, 6, withoute varyauns, *without deviation, straightway*; varyauns, 232, *divergence.*
vathe, 299, *interjection*, ? *what*; cf. whath, 231.
vengeabyl, 100, *revengeful, prone to vengeance.*
venym (adj.), 192, *venomous.*
verament, 14, *truly, verily.*
verdyth (r. w. lyth, *light*), 339, *verdict.*
verray (adj.), 2, *true*; verry (adj.), 60; veray (adv.), 183, *truly.*
vertuis (adj.), 256, *virtuous.*
veruent, 216, *fervent*; veruently, 15, *flamingly.*
veryfye, 114, *confirm as true.*
vesage, 34, *visage, face.*
vesyte (v.), 98; vycyte, 240, *visit.*
vesytacion, 49, *visitation.*
vetaylys, 244, *victuals.*
veyn (adj.), 34, *worthless.*
vmbyl, 260, *humble, lowly.*
vnbegete (pp.), 258, *unbegotten.*
vndowteful, 75, *free from doubt.*

vndryd, 97, *hundred.*
vnhede (r. w. dede), 24, *uncover.*
vnhende, 176, *discourteous, unknightly.*
vnkende (r. w. wynde), 7, *unnatural*; 99 (r. w. ende).
vnknowlage, 113, *lack of knowledge.*
vnlosne (v.), 237, *unloose, untie.*
vnpynne, 306, *unfasten.*
vnstable (*altered from* vnstabyl, r. w. able), 203, *unreliable, inconsistent.*
vnterest, 167, *uttermost.*
vntey (r. w. wey), 224, *untie.*
vnthende (adj.), 32, *poor, meagre.* *See* the.
voydnes, 116, *emptiness.*
vyl (sb.), 263, *for* wyl, *will.*

wace, 265 (r. w. place), *was.*
wake (v. int.), 17, *to exist fully, be active*; 252, *watch* (int.); wakyn the way, 314, *watch* (tr.), *guard.*
wakyn, 161; *awake*; cf. ore he be wakyn, *Torrent of Portugal*, N. E. D.
walkyn, 19, *welken, sky.*
walterid, 317, *weltering, steeped in.*
wan (v.), 226, *won*; wonnyn (pp.), 146.
wantruste, 212, *mistrust, lack of hope.*
war (adj.), 4, be war, *take note*; ware, 187, *informed, aware*; be whare, 54, *beware.*
wardeyn, 91, *guardian.*
ware (sb.), 187, *goods.*
wark (sb.), 307, l. 1046, *work. See* werk.
warly, 310, *warily.*
wasch (pp.), 348, *washed.*
wast (pp.), 27, *wasted.*
watt, 274, *fellow*; cf. *Richard the Redeless*, wattis, *people.*
wawys, 42, *waves.*
way, 111, do way, *leave off.*
wayted, 169, *kept watch*; wayten, 314, *watch, spy.*
weche (v.), 262, 312, *watch.*
weche, þe weche, 12, *which*; þe which, 3; wich (þat), 44; wheche book, 95; whiche, 164.
wede, 25, *garment.*
wedyr (sb.), 307, grett wedyr, *storm.*
wedyr, 113, *whither.*
weldygh, 90, *wieldeth, disposeth*; weldyth, 17.
wele (v.), 188, *will.* *See* wole.
welsom, 27, *wild, unpleasant*; whylsum, 191.
welthis, 95, ryches and welthis.
wem, 4, *stain, pollution.*
wene, 2, *ween, believe*; wend (pt.), 107.
wene, 4, withoutyn wene, *doubt.*
wepone, 265 (st. dir.), *weapons.*
wepyng, 28, wepyng dale, *hell.*
werch (v.), 55 (r. w. cherch), *work*; werke, 19; werkyn, 86; werkyht, 55, *worketh*; werke, 236, l. 158, *worketh.*
werd, 1, *world*; werlde, 26; werdl, 172.
werdly, 10, *worldly*; wurdly, 54.
were, 137 (r. w. to be), *weary.*
werke (v.). *See* werch.
werk (sb.), 16, *work*; werk wylde, 136, *difficult, tiresome business.* *See* wark.
werm, 19, *worm, creeping thing*; 25, *serpent*; worm, 26.
weryd (pt.), 328, awey he weryd þo ffyndys, ? *exorcized.*
weryn, 42; ware, 65; wore, 82; worn, 122; where, 240, *were*; wore, 318, *wert.*
wese, 159, *wise.*
wete (v.), 30, *know*; wetyn, 267; *See* wyst, woot.
weyll (adv.), 37, *well.*
weyth, 223, *weight.*
weyys, 159, this weyys (r. w. reyse), *this way.*
whanhope, 11, *despair.*
whan, 307, *wan, gloomy.*
whar, 277, whar, whar, *a hunting cry* (cf. modern 'Ware fox').
whath, 231, *what.* *See* vathe.
whe, 235, *we.*
whight, 55. *creature, man*; wyght, 55; wythe, 115; wyhgte, 135; whyt, 138; whyht, 176; whith, 213; wytys, 307; whithtys, 327.
whith, 72, *with*; whit, 240; *for usual* w^t^ or w*ith.*
who, 206, *woe*; whoo, 34.
whonde (v.), 115 (r. w. honde), *hesitate.*
whow, 65, *how*; whov, 164.
whyght, 89, *white.*
whyle (sb.), *time*; wheyle, 230. *See* qwyle.
whylsum, 191. *See* welsom.
whysshe, 251, *wish.*
withsytt, 200, *oppose, withstand*; withsett (2 sg.), 200.
wo (adj.), 14, *miserable.*
woke (sb.), 4, *week.*
wole (v. sg.), 17, *will*; wele (sg.), 88; wole (pl.), 292; wyl (pl.), 1; wul (pl.), 67, 332; wolyn (pl.), 291; welyn (pl.), 292; wolne (pl.), 258; wole, 88, *for* wolde.
won (v.), 355 (r. w. on), *wane, cease,*

slacken, or for wonde, *hesitate See.* whonde. *See* Dr. Greg's Note in *Assumption of the Virgin.*
wond, 87, *wand, rod*; whande, 87.
wone (sb.), 17, *habitation, place*; won, 137, *habitation, house.*
wonyn (v.), 149, *dwell*; 146, wonyght, *dwelleth.*
wood (adj.), 41, *mad.*
woolde, 41, *wold, ground*; wolde, 154, 327; 154, on wolde *altered to* ouer mowlde.
woot (v.), 26, *know*; wot, 35. *See* wete.
worde, 183, *world. See* werd.
wore, 88, þerfore now wole I · so 'my staff holde I · þis jurny to wore. *were to*, i. e. *arrived. See* to (adv.), 'expressing motion resulting in arrival', N. E. D.
woundyr (adv.), 21, *wondrously*; 138 (adj.), wonderful.
wrake (sb.), 129, *injury, harm*; 189, 288, 375, *persecution, vengeance, injury*; 86, thi wurde xal I werkyn withoutyn wrake, ?.
wrech (sb.), 194, werke us mech wrake, both wrech and woo, *punishment, destruction*; wreche, 327.
wrecche (sb.), 26, *wretch, outcast.*
wreke (v.), 148, from woe us wreke; 306, from wo I wole hem wreke, *protect* or *rescue*; wrokyn (pp.), out of whoo ȝet art not wrokyn, *rescued* (cf. *Chronicles of Robert of Brunne*, ed. Furnivall, l. 16076, 'Christendam fro Paen wreche', *rescue, defend, vindicate*); wreke (pp.), 171, *avenged*; wrokyn (pp.), 133, *avenged.*
wrete (pp.), 3, *written*; wretyn, 53.
wrethe (sb.), 26, 209, *wrath.*
wrothe (adj.), 2, *wroth.*
wrougth (pp.), 2, *wrought, created*; wrouth, 9; wrowth, 17.
wry (v.), 229, *turn aside.*
wryth (sb.), 136, *see* tymbre wryth.
wrythe (v.), 28, *twist.*
wul (adv.), 330, *well, very* (*usual* wele).
wul (v.), 67, 332. *See* wole.
wurth (adj.), 158, *valuable.*
wyght (adv.), 40, *strongly, actively*; 310, wyth, *swiftly, readily*; 328, l. 10, *swiftly* or (adj.) *strong*; whyt as thought, 279 (adj.), *swift.*
wyght (sb.), 55, *creature. See* whight.
wyghtly, 152, *nimbly.*
wyk (adj.), 25, *wicked*
wyl (adv.), 16, *well. See* wul.
wyndand, 369, wyndand wod (Halliwell prints wynd and wod), ?*raving mad*
wynne, 164, to jherusalem fast now wynne, *make for Jerusalem*; 146, *gain, win*; wonnyn (pp.), 146, *won.*
wys, 73, *for* i-wys.
wysse (v.), 94, *direct, guide.*
wyst (v.), 112, 310, *knew*; 113, I myght wel a wyst, *known. See* wete, woot.
wyte (v.), 115, *blame.*
wyth (adv.), 310. *See* wyght.
wytte, 65, *for* wylte, *wilt.*
wyttys, 155, *wits.*
wytys, 307, *wights. See* whight.

xad (v. inf.), 259, *shed.*
xal (sg.), 1, *shall* (x forms are the standard forms for this text, except in the *Assumption* Play. *See* shalle); xul (pl.), 6; xuln (pl.), 298; xulde (pt.), 3, *should.*
xamefullest, 273, *shamefullest.*

ye, 87 (*for usual* ȝe), ye.
yif, 356, *if.*
yis, 87, *yes.*
yit, 356, *yet.*
yon (adj.), 159 (*for usual* ȝon), *yon.*
yong, 158 (*for usual* ȝonge), *young.*
yow, 160, 188 (*for usual* ȝow), *you.*
yowr, 161 (*for usual* ȝoure), *your.*
ys, 87, *is.*
yt, 87, *it.*

NOTES

Calsydon. p. 244, l. 374. The reference is possibly to the hardness of chalcedony (cf. 'chalcedonius . . . sculpturis resistere dicitur', Bede, *On the Apocalypse*, and 'The calcydone whiche harde is in to grave', *Court of Sapience*); or possibly to the fact that chalcedony was held to typify those who show forth the light that is within them when called upon to give public testimony of their faith (cf. 'Chalcedonius . . . habet fulgorem sub dio non in domo. Quo demonstrantur hi qui . . . cum vel doctrinae vel aliis sanctorum usibus in servitute ad publicum procedere iubentur, mox quid fulgoris intus gesserint ostendunt', Bede). The path was to lead to the Trial and Death; it is the way taken 'for mannys love'.

glete. p. 155, l. 125. Miss H. M. R. Murray suggests that 'glete' may be connected with the verb 'gleit' or 'glete', N. E. D. = to glitter. In this case the meaning of the passage may be: 'He may put (in) us wits to recognize in the flesh, by the gleam (of the star), God's noble son.'

laue. p. 91, l. 275. Halliwell gives 'lave' = *washed.* Miss Murray suggests that 'laue' may be a Northern form of 'loaf'. Cf. 'whyt as payndemayn', *Sir Thopas.*

on a myche. p. 153, l. 73. The word 'myche' has been altered by the reviser to 'mysse'. 'on a myc' is still quite clear, and the full form can be deduced from the rhyme words 'rych' and 'dych'. Miss G. D. Willcock suggests 'miche', *a loaf of bread*, N. E. D. (cf. *Rel. Ant.* ii. 192: 'with-oute wyn and miche', where the word rhymes, as in the text, with 'riche' and 'diche'). The word would here be used with reference to the sacramental wafer, and the missing verb must have been some word equivalent to the 'beleuen' that has replaced it. The meaning is borne out by 'þer-inne sette here sacrementys' in the following line.

PROVERBS, SAYINGS, ALLUSIONS

p. 17. I am fadyr of myth
My sone kepyth ryth
My gost hath lyth
And grace with-alle.

p. 25. I walke as werme with-outyn wede
Awey is schrowde and sho.

p. 28. Schort lykyng xal be longe bought.

p. 29. ȝe must delve and I xal spynne.

p. 31. þat alle ȝour here levyng
May be to his plesyng
And at ȝour hens partyng
To come to good place.

p. 54. By-ware of custome ffor he wyl dere.

p. 58. Of Regalle lyff xal come suche foyson
þat a clene mayde modyr xal be.

p. 64. So xulde euery curat in þis werde wyde
ȝeve a part to his chauncel i-wys
A part to his parochonerys þat to povert slyde
The thryd part to kepe for hym and his

p. 65. And þo þat departe in sorwe god make þer metyng glad.

p. 67. Aftere grett sorwe · evyr gret grace growyht.

p. 70. Aftere grett sorwe · evyr grett gladnes is had.

p. 110. Olde cokwold þi bow is bent
Newly now after þe frensche gyse.

p. 111. Many a man doth bete þe bow
Another man hath þe brydde.

p. 128. But ȝit sum mede and ȝe me take
I wyl with-drawe my gret rough toth
Gold or sylvyr I wol not for-sake
But evyn as alle somnorys doth.

p. 132. In ffeyth I suppose þat þis woman slepte
Withowtyn alle coverte whylle þat it dede snowe.

p. 140. Whan women travayl grace doth growe.

p. 202. þe hare fro þe fforme we xal a-rere.

p. 227. In trost is treson.

p. 252. Mony makyth schapman.

p. 288. Ffor alle his barfot goyng · fro me xal he not skyp.

p. 314. þow þer come both jakke and gylle.

p. 326. Ffor mede doth most in every qwest
And mede is mayster bothe est and west
With mede men may bynde berys.

p. 356. Whoso clyme ouer hie · he hath a foule falle.

p. 357. It wolde cause the comownys to ryse
And rathere the devyl sle hym · than we schulde that abide.

jolyere than þe jay, p. 154; lyth as ro, p. 327; shrynkyd as a snayle, p. 198; whyte as laue, p. 91; whyte as swan, p. 56; wroth as wynde, pp. 7, 315.

HOMERTON
COLLEGE
LIBRARY